A PRACTICAL APPROACH TO SIGNALS AND SYSTEMS

A PRACTICAL APPROACH TO SIGNALS AND SYSTEMS

D. Sundararajan

John Wiley & Sons (Asia) Pte Ltd

Other Wiley Editorial Offices

John Wiley & Sons, Ltd, The Atrium, Southern Gate, Chichester, West Sussex, PO19 8SQ, UK

John Wiley & Sons Inc., 111 River Street, Hoboken, NJ 07030, USA

Jossey-Bass, 989 Market Street, San Francisco, CA 94103-1741, USA

Wiley-VCH Verlag GmbH, Boschstr. 12, D-69469 Weinheim, Germany

John Wiley & Sons Australia Ltd, 42 McDougall Street, Milton, Queensland 4064, Australia

John Wiley & Sons Canada Ltd, 6045 Freemont Blvd, Mississauga, ONT, L5R 4J3, Canada

Wiley also publishes its books in a variety of electronic formats. Some content that appears in print may not be available in electronic books.

Library of Congress Cataloging-in-Publication Data

Sundararajan, D.
   Practical approach to signals and systems / D. Sundararajan.
      p. cm.
   Includes bibliographical references and index.
   ISBN 978-0-470-82353-8 (cloth)
   1. Signal theory (Telecommunication)   2. Signal processing.   3. System analysis. I. Title.
   TKTK5102.9.S796 2008
   621.382'23–dc22
                                                                                2008012023
ISBN 978-0-470-82353-8 (HB)

Typeset in 11/13pt Times by Thomson Digital, Noida, India.
Printed and bound in Singapore by Markono Print Media Pte Ltd, Singapore.
This book is printed on acid-free paper responsibly manufactured from sustainable forestry in which at least two trees are planted for each one used for paper production.

Contents

Preface

The increasing number of applications, requiring a knowledge of the theory of signals and systems, and the rapid developments in digital systems technology and fast numerical algorithms call for a change in the content and approach used in teaching the subject. I believe that a modern signals and systems course should emphasize the practical and computational aspects in presenting the basic theory. This approach to teaching the subject makes the student more effective in subsequent courses. In addition, students are exposed to practical and computational solutions that will be of use in their professional careers. This book is my attempt to adapt the theory of signals and systems to the use of computers as an efficient analysis tool.

A good knowledge of the fundamentals of the analysis of signals and systems is required to specialize in such areas as signal processing, communication, and control. As most of the practical signals are continuous functions of time, and since digital systems are mostly used to process them, the study of both continuous and discrete signals and systems is required. The primary objective of writing this book is to present the fundamentals of time-domain and frequency-domain methods of signal and linear time-invariant system analysis from a practical viewpoint. As discrete signals and systems are more often used in practice and their concepts are relatively easier to understand, for each topic, the discrete version is presented first, followed by the corresponding continuous version. Typical applications of the methods of analysis are also provided. Comprehensive coverage of the transform methods, and emphasis on practical methods of analysis and physical interpretation of the concepts are the key features of this book. The well-documented software, which is a supplement to this book and available on the website (www.wiley.com/go/sundararajan), greatly reduces much of the difficulty in understanding the concepts. Based on this software, a laboratory course can be tailored to suit individual course requirements.

This book is intended to be a textbook for a junior undergraduate level one-semester signals and systems course. This book will also be useful for self-study. Answers to selected exercises, marked ∗, are given at the end of the book. A Solutions manual and slides for instructors are also available on the website (www.wiley.com/go/sundararajan). I assume responsibility for any errors in this book and in the accompanying supplements, and would very much appreciate receiving readers' suggestions and pointing out any errors (email address: d_sundararajan@yahoo.com).

I am grateful to my editor and his team at Wiley for their help and encouragement in completing this project. I thank my family and my friend Dr A. Pedar for their support during this endeavor.

D. Sundararajan

Abbreviations

dc: Constant
DFT: Discrete Fourier transform
DTFT: Discrete-time Fourier transform
FT: Fourier transform
FS: Fourier series
IDFT: Inverse discrete Fourier transform
Im: Imaginary part of a complex number or expression
LTI: Linear time-invariant
Re: Real part of a complex number or expression
ROC: Region of convergence

1

Introduction

In typical applications of science and engineering, we have to process signals, using systems. While the applications vary from communication to control, the basic analysis and design tools are the same. In a signals and systems course, we study these tools: convolution, Fourier analysis, z-transform, and Laplace transform. The use of these tools in the analysis of linear time-invariant (LTI) systems with deterministic signals is presented in this book. While most practical systems are nonlinear to some extent, they can be analyzed, with acceptable accuracy, assuming linearity. In addition, the analysis is much easier with this assumption. A good grounding in LTI system analysis is also essential for further study of nonlinear systems and systems with random signals.

For most practical systems, input and output signals are continuous and these signals can be processed using continuous systems. However, due to advances in digital systems technology and numerical algorithms, it is advantageous to process continuous signals using digital systems (systems using digital devices) by converting the input signal into a digital signal. Therefore, the study of both continuous and digital systems is required. As most practical systems are digital and the concepts are relatively easier to understand, we describe discrete signals and systems first, immediately followed by the corresponding description of continuous signals and systems.

1.1 The Organization of this Book

Four topics are covered in this book. The time-domain analysis of signals and systems is presented in Chapters 2–5. The four versions of the Fourier analysis are described in Chapters 6–9. Generalized Fourier analysis, the z-transform and the Laplace transform, are presented in Chapters 10 and 11. State space analysis is introduced in Chapters 12 and 13.

The amplitude profile of practical signals is usually arbitrary. It is necessary to represent these signals in terms of well-defined basic signals in order to carry out

A Practical Approach to Signals and Systems   D. Sundararajan
© 2008 John Wiley & Sons (Asia) Pte Ltd

efficient signal and system analysis. The impulse and sinusoidal signals are fundamental in signal and system analysis. In Chapter 2, we present discrete signal classifications, basic signals, and signal operations. In Chapter 3, we present continuous
signal classifications, basic signals, and signal operations.

The study of systems involves modeling, analysis, and design. In Chapter 4, we
start with the modeling of a system with the difference equation. The classification
of systems is presented next. Then, the convolution–summation model is introduced.
The zero-input, zero-state, transient, and steady-state responses of a system are derived
from this model. System stability is considered in terms of impulse response. The basic
components of discrete systems are identified. In Chapter 5, we start with the classification of systems. The modeling of a system with the differential equation is presented
next. Then, the convolution-integral model is introduced. The zero-input, zero-state,
transient, and steady-state responses of a system are derived from this model. System stability is considered in terms of impulse response. The basic components of
continuous systems are identified.

Basically, the analysis of signals and systems is carried out using impulse or sinusoidal signals. The impulse signal is used in time-domain analysis, which is presented
in Chapters 4 and 5. Sinusoids (more generally complex exponentials) are used as the
basic signals in frequency-domain analysis. As frequency-domain analysis is generally more efficient, it is most often used. Signals occur usually in the time-domain. In
order to use frequency-domain analysis, signals and systems must be represented in
the frequency-domain. Transforms are used to obtain the frequency-domain representation of a signal or a system from its time-domain representation. All the essential
transforms required in signal and system analysis use the same family of basis signals,
a set of complex exponential signals. However, each transform is more advantageous
to analyze certain types of signal and to carry out certain types of system operations,
since the basis signals consists of a finite or infinite set of complex exponential signals
with different characteristics—continuous or discrete, and the exponent being complex or pure imaginary. The transforms that use the complex exponential with a pure
imaginary exponent come under the heading of Fourier analysis. The other transforms
use exponentials with complex exponents as their basis signals.

There are four versions of Fourier analysis, each primarily applicable to a different
type of signals such as continuous or discrete, and periodic or aperiodic. The discrete
Fourier transform (DFT) is the only one in which both the time- and frequency-domain
representations are in finite and discrete form. Therefore, it can approximate other
versions of Fourier analysis through efficient numerical procedures. In addition, the
physical interpretation of the DFT is much easier. The basis signals of this transform is
a finite set of harmonically related discrete exponentials with pure imaginary exponent.
In Chapter 6, the DFT, its properties, and some of its applications are presented.

Fourier analysis of a continuous periodic signal, which is a generalization of the
DFT, is called the Fourier series (FS). The FS uses an infinite set of harmonically
related continuous exponentials with pure imaginary exponent as the basis signals.

This transform is useful in frequency-domain analysis and design of periodic signals and systems with continuous periodic signals. In Chapter 7, the FS, its properties, and some of its applications are presented.

Fourier analysis of a discrete aperiodic signal, which is also a generalization of the DFT, is called the discrete-time Fourier transform (DTFT). The DTFT uses a continuum of discrete exponentials, with pure imaginary exponent, over a finite frequency range as the basis signals. This transform is useful in frequency-domain analysis and design of discrete signals and systems. In Chapter 8, the DTFT, its properties, and some of its applications are presented.

Fourier analysis of a continuous aperiodic signal, which can be considered as a generalization of the FS or the DTFT, is called the Fourier transform (FT). The FT uses a continuum of continuous exponentials, with pure imaginary exponent, over an infinite frequency range as the basis signals. This transform is useful in frequency-domain analysis and design of continuous signals and systems. In addition, as the most general version of Fourier analysis, it can represent all types of signals and is very useful to analyze a system with different types of signals, such as continuous or discrete, and periodic or aperiodic. In Chapter 9, the FT, its properties, and some of its applications are presented.

Generalization of Fourier analysis for discrete signals results in the z-transform. This transform uses a continuum of discrete exponentials, with complex exponent, over a finite frequency range of oscillation as the basis signals. With a much larger set of basis signals, this transform is required for the design, and transient and stability analysis of discrete systems. In Chapter 10, the z-transform is derived from the DTFT and, its properties and some of its applications are presented. Procedures for obtaining the forward and inverse z-transforms are described.

Generalization of Fourier analysis for continuous signals results in the Laplace transform. This transform uses a continuum of continuous exponentials, with complex exponent, over an infinite frequency range of oscillation as the basis signals. With a much larger set of basis signals, this transform is required for the design, and transient and stability analysis of continuous systems. In Chapter 11, the Laplace transform is derived from the FT and, its properties and some of its applications are presented. Procedures for obtaining the forward and inverse Laplace transforms are described.

In Chapter 12, state-space analysis of discrete systems is presented. This type of analysis is more general in that it includes the internal description of a system in contrast to the input–output description of other types of analysis. In addition, this method is easier to extend to system analysis with multiple inputs and outputs, and nonlinear and time-varying system analysis. In Chapter 13, state-space analysis of continuous systems is presented.

In Appendix A, transform pairs and properties are listed. In Appendix B, useful mathematical formulas are given.

The basic problem in the study of systems is how to analyze systems with arbitrary input signals. The solution, in the case of linear time-invariant (LTI) systems, is to decompose the signal in terms of basic signals, such as the impulse or the sinusoid. Then, with knowledge of the response of a system to these basic signals, the response of the system to any arbitrary signal that we shall ever encounter in practice, can be obtained. Therefore, the study of the response of systems to the basic signals, along with the methods of decomposition of arbitrary signals in terms of the basic signals, constitute the study of the analysis of systems with arbitrary input signals.

2

Discrete Signals

A signal represents some information. Systems carry out tasks or produce output signals in response to input signals. A control system may set the speed of a motor in accordance with an input signal. In a room-temperature control system, the power to the heating system is regulated with respect to the room temperature. While signals may be electrical, mechanical, or of any other form, they are usually converted to electrical form for processing convenience. A speech signal is converted from a pressure signal to an electrical signal in a microphone. Signals, in almost all practical systems, have arbitrary amplitude profile. These signals must be represented in terms of simple and well-defined mathematical signals for ease of representation and processing. The response of a system is also represented in terms of these simple signals. In Section 2.1, signals are classified according to some properties. Commonly used basic discrete signals are described in Section 2.2. Discrete signal operations are presented in Section 2.3.

2.1 Classification of Signals

Signals are classified into different types and, the representation and processing of a signal depends on its type.

2.1.1 Continuous, Discrete and Digital Signals

A continuous signal is specified at every value of its independent variable. For example, the temperature of a room is a continuous signal. One cycle of the continuous complex exponential signal, $x(t) = e^{j(\frac{2\pi}{16}t + \frac{\pi}{3})}$, is shown in Figure 2.1(a). We denote a continuous signal, using the independent variable t, as $x(t)$. We call this representation the time-domain representation, although the independent variable is not time for some signals. Using Euler's identity, the signal can be expressed, in terms of cosine and

A Practical Approach to Signals and Systems   D. Sundararajan
© 2008 John Wiley & Sons (Asia) Pte Ltd

(a)   (b)

Figure 2.1 (a) The continuous complex exponential signal, $x(t) = e^{j(\frac{2\pi}{16}t+\frac{\pi}{3})}$; (b) the discrete complex exponential signal, $x(n) = e^{j(\frac{2\pi}{16}n+\frac{\pi}{3})}$

sine signals, as

$$x(t) = e^{j(\frac{2\pi}{16}t+\frac{\pi}{3})} = \cos\left(\frac{2\pi}{16}t + \frac{\pi}{3}\right) + j\sin\left(\frac{2\pi}{16}t + \frac{\pi}{3}\right)$$

The real part of $x(t)$ is the real sinusoid $\cos(\frac{2\pi}{16}t + \frac{\pi}{3})$ and the imaginary part is the real sinusoid $\sin(\frac{2\pi}{16}t + \frac{\pi}{3})$, as any complex signal is an ordered pair of real signals. While practical signals are real-valued with arbitrary amplitude profile, the mathematically well-defined complex exponential is predominantly used in signal and system analysis.

A discrete signal is specified only at discrete values of its independent variable. For example, a signal $x(t)$ is represented only at $t = nT_s$ as $x(nT_s)$, where T_s is the sampling interval and n is an integer. The discrete signal is usually denoted as $x(n)$, suppressing T_s in the argument of $x(nT_s)$. The important advantage of discrete signals is that they can be stored and processed efficiently using digital devices and fast numerical algorithms. As most practical signals are continuous signals, the discrete signal is often obtained by sampling the continuous signal. However, signals such as yearly population of a country and monthly sales of a company are inherently discrete signals. Whether a discrete signal arises inherently or by sampling, it is represented as a sequence of numbers $\{x(n), -\infty < n < \infty\}$, where the independent variable n is an integer. Although $x(n)$ represents a single sample, it is also used to denote the sequence instead of $\{x(n)\}$. One cycle of the discrete complex exponential signal, $x(n) = e^{j(\frac{2\pi}{16}n+\frac{\pi}{3})}$, is shown in Figure 2.1(b). This signal is obtained by sampling the signal (replacing t by nT_s) in Figure 2.1(a) with $T_s = 1$ s. In this book, we assume that the sampling interval, T_s, is a constant. In sampling a signal, the sampling interval, which depends on the frequency content of the signal, is an important parameter. The sampling interval is required again to convert the discrete signal back to its corresponding continuous form. However, when the signal is in discrete form, most of the processing is independent of the sampling interval. For example, summing of a set of samples of a signal is independent of the sampling interval.

When the sample values of a discrete signal are quantized, it becomes a digital signal. That is, both the dependent and independent variables of a digital signal are in discrete form. This form is actually used to process signals using digital devices, such as a digital computer.

2.1.2 Periodic and Aperiodic Signals

The smallest positive integer $N > 0$ satisfying the condition $x(n + N) = x(n)$, for all n, is the period of the periodic signal $x(n)$. Over the interval $-\infty < n < \infty$, a periodic signal repeats its values in any interval equal to its period, at intervals of its period. Cosine and sine waves, and the complex exponential, shown in Figure 2.1, are typical examples of a periodic signal. A signal with constant value (dc) is periodic with any period. In Fourier analysis, it is considered as $A \cos(\omega n)$ or $Ae^{j\omega n}$ with the frequency ω equal to zero (period equal to ∞).

When the period of a periodic signal approaches infinity, there is no repetition of a pattern and it degenerates into an aperiodic signal. Typical aperiodic signals are shown in Figure 2.3.

It is easier to decompose an arbitrary signal in terms of some periodic signals, such as complex exponentials, and the input–output relationship of LTI systems becomes a multiplication operation for this type of input signal. For these reasons, most of the analysis of practical signals, which are mostly aperiodic having arbitrary amplitude profile, is carried out using periodic basic signals.

2.1.3 Energy and Power Signals

The power or energy of a signal are also as important as its amplitude in its character-ization. This measure involves the amplitude and the duration of the signal. Devices, such as amplifiers, transmitters, and motors, are specified by their output power. In signal processing systems, the desired signal is usually mixed up with a certain amount of noise. The quality of these systems is indicated by the signal-to-noise power ratio. Note that noise signals, which are typically of random type, are usually characterized by their average power. In the most common signal approximation method, Fourier analysis, the goodness of the approximation improves as more and more frequency components are used to represent a signal. The quality of the approximation is mea-sured in terms of the square error, which is an indicator of the difference between the energy or power of a signal and that of its approximate version.

The instantaneous power dissipated in a resistor of 1Ω is $x^2(t)$, where $x(t)$ may be the voltage across it or the current through it. By integrating the power over the interval in which the power is applied, we get the energy dissipated. Similarly, the sum of the squared magnitude of the values of a discrete signal $x(n)$ is an indicator of its energy and is given as

$$E = \sum_{n=-\infty}^{\infty} |x(n)|^2$$

The use of the magnitude $|x(n)|$ makes the expression applicable to complex signals as well. Due to the squaring operation, the energy of a signal $2x(n)$, with double the amplitude, is four times that of $x(n)$. Aperiodic signals with finite energy are called energy signals. The energy of $x(n) = 4(0.5)^n$, $n \geq 0$ is

$$E = \sum_{n=0}^{\infty} |4(0.5)^n|^2 = \frac{16}{1 - 0.25} = \frac{64}{3}$$

If the energy of a signal is infinite, then it may be possible to characterize it in terms of its average power. The average power is defined as

$$P = \lim_{N \to \infty} \frac{1}{2N + 1} \sum_{n=-N}^{N} |x(n)|^2$$

For a periodic signal with period N, the average power can be determined as

$$P = \frac{1}{N} \sum_{n=0}^{N-1} |x(n)|^2$$

Signals, periodic or aperiodic, with finite average power are called power signals. Cosine and sine waveforms are typical examples of power signals. The average power of the cosine wave $2\cos(\frac{2\pi}{4}n)$ is

$$P = \frac{1}{4} \sum_{n=0}^{3} |x(n)|^2 = \frac{1}{4}(2^2 + 0^2 + (-2)^2 + 0^2) = 2$$

A signal is an energy signal or a power signal, since the average power of an energy signal is zero while that of a power signal is finite. Signals with infinite average power and infinite energy, such as $x(n) = n$, $0 \leq n < \infty$, are neither power signals nor energy signals. The measures of signal power and energy are indicators of the signal size, since the actual energy or power depends on the load.

2.1.4 Even- and Odd-symmetric Signals

The storage and processing requirements of a signal can be reduced by exploiting its symmetry. A signal $x(n)$ is even-symmetric, if $x(-n) = x(n)$ for all n. The signal is symmetrical about the vertical axis at the origin. The cosine waveform, shown in Figure 2.2(b), is an example of an even-symmetric signal. A signal $x(n)$ is odd-symmetric, if $x(-n) = -x(n)$ for all n. The signal is asymmetrical

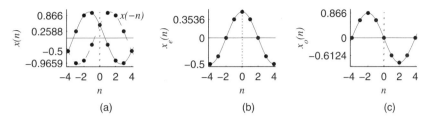

Figure 2.2 (a) The sinusoid $x(n) = \cos(\frac{2\pi}{8}n + \frac{\pi}{3})$ and its time-reversed version $x(-n)$; (b) its even component $x_e(n) = \frac{1}{2}\cos(\frac{2\pi}{8}n)$; (c) its odd component $x_o(n) = -\frac{\sqrt{3}}{2}\sin(\frac{2\pi}{8}n)$

about the vertical axis at the origin. For an odd-symmetric signal, $x(0) = 0$. The sine waveform, shown in Figure 2.2(c), is an example of an odd-symmetric signal.

The sum $(x(n) + y(n))$ of two odd-symmetric signals, $x(n)$ and $y(n)$, is an odd-symmetric signal, since $x(-n) + y(-n) = -x(n) - y(n) = -(x(n) + y(n))$. For example, the sum of two sine signals is an odd-symmetric signal. The sum $(x(n) + y(n))$ of two even-symmetric signals, $x(n)$ and $y(n)$, is an even-symmetric signal, since $x(-n) + y(-n) = (x(n) + y(n))$. For example, the sum of two cosine signals is an even-symmetric signal. The sum $(x(n) + y(n))$ of an odd-symmetric signal $x(n)$ and an even-symmetric signal $y(n)$ is neither even-symmetric nor odd-symmetric, since $x(-n) + y(-n) = -x(n) + y(n) = -(x(n) - y(n))$. For example, the sum of cosine and sine signals with nonzero amplitudes is neither even-symmetric nor odd-symmetric.

Since $x(n)y(n) = (-x(-n))(-y(-n)) = x(-n)y(-n)$, the product of two odd-symmetric or two even-symmetric signals is an even-symmetric signal. The product $z(n) = x(n)y(n)$ of an odd-symmetric signal $y(n)$ and an even-symmetric signal $x(n)$ is an odd-symmetric signal, since $z(-n) = x(-n)y(-n) = x(n)(-y(n)) = -z(n)$.

An arbitrary signal $x(n)$ can always be decomposed in terms of its even-symmetric and odd-symmetric components, $x_e(n)$ and $x_o(n)$, respectively. That is, $x(n) = x_e(n) + x_o(n)$. Replacing n by $-n$, we get $x(-n) = x_e(-n) + x_o(-n) = x_e(n) - x_o(n)$. Solving for $x_e(n)$ and $x_o(n)$, we get

$$x_e(n) = \frac{x(n) + x(-n)}{2} \quad \text{and} \quad x_o(n) = \frac{x(n) - x(-n)}{2}$$

As the sum of an odd-symmetric signal $x_o(n)$, over symmetric limits, is zero,

$$\sum_{n=-N}^{N} x_o(n) = 0 \qquad \sum_{n=-N}^{N} x(n) = \sum_{n=-N}^{N} x_e(n) = x_e(0) + 2\sum_{n=1}^{N} x_e(n)$$

For example, the even-symmetric component of $x(n) = \cos(\frac{2\pi}{8}n + \frac{\pi}{3})$ is

$$x_e(n) = \frac{x(n) + x(-n)}{2} = \frac{\cos\left(\frac{2\pi}{8}n + \frac{\pi}{3}\right) + \cos\left(\frac{2\pi}{8}(-n) + \frac{\pi}{3}\right)}{2}$$

$$= \frac{2\cos\left(\frac{2\pi}{8}n\right)\cos\left(\frac{\pi}{3}\right)}{2} = \frac{\cos\left(\frac{2\pi}{8}n\right)}{2}$$

The odd-symmetric component is

$$x_o(n) = \frac{x(n) - x(-n)}{2} = \frac{\cos\left(\frac{2\pi}{8}n + \frac{\pi}{3}\right) - \cos\left(\frac{2\pi}{8}(-n) + \frac{\pi}{3}\right)}{2}$$

$$= \frac{-2\sin\left(\frac{2\pi}{8}n\right)\sin\left(\frac{\pi}{3}\right)}{2} = -\frac{\sqrt{3}}{2}\sin\left(\frac{2\pi}{8}n\right)$$

The sinusoid $x(n)$ and its time-reversed version $x(-n)$, its even component, and its odd component are shown, respectively, in Figures 2.2(a–c). As the even and odd components of a sinusoid are, respectively, cosine and sine functions of the same frequency as that of the sinusoid, these results can also be obtained by expanding the expression characterizing the sinusoid.

If a continuous signal is sampled with an adequate sampling rate, the samples uniquely correspond to that signal. Assuming that the sampling rate is adequate, in Figure 2.2 (and in other figures in this book), we have shown the corresponding continuous waveform only for clarity. It should be remembered that a discrete signal is represented only by its sample values.

2.1.5 Causal and Noncausal Signals

Most signals, in practice, occur at some finite time instant, usually chosen as $n = 0$, and are considered identically zero before this instant. These signals, with $x(n) = 0$ for $n < 0$, are called causal signals. Signals, with $x(n) \neq 0$ for $n < 0$, are called noncausal signals. Sine and cosine signals, shown in Figures 2.1 and 2.2, are noncausal signals. Typical causal signals are shown in Figure 2.3.

2.1.6 Deterministic and Random Signals

Signals such as $x(n) = \sin(\frac{2\pi}{8}n)$, whose values are known for any value of n, are called deterministic signals. Signals such as those generated by thermal noise in conductors or speech signals, whose future values are not exactly known, are called random signals. Despite the fact that rainfall record is available for several years in the past, the amount of future rainfall at a place cannot be exactly predicted. This type of signal is characterized by a probability model or a statistical model. The study of random

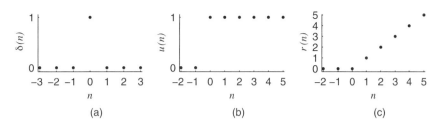

Figure 2.3   (a) The unit-impulse signal, $\delta(n)$; (b) the unit-step signal, $u(n)$; (c) the unit-ramp signal, $r(n)$

signals is important in practice, since all practical signals are random to some extent. However, the analysis of systems is much simpler, mathematically, with deterministic signals. The input–output relationship of a system remains the same whether the input signal is random or deterministic. The time-domain and frequency-domain methods of system analysis are common to both types of signals. The key difference is to find a suitable mathematical model for random signals. In this book, we confine ourselves to the study of deterministic signals.

2.2 Basic Signals

As we have already mentioned, most practical signals have arbitrary amplitude profile. These signals are, for processing convenience, decomposed in terms of mathematically well-defined and simple signals. These simple signals, such as the sinusoid with infinite duration, are not practical signals. However, they can be approximated to a desired accuracy.

2.2.1 Unit-impulse Signal

The unit-impulse signal, shown in Figure 2.3(a), is defined as

$$\delta(n) = \begin{cases} 1 & \text{for } n = 0 \\ 0 & \text{for } n \neq 0 \end{cases}$$

The unit-impulse signal is an all-zero sequence except that it has a value of one when its argument is equal to zero. A time-shifted unit-impulse signal $\delta(n - m)$, with argument $(n - m)$, has its only nonzero value at $n = m$. Therefore, $\sum_{n=-\infty}^{\infty} x(n)\delta(n - m)$ $= x(m)$ is called the sampling or sifting property of the impulse. For example,

$$\sum_{n=-\infty}^{\infty} 2^n \delta(n) = 1 \quad \sum_{n=-2}^{0} 2^n \delta(n - 1) = 0 \quad \sum_{n=-2}^{0} 2^n \delta(-n - 1) = 0.5$$

$$\sum_{n=-2}^{0} 2^n \delta(n + 1) = 0.5 \quad \sum_{n=-\infty}^{\infty} 2^n \delta(n + 2) = 0.25 \quad \sum_{n=-\infty}^{\infty} 2^n \delta(n - 3) = 8$$

In the second summation, the argument $n - 1$ of the impulse never becomes zero within the limits of the summation.

The decomposition of an arbitrary signal in terms of scaled and shifted impulses is a major application of this signal. Consider the product of a signal with a shifted impulse $x(n)\delta(n - m) = x(m)\delta(n - m)$. Summing both sides with respect to m, we get

$$\sum_{m=-\infty}^{\infty} x(n)\delta(n - m) = x(n) \sum_{m=-\infty}^{\infty} \delta(n - m) = x(n) = \sum_{m=-\infty}^{\infty} x(m)\delta(n - m)$$

The general term $x(m)\delta(n - m)$ of the last sum, which is one of the constituent impulses of $x(n)$, is a shifted impulse $\delta(n - m)$ located at $n = m$ with value $x(m)$. The summation operation sums all these impulses to form $x(n)$. Therefore, the signal $x(n)$ is represented by the sum of scaled and shifted impulses with the value of the impulse at any n being $x(n)$. The unit-impulse is the basis function and $x(n)$ is its coefficient. As the value of the sum is nonzero only at $n = m$, the sum is effective only at that point. By varying the value of n, we can sift out all the values of $x(n)$. For example, consider the signal $x(-2) = 2, x(0) = 3, x(2) = -4, x(3) = 1$, and $x(n) = 0$ otherwise. This signal can be expressed, in terms of impulses, as

$$x(n) = 2\delta(n + 2) + 3\delta(n) - 4\delta(n - 2) + \delta(n - 3)$$

With $n = 2$, for instance,

$$x(2) = 2\delta(4) + 3\delta(2) - 4\delta(0) + \delta(-1) = -4$$

2.2.2 Unit-step Signal

The unit-step signal, shown in Figure 2.3(b), is defined as

$$u(n) = \begin{cases} 1 & \text{for } n \geq 0 \\ 0 & \text{for } n < 0 \end{cases}$$

The unit-step signal is an all-one sequence for positive values of its argument and is an all-zero sequence for negative values of its argument. The causal form of a signal $x(n)$, $x(n)$ is zero for $n < 0$, is obtained by multiplying it by the unit-step signal as $x(n)u(n)$. For example, $\sin(\frac{2\pi}{6}n)$ has nonzero values in the range $-\infty < n < \infty$, whereas the values of $\sin(\frac{2\pi}{6}n)u(n)$ are zero for $n < 0$ and $\sin(\frac{2\pi}{6}n)$ for $n \geq 0$. A shifted unit-step signal, for example $u(n - 1)$, is $u(n)$ shifted by one sample interval to the right (the first nonzero value occurs at $n = 1$). Using scaled and shifted unit-step signals, any signal, described differently over different intervals, can be specified, for easier mathematical analysis, by a single expression, valid for all n. For example, a pulse signal with its only nonzero values defined as $x(-1) = 2, x(0) = 2, x(1) = -3$, and $x(2) = -3$ can be expressed as $x(n) = 2u(n+1) - 5u(n-1) + 3u(n-3)$.

2.2.3 Unit-ramp Signal

Another signal that is often used in the analysis of systems is the unit-ramp signal, shown in Figure 2.3(c). It is defined as

$$r(n) = \begin{cases} n & \text{for } n \geq 0 \\ 0 & \text{for } n < 0 \end{cases}$$

The unit-ramp signal increases linearly for positive values of its argument and is an all-zero sequence for negative values of its argument.

The three signals, the unit-impulse, the unit-step, and the unit-ramp, are closely related. The unit-impulse signal $\delta(n)$ is equal to $u(n) - u(n-1)$. The unit-step signal $u(n)$ is equal to $\sum_{l=-\infty}^{n} \delta(l)$. The shifted unit-step signal $u(n-1)$ is equal to $r(n) - r(n-1)$. The unit-ramp signal $r(n)$ is equal to $\sum_{l=-\infty}^{n} u(l-1)$.

2.2.4 Sinusoids and Exponentials

The sinusoidal waveform or sinusoid is the well-known trigonometric sine and cosine functions, with arbitrary shift along the horizontal axis. The sinusoidal waveforms are oscillatory, with peaks occurring at equal distance from the horizontal axis. The waveforms have two zero-crossings in each cycle. As the sinusoidal waveforms of a particular frequency and amplitude have the same shape with the peaks occurring at different instants, we have to define a reference position to distinguish the innumerable number of different sinusoids. Let the occurrence of the positive peak at the origin be the reference position. Then, as the cosine wave has its positive peak at that point, it becomes the reference waveform and is characterized by a phase of zero radians. The other sinusoidal waveforms can be obtained by shifting the cosine waveform to the right or left. A shift to the right is considered as negative and a shift to the left is positive. The phase of the sine wave is $-\pi/2$ radians, as we get the sine wave by shifting a cosine wave to the right by $\pi/2$ radians. The other sinusoidal waveforms have arbitrary phases. The sine and cosine waves are important special cases of sinusoidal waveforms.

2.2.4.1 The Polar Form of Sinusoids

The polar form specifies a sinusoid, in terms of its amplitude and phase, as

$$x(n) = A\cos(\omega n + \theta) \qquad n = -\infty, \ldots, -1, 0, 1, \ldots, \infty$$

where A, ω, and θ are, respectively, the amplitude, the angular frequency, and the phase. The amplitude A is the distance of either peak of the waveform from the horizontal axis ($A = 1$ for the waves shown in Figure 2.1). A discrete sinusoid has to complete an integral number of cycles (say k, where $k > 0$ is an integer) over an integral number of sample points, called its period (denoted by N, where $N > 0$ is an integer), if it is periodic. Then, as

$$\cos(\omega(n + N) + \theta) = \cos(\omega n + \omega N + \theta) = \cos(\omega n + \theta) = \cos(\omega n + \theta + 2k\pi)$$

$N = 2k\pi/\omega$. Note that k is the smallest integer that will make $2k\pi/\omega$ an integer. The cyclic frequency, denoted by f, of a sinusoid is the number of cycles per sample and is equal to the number of cycles the sinusoid makes in a period divided by the period, $f = k/N = \omega/2\pi$ cycles per sample. Therefore, the cyclic frequency of a discrete periodic sinusoid is a rational number. The angular frequency (the number of radians per sample) of a sinusoid is 2π times its cyclic frequency, that is $\omega = 2\pi f$ radians per sample.

The angular frequency of the sinusoids, shown in Figure 2.1(b), is $\omega = \pi/8$ radians per sample. The period of the discrete sinusoids is $N = 2k\pi/\omega = 16$ samples, with $k = 1$. The cyclic frequency of the sinusoid $\sin((2\sqrt{2}\pi/16)n + \pi/3)$ is $\sqrt{2}/16$. As it is an irrational number, the sinusoid is not periodic. The cyclic frequency of the sinusoids in Figure 2.1(b) is $f = k/N = 1/16$ cycles per sample. The phase of the sinusoid $\cos((2\pi/16)n + \pi/3)$ in Figure 2.1(b) is $\theta = \pi/3$ radians. As it repeats a pattern over its period, the sinusoid remains the same by a shift of an integral number of its period. A phase-shifted sine wave can be expressed in terms of a phase-shifted cosine wave as $A \sin(\omega n + \theta) = A \cos(\omega n + (\theta - \frac{\pi}{2}))$. The phase of the sinusoid

$$\sin\left(\frac{2\pi}{16}n + \frac{\pi}{3}\right) = \cos\left(\frac{2\pi}{16}n + \left(\frac{\pi}{3} - \frac{\pi}{2}\right)\right) = \cos\left(\frac{2\pi}{16}n - \frac{\pi}{6}\right)$$

in Figure 2.1(b) is $-\pi/6$ radians. A phase-shifted cosine wave can be expressed in terms of a phase-shifted sine wave as $A \cos(\omega n + \theta) = A \sin(\omega n + (\theta + \pi/2))$.

2.2.4.2 The Rectangular Form of Sinusoids

An arbitrary sinusoid is neither even- nor odd-symmetric. The even and odd components of a sinusoid are, respectively, cosine and sine waveforms. That is, a sinusoid is a linear combination of cosine and sine waveforms of the same frequency as that of the sinusoid. Expression of a sinusoid in terms of its cosine and sine components is called its rectangular form and is given as

$$A \cos(\omega n + \theta) = A \cos(\theta) \cos(\omega n) - A \sin(\theta) \sin(\omega n) = C \cos(\omega n) + D \sin(\omega n)$$

where $C = A \cos \theta$ and $D = -A \sin \theta$. The inverse relation is $A = \sqrt{C^2 + D^2}$ and $\theta = \cos^{-1}(C/A) = \sin^{-1}(-D/A)$. For example,

$$2 \cos\left(\frac{2\pi}{16}n + \frac{\pi}{3}\right) = \cos\left(\frac{2\pi}{16}n\right) - \sqrt{3} \sin\left(\frac{2\pi}{16}n\right)$$

$$\frac{3}{\sqrt{2}} \cos\left(\frac{2\pi}{16}n\right) + \frac{3}{\sqrt{2}} \sin\left(\frac{2\pi}{16}n\right) = 3 \cos\left(\frac{2\pi}{16}n - \frac{\pi}{4}\right)$$

2.2.4.3 The Sum of Sinusoids of the Same Frequency

The sum of sinusoids of arbitrary amplitudes and phases, but with the same frequency, is also a sinusoid of the same frequency. Let

$$x_1(n) = A_1 \cos(\omega n + \theta_1) \qquad \text{and} \qquad x_2(n) = A_2 \cos(\omega n + \theta_2)$$

Then,

$$\begin{aligned}
x(n) = x_1(n) + x_2(n) &= A_1 \cos(\omega n + \theta_1) + A_2 \cos(\omega n + \theta_2) \\
&= \cos(\omega n)(A_1 \cos(\theta_1) + A_2 \cos(\theta_2)) - \sin(\omega n)(A_1 \sin(\theta_1) + A_2 \sin(\theta_2)) \\
&= A \cos(\omega n + \theta) = \cos(\omega n)(A \cos(\theta)) - \sin(\omega n)(A \sin(\theta))
\end{aligned}$$

Solving for A and θ, we get

$$A = \sqrt{A_1^2 + A_2^2 + 2A_1 A_2 \cos(\theta_1 - \theta_2)}$$

$$\theta = \tan^{-1} \frac{A_1 \sin(\theta_1) + A_2 \sin(\theta_2)}{A_1 \cos(\theta_1) + A_2 \cos(\theta_2)}$$

Any number of sinusoids can be combined into a single sinusoid by repeatedly using the formulas. Note that the formula for the rectangular form of the sinusoid is a special case of the sum of two sinusoids, one sinusoid being the cosine and the other being the sine.

Example 2.1. Determine the sum of the two sinusoids $x_1(n) = 2 \cos(\frac{2\pi}{8}n + \frac{\pi}{3})$ and $x_2(n) = -3 \cos(\frac{2\pi}{8}n - \frac{\pi}{6})$.

Solution

As

$$x_2(n) = -3\cos\left(\frac{2\pi}{8}n - \frac{\pi}{6}\right) = 3\cos\left(\frac{2\pi}{8}n - \frac{\pi}{6} + \pi\right) = 3\cos\left(\frac{2\pi}{8}n + \frac{5\pi}{6}\right)$$

$$A_1 = 2 \quad A_2 = 3 \quad \theta_1 = \frac{\pi}{3} \text{ and } \theta_2 = \frac{5\pi}{6}$$

Substituting the numerical values in the equations, we get

$$A = \sqrt{2^2 + 3^2 + 2(2)(3)\cos\left(\frac{\pi}{3} - \frac{5\pi}{6}\right)} = \sqrt{13} = 3.6056$$

$$\theta = \tan^{-1}\frac{2\sin(\frac{\pi}{3}) + 3\sin(\frac{5\pi}{6})}{2\cos(\frac{\pi}{3}) + 3\cos(\frac{5\pi}{6})} = 2.03 \text{ radians}$$

The waveforms of the two sinusoids and their sum, $x(n) = 3.6056\cos(\frac{2\pi}{8}n + 2.03)$, are shown, respectively, in Figures 2.4(a), (b), and (c). □

2.2.4.4 Exponentials

A constant a raised to the power of a variable n, $x(n) = a^n$ is the exponential function. We are more familiar with the exponential of the form e^{-2t} with base e and this form is used in the analysis of continuous signals and systems. The exponential e^{sn} is the same as a^n, where $s = \log_e a$ and $a = e^s$. For example, $e^{-0.2231n} = (0.8)^n$ is a decaying discrete exponential. As both the forms are used in the analysis of discrete signals and systems, it is necessary to get used to both of them.

With base e, the most general form of the continuous exponential is Pe^{st}, where P or s or both may be complex-valued. Let $s = \sigma + j\omega$. Then, $e^{st} = e^{(\sigma+j\omega)t} = e^{\sigma t}e^{j\omega t}$. Exponential $e^{j\omega t} = \cos(\omega t) + j\sin(\omega t)$ is a constant-amplitude oscillating signal with the frequency of oscillation in the range $0 \le \omega \le \infty$. When the real part of s is positive $(\sigma > 0)$, e^{st} is a growing exponential. When $\sigma < 0$, e^{st} is a decaying exponential. When $\sigma = 0$, e^{st} oscillates with constant amplitude. When $s = 0$, e^{st} is a constant signal.

Figure 2.4 (a) The sinusoid $x_1(n) = 2\cos(\frac{2\pi}{8}n + \frac{\pi}{3})$; (b) the sinusoid $x_2(n) = 3\cos(\frac{2\pi}{8}n + \frac{5\pi}{6})$; (c) the sum of $x_1(n)$ and $x_2(n)$, $x(n) = 3.6056\cos(\frac{2\pi}{8}n + 2.03)$

With base a, the most general form of the discrete exponential is Pa^n, where P or a or both may be complex-valued. Let $a = re^{j\omega}$. Then, $a^n = r^n e^{j\omega n}$. Exponential $e^{j\omega n} = \cos(\omega n) + j\sin(\omega n)$ is a constant-amplitude oscillating signal with the frequency of oscillation in the range $0 \leq \omega \leq \pi$, since $e^{\pm j\omega n} = e^{j(2\pi \pm \omega)n} = e^{j(4\pi \pm \omega)n} = \cdots$. When $|a| = r > 1$, a^n is a growing exponential. When $|a| = r < 1$, a^n is a decaying exponential. When $|a| = r = 1$, a^n is a constant-amplitude signal.

2.2.4.5 The Complex Sinusoids

In practice, the real sinusoid $A\cos(\omega n + \theta)$ is most often used and is easy to visualize. At a specific frequency, a sinusoid is characterized by two real-valued quantities, the amplitude and the phase. These two values can be combined into a complex constant that is associated with a complex sinusoid. Then, we get a single waveform with a single coefficient, although both of them are complex. Because of its compact form and ease of manipulation, the complex sinusoid is used in almost all theoretical analysis. The complex sinusoid is given as

$$x(n) = Ae^{j(\omega n + \theta)} = Ae^{j\theta}e^{j\omega n} \qquad n = -\infty, \ldots, -1, 0, 1, \ldots, \infty$$

The term $e^{j\omega n}$ is the complex sinusoid with unit magnitude and zero phase. Its complex (amplitude) coefficient is $Ae^{j\theta}$. The amplitude and phase of the sinusoid are represented by the single complex number $Ae^{j\theta}$. The complex sinusoid is a functionally equivalent mathematical representation of a real sinusoid. By adding its complex conjugate, $Ae^{-j(\omega n + \theta)}$, and dividing by two, due to Euler's identity, we get

$$x(n) = \frac{A}{2}\left(e^{j(\omega n + \theta)} + e^{-j(\omega n + \theta)}\right) = A\cos(\omega n + \theta)$$

The use of two complex sinusoids to represent a single real sinusoid requires four real quantities instead of two. This redundancy in terms of storage and operations can be avoided. Figure 2.1(b) shows the complex sinusoid $e^{j(\frac{2\pi}{16}n)}$ with complex coefficient $1e^{j\frac{\pi}{3}}$.

2.2.4.6 Exponentially Varying Amplitude Sinusoids

An exponentially varying amplitude sinusoid, $Ar^n\cos(\omega n + \theta)$, is obtained by multiplying a sinusoidal sequence, $A\cos(\omega n + \theta)$, by a real exponential sequence, r^n. The more familiar constant amplitude sinusoid results when the base of the real exponential r is equal to one. If ω is equal to zero, then we get real exponential sequences. Sinusoid, $x(n) = (0.9)^n \cos(\frac{2\pi}{8}n)$, with exponentially decreasing amplitude is shown in Figure 2.5(a). The amplitude of the sinusoid $\cos(\frac{2\pi}{8}n)$ is constrained by the exponential $(0.9)^n$. When the value of the cosine function is equal to one, the waveform

Figure 2.5 (a) Exponentially decreasing amplitude sinusoid, $x(n) = (0.9)^n \cos(\frac{2\pi}{8}n)$; (b) exponentially increasing amplitude sinusoid, $x(n) = (1.1)^n \cos(\frac{2\pi}{8}n)$

reduces to $(0.9)^n$. Therefore, the graph of the function $(0.9)^n$ is the envelope of the positive peaks of the waveform, as shown in Figure 2.5(a). Similarly, the graph of the function $-(0.9)^n$ is the envelope of the negative peaks of the waveform. Sinusoid, $x(n) = (1.1)^n \cos(\frac{2\pi}{8}n)$, with exponentially increasing amplitude is shown in Figure 2.5(b).

The complex exponential representation of an exponentially varying amplitude sinusoid is given as

$$x(n) = \frac{A}{2}r^n \left(e^{j(\omega n + \theta)} + e^{-j(\omega n + \theta)}\right) = Ar^n \cos(\omega n + \theta)$$

2.2.4.7 The Sampling Theorem and the Aliasing Effect

As we have already mentioned, most practical signals are continuous signals. However, digital signal processing is so advantageous that we prefer to convert the continuous signals into digital form and then process it. This process involves sampling the signal in time and in amplitude. The sampling in time involves observing the signal only at discrete instants of time. By sampling a signal, we are reducing the number of samples from infinite (of the continuous signal over any finite duration) to finite (of the corresponding discrete signal over the same duration). This reduction in the number of samples restricts the ability to represent rapid time variations of a signal and, consequently, reduces the effective frequency range of discrete signals. Note that high-frequency components of a signal provide its rapid variations. As practical signals have negligible spectral values beyond some finite frequency range, the representation of a continuous signal by a finite set of samples is possible, satisfying a required accuracy. Therefore, we should be able to determine the sampling interval required for a specific signal.

The sampling theorem states that a continuous signal $x(t)$ can be uniquely determined from its sampled version $x(n)$ if the sampling interval T_s is less than $1/2f_m$, where f_m is the cyclic frequency of the highest-frequency component of $x(t)$. The implies that there are more than two samples per cycle of the highest-frequency component. That is, a sinusoid, which completes f cycles, has a distinct set of $2f + 1$ sample values. A cosine wave, however, can be represented with $2f$

samples. For example, the cyclic frequency of the sinusoid $x(t) = \cos(3(2\pi)t - \frac{\pi}{3})$ is $f = \frac{3(2\pi)}{2\pi} = 3$ Hz and, therefore, $T_s < \frac{1}{2(3)} = \frac{1}{6}$ s and the minimum sampling frequency is $f_s = \frac{1}{T_s} = 2f + 1 = 6 + 1 = 7$ samples per second. In practice, due to nonideal response of physical devices, the sampling frequency used is typically more than twice the theoretical minimum.

Given a sampling interval T_s the cyclic frequency f_m of the highest-frequency component of $x(t)$, for the unambiguous representation of its sampled version, must be less than $1/2T_s$. The corresponding angular frequency ω_m is equal to $2\pi f_m < \pi/T_s$ radians per second. Therefore, the frequency range of the frequency components of the signal $x(t)$, for the unambiguous representation of its sampled version, must be $0 \leq \omega < \pi/T_s$.

To find out why the frequency range is limited, due to sampling of a signal, consider the sinusoid $x(t) = \cos(\omega_0 t + \theta)$ with $0 \leq \omega_0 < \pi/T_s$. The sampled version of $x(t)$ is $x(n) = \cos(\omega_0 n T_s + \theta)$. Now, consider the sinusoid $y(t) = \cos((\omega_0 + 2\pi m/T_s)t + \theta)$, where m is any positive integer. The sampled version of $y(t)$ is identical with that of $x(t)$, as

$$y(n) = \cos\left(\left(\omega_0 + \frac{2\pi m}{T_s}\right)n T_s + \theta\right)$$
$$= \cos(\omega_0 n T_s + 2\pi n m + \theta)$$
$$= \cos(\omega_0 n T_s + \theta) = x(n)$$

Therefore, the effective frequency range is limited to $2\pi/T_s$.
Now, consider the sinusoid

$$z(t) = \cos\left(\left(\frac{2\pi m}{T_s} - \omega_0\right)t - \theta\right)$$

where m is any positive integer. The sampled version of $z(t)$ is identical with that of $x(t)$, as

$$z(n) = \cos\left(\left(\frac{2\pi m}{T_s} - \omega_0\right)n T_s - \theta\right)$$
$$= \cos(2\pi n m - \omega_0 n T_s - \theta)$$
$$= \cos(\omega_0 n T_s + \theta) = x(n)$$

We conclude that it is impossible to differentiate between the sampled versions of two continuous sinusoids with the sum or difference of their angular frequencies equal to an integral multiple of $2\pi/T_s$. Therefore, the effective frequency range is further limited to π/T_s, as given by the sampling theorem. The frequency π/T_s is called the folding frequency, since higher frequencies are folded back and forth into the frequency range from zero to π/T_s.

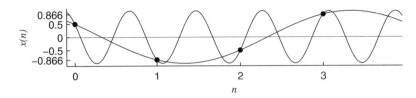

Figure 2.6 The continuous sinusoids $x(t) = \cos(2\pi t + \frac{\pi}{3})$ and $x(t) = \cos(5(2\pi)t + \frac{\pi}{3})$, and their sampled versions, with the sampling interval $T_s = \frac{1}{4}$ seconds, $x(n) = \cos(\frac{2\pi}{4}n + \frac{\pi}{3})$ and $x(n) = \cos(5\frac{2\pi}{4}n + \frac{\pi}{3}) = \cos(\frac{2\pi}{4}n + \frac{\pi}{3})$

Consider the continuous sinusoids $x(t) = \cos(2\pi t + \frac{\pi}{3})$ and $x(t) = \cos(5(2\pi)t + \frac{\pi}{3})$, and their sampled versions, obtained from the corresponding continuous sinusoids by replacing t by $nT_s = n\frac{1}{4}$ with the sampling interval $T_s = \frac{1}{4}$ s, $x(n) = \cos(\frac{2\pi}{4}n + \frac{\pi}{3})$ and $x(n) = \cos(5\frac{2\pi}{4}n + \frac{\pi}{3})$, shown in Figure 2.6. We can easily distinguish one continuous sinusoid from the other, as they are clearly different. However, the set of sample values, shown by dots, of the two discrete sinusoids are the same and it is impossible to differentiate them. The sample values of both the sinusoids are the same, since

$$\cos\left(5\frac{2\pi}{4}n + \frac{\pi}{3}\right) = \cos\left((4+1)\frac{2\pi}{4}n + \frac{\pi}{3}\right) = \cos\left(\frac{2\pi}{4}n + \frac{\pi}{3}\right)$$

With the sampling interval $T_s = \frac{1}{4}$ s, the effective frequency range is limited to $\pi/T_s = 4\pi$. Therefore, the continuous sinusoid $\cos(5(2\pi)t + \frac{\pi}{3})$, with its angular frequency 10π greater than the folding frequency 4π, appears as or impersonates a lower-frequency discrete sinusoid. The impersonation of high-frequency continuous sinusoids by low-frequency discrete sinusoids, due to an insufficient number of samples in a cycle (the sampling interval is not short enough), is called the aliasing effect.

As only scaling of the frequency axis is required for any other sampling interval, most of the analysis of discrete signals is carried out assuming that the sampling interval is 1 s. The effective frequency range becomes $0-\pi$ and it is referred to as half the fundamental range. Low frequencies are those near zero and high frequencies are those near π. The range, 0 to 2π or $-\pi$ to π, is called the fundamental range of frequencies.

2.3 Signal Operations

In addition to the arithmetic operations, time shifting, time reversal, and time scaling operations are also commonly used in the analysis of discrete signals. The three operations described in this section are with respect to the independent variable, n.

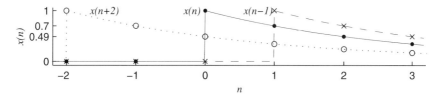

Figure 2.7 The exponential signal $x(n) = (0.7)^n u(n)$, the right-shifted signal, $x(n-1) = (0.7)^{(n-1)} u(n-1)$, and the left-shifted signal, $x(n+2) = (0.7)^{(n+2)} u(n+2)$

2.3.1 Time Shifting

By replacing n by $n + N$, where N is an integer, we get the shifted version, $x(n + N)$, of the signal $x(n)$. The value of $x(n)$ at $n = n_0$ occurs at $n = n_0 - N$ in $x(n + N)$. The exponential signal $x(n) = (0.7)^n u(n)$ is shown in Figure 2.7 by dots. The signal $x(n - 1)$, shown in Figure 2.7 by crosses, is the signal $x(n)$ shifted by one sample interval to the right (delayed by one sample interval, as the sample values of $x(n)$ occur one sample interval later). For example, the first nonzero sample value occurs at $n = 1$ as $(0.7)^{1-1} u(1 - 1) = (0.7)^0 u(0) = 1$. That is, the value of the function $x(n)$ at n_0 occurs in the shifted signal one sample interval later at $n_0 + 1$. The signal $x(n + 2)$, shown in Figure 2.7 by unfilled circles, is the signal $x(n)$ shifted by two sample intervals to the left (advanced by two sample intervals, as the sample values of $x(n)$ occur two sample intervals earlier). For example, the first nonzero sample value occurs at $n = -2$ as $(0.7)^{-2+2} u(-2 + 2) = (0.7)^0 u(0) = 1$. That is, the value of the function $x(n)$ at n_0 occurs in the shifted signal two sample intervals earlier at $n_0 - 2$.

2.3.2 Time Reversal

Forming the mirror image of a signal about the vertical axis at the origin is the time reversal or folding operation. This is achieved by replacing the independent variable n in $x(n)$ by $-n$ and we get $x(-n)$. The value of $x(n)$ at $n = n_0$ occurs at $n = -n_0$ in $x(-n)$. The exponential signal $x(n) = (0.7)^n u(n)$ is shown in Figure 2.8 by dots. The folded signal $x(-n)$ is shown in Figure 2.8 by crosses. Consider the folded and shifted signal $x(-n + 2) = x(-(n - 2))$, shown in Figure 2.8 by unfilled circles. This

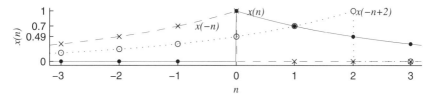

Figure 2.8 The exponential signal $x(n) = (0.7)^n u(n)$, the folded signal, $x(-n) = (0.7)^{-n} u(-n)$, and the shifted and folded signal, $x(-n + 2) = (0.7)^{(-n+2)} u(-n + 2)$

signal can be formed by first folding $x(n)$ to get $x(-n)$ and then shifting it to the right by two sample intervals (n is replaced by $n-2$). This signal can also be formed by first shifting $x(n)$ to the left by two sample intervals to get $x(n+2)$ and then folding it about the vertical axis (n is replaced by $-n$). That is, the value of the function $x(n)$ at n_0 occurs in the reversed and shifted signal at $-(n_0-2)$.

2.3.3 Time Scaling

Replacing the independent variable n in $x(n)$ by an or n/a results in the time-scaled signal $x(an)$ (time-compressed version of $x(n)$) or $x(n/a)$ (time expanded version of $x(n)$), with $a \neq 0$ being an integer. The value of $x(n)$ at $n = n_0$ occurs at $n = n_0/a$ (n_0 being an integral multiple of a) in $x(an)$ and at $n = an_0$ in $x(n/a)$. Consider the signal $x(n) = (0.8)^n u(n)$, shown in Figure 2.9 by dots. The time-compressed version with $a = 2$, $y(n) = x(2n)$, is shown in Figure 2.9 by crosses. The values of the signal $y(n) = x(2n)$ are the even-indexed values of $x(n)$. That is, $y(0) = x(0)$, $y(1) = x(2)$, $y(2) = x(4)$, and so on. The odd-indexed values of $x(n)$ are lost in the time compression operation. In general, $x(an)$ is composed only of every ath sample of $x(n)$.

The time-expanded version with $a = 2$, $y(n) = x(n/2)$, is shown in Figure 2.9 by unfilled circles. The values of the time-expanded signal are defined from that of $x(n)$ only for the even-indexed values of $y(n)$. That is, $y(0) = x(0)$, $y(2) = x(1)$, $y(4) = x(2)$, and so on. Odd-indexed values of $y(n)$ are assigned the value zero. In general, $y(n) = x(n/a)$ is defined only for $n = 0, \pm a, \pm 2a, \pm 3a, \ldots$, and the rest of the values of $y(n)$ are undefined. Interpolation by assigning the value zero is often used in practice. Of course, the undefined values can also be defined using a suitable interpolation formula.

In general, the three operations described on a signal $x(n)$ can be expressed as $y(n) = x(an - b)$ or $y(n) = x[(n/a) - b]$. The signal $y(n)$ can be generated by replacing n by $(an - b)$ or $[(n/a) - b]$ in $x(n)$. However, it is instructive to consider it as the result of a sequence of two steps: (i) first shifting the signal $x(n)$ by b to get $x(n - b)$, and then (ii) time scaling (replace n by an or n/a) the shifted signal by a to get $y(n) = x(an - b)$ or $y(n) = x[(n/a) - b]$. Note that the time reversal operation is a special case of the time scaling operation with $a = -1$.

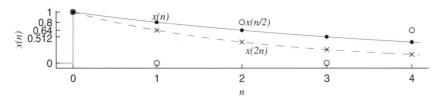

Figure 2.9   The exponential $x(n) = (0.8)^n u(n)$, $x(2n)$, and $x(\frac{n}{2})$

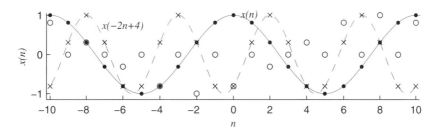

Figure 2.10 The sinusoid, $x(n) = \cos(\frac{\pi}{5}n)$, $x(-2n + 4)$, and $x(-\frac{n}{2} + 4)$

Let $x(n) = \cos(\frac{\pi}{5}n)$, shown in Figure 2.10 by dots. It is required to find $x(-2n + 4)$. The shifted signal is $x(n + 4) = \cos(\frac{\pi}{5}(n + 4))$. Now scaling this signal by -2 yields the signal $x(-2n + 4) = \cos(\frac{\pi}{5}(-2n + 4))$, shown in Figure 2.10 by crosses. The value of the function $x(n)$ at an even n_0 occurs in the scaled and shifted signal at $-(n_0 - 4)/2$.

Let us find $x(-\frac{n}{2} + 4)$. Scaling the shifted signal by $-\frac{1}{2}$ yields the signal $x(-\frac{n}{2} + 4)$ $= \cos(\frac{\pi}{5}(-\frac{n}{2} + 4))$, shown in Figure 2.10 by unfilled circles. The value of the function $x(n)$ at n_0 occurs in the scaled and shifted signal at $-2(n_0 - 4)$.

2.4 Summary

- In this chapter, signal classifications, basic discrete signals, and signal operations have been presented.
- Storage and processing requirements of a signal depend on its type.
- As practical signals have arbitrary amplitude profile, these signals are usually decomposed and processed in terms of basic signals, such as the sinusoid or the impulse.
- In addition to arithmetic operations, time shifting, time reversal, and time scaling operations are also commonly used in the analysis of discrete signals.

Further Reading

1. Sundararajan, D., *Discrete Fourier Transform, Theory, Algorithms, and Applications*, World Scientific, Singapore, 2001.
2. Sundararajan, D., *Digital Signal Processing, Theory and Practice*, World Scientific, Singapore, 2003.
3. Lathi, B. P., *Linear Systems and Signals*, Oxford University Press, New York, 2004.

Exercises

2.1   Is $x(n)$ an energy signal, a power signal, or neither? If it is an energy signal, find its energy. If it is a power signal, find its average power.

2.1.1 $x(0) = 2$, $x(-1) = 2$, $x(-2) = -2$, $x(-3) = -2$, and $x(n) = 0$ otherwise.

*2.1.2 $x(n) = 2(0.8)^n u(n)$.

2.1.3 $x(n) = 2^n$.

2.1.4 $x(n) = Ce^{j(\frac{6\pi n}{8})}$.

2.1.5 $x(n) = 3\cos\left(\frac{\pi n}{2} + \frac{\pi}{4}\right)$.

2.1.6 $x(n) = u(n)$.

2.1.7 $x(n) = 2$.

2.1.8 $x(n) = \frac{2}{n} u(n-1)$.

2.1.9 $x(n) = n$.

2.2   Is $x(n)$ even-symmetric, odd-symmetric, or neither? List the values of $x(n)$ for $n = -3, -2, -1, 0, 1, 2, 3$.

2.2.1 $x(n) = 2\sin\left(\frac{\pi}{5}n - \frac{\pi}{3}\right)$.

2.2.2 $x(n) = \sin\left(\frac{\pi}{5}n\right)$.

2.2.3 $x(n) = 2\cos\left(\frac{\pi}{5}n\right)$.

2.2.4 $x(n) = 3$.

2.2.5 $x(n) = n$.

2.2.6 $x(n) = \frac{2\sin(\frac{\pi}{3}n)}{n}$.

2.2.7 $x(n) = \frac{2\sin^2(\frac{\pi}{3}n)}{n}$.

2.2.8 $x(0) = 0$ and $x(n) = \frac{(-1)^n}{n}$ otherwise.

2.2.9 $x(n) = 3\delta(n)$

2.3   Find the even and odd components of the signal. List the values of the signal and its components for $n = -3, -2, -1, 0, 1, 2, 3$. Verify that the values of the components add up to the values of the signal. Verify that the sum of the values of the even component and that of the signal are equal.

2.3.1 $x(0) = 1$, $x(1) = 1$, $x(2) = -1$, $x(3) = -1$, and $x(n) = 0$ otherwise.

2.3.2 $x(n) = 3\cos\left(\frac{\pi}{5}n + \frac{\pi}{6}\right)$

*2.3.3 $x(n) = (0.4)^n u(n)$

2.3.4 $x(n) = u(n+1)$

2.3.5 $x(n) = e^{-j(\frac{\pi}{3}n)}$

2.3.6 $x(n) = n\,u(n)$

2.4   Evaluate the summation.

2.4.1 $\sum_{n=0}^{\infty} \delta(n)(0.5)^n u(n)$.

*2.4.2 $\sum_{n=0}^{\infty} \delta(n+1)(0.5)^n$.

2.4.3 $\sum_{n=0}^{\infty} \delta(n-2)(0.5)^n u(n)$.

2.4.4 $\sum_{n=-\infty}^{\infty} \delta(n+1)(0.5)^n$.

2.5   Express the signal in terms of scaled and shifted impulses.

2.5.1 $x(0) = 2$, $x(1) = 3$, $x(2) = -1$, $x(3) = -4$, and $x(n) = 0$ otherwise.

2.5.2 $x(0) = 5$, $x(-1) = 3$, $x(2) = -7$, $x(-3) = -4$, and $x(n) = 0$ otherwise.

2.6   If the waveform is periodic, what is its period?

2.6.1 $x(n) = 4\cos(0.7\pi n)$.

2.6.2 $x(n) = 2\cos(\sqrt{2}n)$.

2.6.3 $x(n) = 43 + 2\cos\left(\frac{2\pi}{7}n\right)$.

2.6.4 $x(n) = 2\cos\left(\frac{\pi}{5\sqrt{2}}n\right)$.

*2.6.5 $x(n) = 4\cos\left(\frac{4\pi}{9}n\right)$.

2.7   Find the rectangular form of the sinusoid. List the sample values of one cycle, starting from $n = 0$, of the sinusoid.

2.7.1 $x(n) = -2\sin\left(\frac{\pi}{6}n - \frac{\pi}{3}\right)$.

2.7.2 $x(n) = -2\cos\left(\frac{\pi}{6}n - \frac{\pi}{4}\right)$.

2.7.3 $x(n) = \cos\left(\frac{\pi}{6}n\right)$.

2.7.4 $x(n) = 3\sin\left(\frac{\pi}{6}n + \frac{\pi}{3}\right)$.

2.7.5 $x(n) = -\sin\left(\frac{\pi}{6}n\right)$.

*2.7.6 $x(n) = 4\cos\left(\frac{\pi}{6}n - \frac{\pi}{6}\right)$.

2.8   Find the polar form of the sinusoid. List the sample values of one cycle, starting from $n = 0$, of the sinusoid.

2.8.1 $x(n) = -2\sin\left(\frac{\pi}{6}n\right)$.

2.8.2 $x(n) = -2\cos\left(\frac{\pi}{6}n\right) - 2\sin\left(\frac{\pi}{6}n\right)$.

*2.8.3 $x(n) = 3\cos\left(\frac{\pi}{6}n\right) + \sqrt{3}\sin\left(\frac{\pi}{6}n\right)$.

2.8.4 $x(n) = -3\cos\left(\frac{\pi}{6}n\right)$.

2.8.5 $x(n) = \sqrt{3}\cos\left(\frac{\pi}{6}n\right) - \sin\left(\frac{\pi}{6}n\right)$.

2.9   Given $x_1(n) = A_1 e^{j(\omega n + \theta_1)}$ and $x_2(n) = A_2 e^{j(\omega n + \theta_2)}$, derive expressions for A and θ of the complex sinusoid $x(n) = x_1(n) + x_2(n) = A e^{j(\omega n + \theta)}$ in terms of those of $x_1(n)$ and $x_2(n)$.

2.10 Given the complex sinusoids $x_1(n) = A_1 e^{j(\omega n + \theta_1)}$ and $x_2(n) = A_2 e^{j(\omega n + \theta_2)}$, find the complex sinusoid $x(n) = x_1(n) + x_2(n) = A e^{j(\omega n + \theta)}$, using the formulas derived in Exercise 2.9. Find the sample values of one cycle, starting from $n = 0$, of the complex sinusoids $x_1(n)$ and $x_2(n)$ and verify that the sample values of $x_1(n) + x_2(n)$ are the same as those of $x(n)$.

2.10.1 $x_1(n) = -2e^{j\left(\frac{\pi}{3}n + \frac{\pi}{3}\right)}$, $x_2(n) = 3e^{j\left(\frac{\pi}{3}n - \frac{\pi}{6}\right)}$.

2.10.2 $x_1(n) = 3e^{-j\left(\frac{\pi}{3}n + \frac{\pi}{3}\right)}$, $x_2(n) = 2e^{-j\left(\frac{\pi}{3}n - \frac{\pi}{3}\right)}$.

2.10.3 $x_1(n) = 2e^{j\left(\frac{\pi}{3}n\right)}$, $x_2(n) = 3e^{j\left(\frac{\pi}{3}n\right)}$.

2.10.4 $x_1(n) = e^{j\left(\frac{\pi}{3}n - \frac{\pi}{2}\right)}$, $x_2(n) = e^{j\left(\frac{\pi}{3}n\right)}$.

*2.10.5 $x_1(n) = 2e^{j\left(\frac{\pi}{3}n + \frac{\pi}{6}\right)}$, $x_2(n) = 4e^{j\left(\frac{\pi}{3}n + \frac{\pi}{4}\right)}$.

2.11 Find the corresponding exponential of the form a^n. List the values of the exponential for $n = 0, 1, 2, 3, 4, 5$.

2.11.1 $x(n) = e^{0.6931n}$.

2.11.2 $x(n) = e^n$.

*2.11.3 $x(n) = e^{-0.6931n}$.

2.11.4 $x(n) = e^{-0.3567n}$.

2.12 Give the sample values of the exponentially varying amplitude sinusoid for $n = -2, -1, 0, 1, 2, 3, 4$.

2.12.1 $x(n) = (0.8)^n \sin\left(\frac{2\pi}{6}n - \frac{\pi}{6}\right)$.

2.12.2 $x(n) = (-0.6)^n \cos\left(\frac{2\pi}{6}n + \frac{\pi}{3}\right)$.

2.12.3 $x(n) = (1.1)^n \sin\left(\frac{2\pi}{6}n - \frac{\pi}{4}\right)$.

2.12.4 $x(n) = (-1.2)^n \cos\left(\frac{2\pi}{6}n + \frac{\pi}{6}\right)$.

2.12.5 $x(n) = (0.7)^n \cos(\pi n)$.

2.13 Find the next three higher frequency sinusoids with the same set of sample values as that of $x(n)$.

2.13.1 $x(n) = 2\cos\left(2\frac{2\pi}{9}n + \frac{\pi}{6}\right)$.

2.13.2 $x(n) = 4\sin\left(3\frac{2\pi}{7}n - \frac{\pi}{3}\right)$.

2.13.3 $x(n) = \cos\left(4\frac{2\pi}{9}n - \frac{\pi}{6}\right)$.

*2.13.4 $x(n) = 3\sin\left(3\frac{2\pi}{8}n - \frac{\pi}{3}\right)$.

2.13.5 $x(n) = 3\cos(\pi n)$.

2.13.6 $x(n) = 5\cos(0n)$.

2.14 Find the minimum sampling rate required to represent the continuous signal unambiguously.

2.14.1 $x(t) = 3\cos(10\pi t)$.

2.14.2 $x(t) = 3\cos\left(10\pi t + \frac{\pi}{3}\right)$.

*2.14.3 $x(t) = 2\sin(10\pi t)$.

2.14.4 $x(t) = 2\sin\left(10\pi t - \frac{\pi}{6}\right)$.

2.15 The sinusoid $x(n)$ and the value k are specified. Express the sinusoid $x(n + k)$ in polar form. List the sample values of one cycle, starting from $n = 0$, of the sinusoids $x(n)$ and $x(n + k)$.

2.15.1 $x(n) = 2\cos\left(\frac{2\pi}{6}n - \frac{\pi}{3}\right)$, $k = 2$.

2.15.2 $x(n) = -3\sin\left(\frac{2\pi}{6}n + \frac{\pi}{6}\right)$, $k = -1$.

2.15.3 $x(n) = \cos\left(\frac{2\pi}{6}n - \frac{\pi}{6}\right)$, $k = 3$.

2.15.4 $x(n) = -\sin\left(\frac{2\pi}{6}n + \frac{\pi}{3}\right)$, $k = 6$.

*2.15.5 $x(n) = \cos\left(\frac{2\pi}{6}n + \frac{\pi}{2}\right)$, $k = -7$.

2.15.6 $x(n) = \sin\left(\frac{2\pi}{6}n + \frac{2\pi}{3}\right)$, $k = 15$.

2.16 The sinusoid $x(n)$ and the value k are specified. Express the sinusoid $x(-n + k)$ in polar form. List the sample values of one cycle, starting from $n = 0$, of the sinusoids $x(n)$ and $x(-n + k)$.

2.16.1 $x(n) = \sin\left(\frac{2\pi}{6}n - \frac{\pi}{3}\right), k = 0$.

2.16.2 $x(n) = \sin\left(\frac{2\pi}{6}n + \frac{\pi}{3}\right), k = -2$.

*2.16.3 $x(n) = \cos\left(\frac{2\pi}{6}n - \frac{\pi}{6}\right), k = 1$.

2.16.4 $x(n) = \sin\left(\frac{2\pi}{6}n + \frac{\pi}{2}\right), k = -3$.

2.16.5 $x(n) = \cos\left(\frac{2\pi}{6}n - \frac{\pi}{2}\right), k = 6$.

2.16.6 $x(n) = \sin\left(\frac{2\pi}{6}n + \frac{\pi}{6}\right), k = 7$.

2.16.7 $x(n) = \cos\left(\frac{2\pi}{6}n + \frac{\pi}{3}\right), k = 14$.

2.17 The sinusoid $x(n)$ and the values k and a are specified. List the sample values of one cycle, starting from $n = 0$, of the sinusoid $x(n)$ and $x(an + k)$. Assume interpolation using zero-valued samples, if necessary.

2.17.1 $x(n) = -\sin\left(\frac{2\pi}{6}n + \frac{\pi}{3}\right), a = -2, k = 0$.

2.17.2 $x(n) = 2\cos\left(\frac{2\pi}{6}n - \frac{\pi}{6}\right), a = \frac{1}{2}, k = -2$.

*2.17.3 $x(n) = \sin\left(\frac{2\pi}{6}n + \frac{\pi}{6}\right), a = -1, k = 1$.

2.17.4 $x(n) = 3\cos\left(\frac{2\pi}{6}n + \frac{\pi}{3}\right), a = \frac{1}{3}, k = 6$.

2.17.5 $x(n) = \sin\left(\frac{2\pi}{6}n - \frac{\pi}{2}\right), a = -3, k = 7$.

2.17.6 $x(n) = \cos\left(\frac{2\pi}{6}n - \frac{\pi}{6}\right), a = -1, k = 15$.

2.18 The waveform $x(n)$ and the values k and a are specified. List the sample values with indices $n = -3, -2, -1, 0, 1, 2, 3$ of the waveforms $x(n)$ and $x(an + k)$. Assume interpolation using zero-valued samples, if necessary.

2.18.1 $x(0) = 2, x(1) = 3, x(2) = -4, x(3) = 1$, and $x(n) = 0$ otherwise. $a = -2, k = 2$.

2.18.2 $x(0) = 2, x(1) = 3, x(2) = -4, x(3) = 1$, and $x(n) = 0$ otherwise. $a = -\frac{1}{2}, k = 1$.

2.18.3 $x(n) = (0.8)^n$. $a = -3, k = -1$.

2.18.4 $x(n) = (0.8)^n$. $a = \frac{1}{3}, k = 2$.

2.18.5 $x(n) = (1.1)^n$. $a = 2, k = 2$.

2.18.6 $x(n) = (1.1)^n$. $a = -\frac{1}{2}, k = 1$.

2.18.7 $x(n) = -2\sin\left(\frac{2\pi}{6}n + \frac{\pi}{6}\right)u(n)$. $a = \frac{1}{2}, k = 3$.

*2.18.8 $x(n) = -2\sin\left(\frac{2\pi}{6}n + \frac{\pi}{6}\right)u(n)$. $a = -2, k = 2$.

2.18.9 $x(n) = (0.7)^n\cos\left(\frac{2\pi}{6}n - \frac{\pi}{3}\right)u(n)$. $a = \frac{1}{3}, k = 3$.

2.18.10 $x(n) = (0.7)^n\cos\left(\frac{2\pi}{6}n - \frac{\pi}{3}\right)u(n)$. $a = -2, k = 2$.

3

Continuous Signals

While the analysis of continuous signals remains essentially the same as that for the discrete signals, there are differences due to the continuous nature. For example, the summation operation on a discrete signal corresponds to the integration operation on a continuous signal, the difference operation corresponds to the derivative, and the continuous impulse signal is defined in terms of area, in contrast to the discrete impulse signal defined by its amplitude. In this chapter, signal classifications, basic signals, and signal operations of continuous signals are described in Sections 3.1, 3.2, and 3.3, respectively.

3.1 Classification of Signals

Signals are classified into different types and the representation and analysis of a signal depends on its type.

3.1.1 Continuous Signals

A continuous signal $x(t)$ is specified at every value of its independent variable t. Figures 3.1(a) and (b) show, respectively, the damped continuous real exponential, $x(t) = e^{-0.1t}u(t)$, and three cycles of the damped complex exponential, $x(t) = e^{(-0.1+j(3\frac{2\pi}{16}))t}$. As the value of the exponential is decreasing with time, it is called a damped or decaying exponential, characterized by the negative constant, -0.1, in its exponent. An exponential e^{at}, where a is a positive constant, is an example of a growing exponential, as its value is increasing with time. We denote a continuous signal, using the independent variable t, as $x(t)$. We call this representation the time-domain representation, although the independent variable is not time for some signals. While most signals, in practical applications, are real-valued, complex-valued signals are often used in analysis. A complex-valued or complex signal is an ordered pair of

A Practical Approach to Signals and Systems   D. Sundararajan
© 2008 John Wiley & Sons (Asia) Pte Ltd

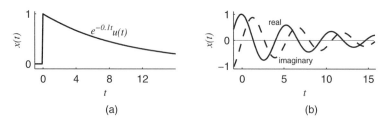

Figure 3.1   (a) The damped continuous real exponential, $x(t) = e^{-0.1t}u(t)$; (b) the damped continuous complex exponential, $x(t) = e^{(-0.1+j(3\frac{2\pi}{16}))t}$

real signals. The damped complex exponential signal, shown in Figure 3.1(b), can be expressed, using Euler's identity, in terms of damped cosine and sine signals as

$$x(t) = e^{\left(-0.1+j(3\frac{2\pi}{16})\right)t} = e^{(-0.1t)}\cos\left(3\frac{2\pi}{16}t\right) + je^{(-0.1t)}\sin\left(3\frac{2\pi}{16}t\right)$$

The real and imaginary parts of $x(t)$ are, respectively, $e^{(-0.1t)}\cos(3\frac{2\pi}{16}t)$ (shown by the solid line in Figure 3.1(b)) and $e^{(-0.1t)}\sin(3\frac{2\pi}{16}t)$ (shown by the dashed line in Figure 3.1(b)).

3.1.2 Periodic and Aperiodic Signals

The smallest positive number $T > 0$ satisfying the condition $x(t + T) = x(t)$, for all t, is the fundamental period of the continuous periodic signal $x(t)$. The reciprocal of the fundamental period is the fundamental cyclic frequency, $f = 1/T$ Hz (cycles per second). The fundamental angular frequency is $\omega = 2\pi f = 2\pi/T$ radians per second. Over the interval $-\infty < t < \infty$, a periodic signal repeats its values over any interval equal to its period, at intervals of its period. Cosine and sine waves are typical examples of a periodic signal. A signal with constant value (dc) is periodic with any period. In Fourier analysis, it is considered as $A\cos(\omega t)$ or $Ae^{j\omega t}$ with the frequency ω equal to zero (period equal to ∞).

When the period of a periodic signal approaches infinity, it degenerates into an aperiodic signal. The exponential signal, shown in Figure 3.1(a), is an aperiodic signal.

It is easier to decompose an arbitrary signal in terms of some periodic signals, such as complex exponentials, and the input–output relationship of a LTI system becomes a multiplication operation for this type of input signals. For these reasons, most of the analysis of practical signals, which are mostly aperiodic, having arbitrary amplitude profile, is carried out using periodic basic signals.

3.1.3 Energy and Power Signals

The energy of a signal $x(t)$ is expressed as the integral of the squared magnitude of its values as

$$E = \int_{-\infty}^{\infty} |x(t)|^2 dt$$

Aperiodic signals with finite energy are called energy signals. The energy of $x(t) = 3e^{-t}, \ t \geq 0$ is

$$E = \int_0^\infty |3e^{-t}|^2 dt = \frac{9}{2}$$

If the energy of a signal is infinite, then it may be possible to characterize it in terms of its average power. The average power is defined as

$$P = \lim_{T \to \infty} \frac{1}{T} \int_{-\frac{T}{2}}^{\frac{T}{2}} |x(t)|^2 dt$$

For periodic signals, the average power can be computed over one period as

$$P = \frac{1}{T} \int_{-\frac{T}{2}}^{\frac{T}{2}} |x(t)|^2 dt$$

where T is the period. Signals, periodic or aperiodic, with finite average power are called power signals. Cosine and sine waveforms are typical examples of power signals. The average power of the cosine wave $3\cos(\frac{\pi}{8}t)$ is

$$P = \frac{1}{16} \int_{-8}^{8} \left| 3\cos\left(\frac{\pi}{8}t\right) \right|^2 dt = \frac{9}{32} \int_{-8}^{8} \left(1 + \cos\left(2\frac{\pi}{8}t\right) \right) dt = \frac{9}{2}$$

A signal is an energy signal or a power signal, since the average power of an energy signal is zero while that of a power signal is finite. Signals with infinite power and infinite energy, such as $x(t) = t, \ t \geq 0$, are neither power signals nor energy signals. The measures of signal power and energy are indicators of the signal size, since the actual energy or power depends on the load.

3.1.4 Even- and Odd-symmetric Signals

The analysis of a signal can be simplified by exploiting its symmetry. A signal $x(t)$ is even-symmetric, if $x(-t) = x(t)$ for all t. The signal is symmetrical about the vertical axis at the origin. The cosine waveform is an example of an even-symmetric signal. A signal $x(t)$ is odd-symmetric, if $x(-t) = -x(t)$ for all t. The signal is asymmetrical about the vertical axis at the origin. For an odd-symmetric signal, $x(0) = 0$. The sine waveform is an example of an odd-symmetric signal.

The sum $(x(t) + y(t))$ of two odd-symmetric signals, $x(t)$ and $y(t)$, is an odd-symmetric signal, since $x(-t) + y(-t) = -x(t) - y(t) = -(x(t) + y(t))$. For example, the sum of two sine signals is an odd-symmetric signal. The sum $(x(t) + y(t))$ of two even-symmetric signals, $x(t)$ and $y(t)$, is an even-symmetric signal, since $x(-t) + y(-t) = (x(t) + y(t))$. For example, the sum of two cosine signals is an even-symmetric signal. The sum $(x(t) + y(t))$ of an odd-symmetric signal $x(t)$ and an even-symmetric signal $y(t)$ is neither even-symmetric nor odd-symmetric, since $x(-t) + y(-t) = -x(t) + y(t) = -(x(t) - y(t))$. For example, the sum of cosine and sine signals with nonzero amplitudes is neither even-symmetric nor odd-symmetric.

Since $x(t)y(t) = (-x(-t))(-y(-t)) = x(-t)y(-t)$, the product of two odd-symmetric or two even-symmetric signals is an even-symmetric signal. The product $z(t) = x(t)y(t)$ of an odd-symmetric signal $y(t)$ and an even-symmetric signal $x(t)$ is an odd-symmetric signal, since $z(-t) = x(-t)y(-t) = x(t)(-y(t)) = -z(t)$.

An arbitrary signal $x(t)$ can be decomposed in terms of its even-symmetric and odd-symmetric components, $x_e(t)$ and $x_o(t)$, respectively. That is, $x(t) = x_e(t) + x_o(t)$. Replacing t by $-t$, we get $x(-t) = x_e(-t) + x_o(-t) = x_e(t) - x_o(t)$. Solving for $x_e(t)$ and $x_o(t)$, we get

$$x_e(t) = \frac{x(t) + x(-t)}{2} \quad \text{and} \quad x_o(t) = \frac{x(t) - x(-t)}{2}$$

As the integral of an odd-symmetric signal $x_o(t)$, over symmetric limits, is zero,

$$\int_{-t_0}^{t_0} x_o(t)\, dt = 0 \qquad \int_{-t_0}^{t_0} x(t)\, dt = \int_{-t_0}^{t_0} x_e(t)\, dt = 2 \int_{0}^{t_0} x_e(t)\, dt$$

For example, the even-symmetric component of $x(t) = e^{j(\frac{2\pi}{16}t + \frac{\pi}{3})}$ is

$$x_e(t) = \frac{x(t) + x(-t)}{2} = e^{j(\frac{\pi}{3})}\frac{e^{j(\frac{2\pi}{16}t)} + e^{j(\frac{2\pi}{16}(-t))}}{2} = e^{j(\frac{\pi}{3})}\cos\left(\frac{2\pi}{16}t\right)$$

The odd-symmetric component is

$$x_o(t) = \frac{x(t) - x(-t)}{2} = e^{j(\frac{\pi}{3})}\frac{e^{j(\frac{2\pi}{16}t)} - e^{j(\frac{2\pi}{16}(-t))}}{2} = je^{j(\frac{\pi}{3})}\sin\left(\frac{2\pi}{16}t\right)$$

The complex exponential, its even component, and its odd component are shown, respectively, in Figures 3.2(a), (b), and (c).

Figure 3.2 (a) The complex exponential $x(t) = e^{j(\frac{2\pi}{16}t + \frac{\pi}{3})}$; (b) its even component $x_e(t) = e^{j(\frac{\pi}{3})}\cos(\frac{2\pi}{16}t)$; (c) its odd component $x_o(t) = je^{j(\frac{\pi}{3})}\sin(\frac{2\pi}{16}t)$

3.1.5 Causal and Noncausal Signals

Most signals, in practice, occur at some finite time instant, usually chosen as $t = 0$, and are considered identically zero before this instant. These signals, with $x(t) = 0$ for $t < 0$, are called causal signals (for example, the exponential shown in Figure 3.1a). Signals, with $x(t) \neq 0$ for $t < 0$, are called noncausal signals (for example, the complex exponential shown in Figure 3.1b).

3.2 Basic Signals

While the input signal to a system, in practice, is arbitrary, some mathematically well-defined and simple signals are used for testing systems and decomposition of the arbitrary signals for analysis. These signals, for example, the sinusoid with infinite duration and the impulse with infinite bandwidth, are mathematical idealizations and are not practical signals. However, they are convenient in the analysis of signals and systems. In practice, they can be approximated to a desired accuracy.

3.2.1 Unit-step Signal

A system is usually turned on by closing a switch. While practical switches have finite switching time, in theoretical analysis, zero switching time is assumed for convenience. This implies that the input signal is applied instantaneously. A function representing such a signal does not exist in the normal function theory, since the derivative of a function, at a discontinuity, is not defined. As this type of function is required frequently in the analysis of systems, we define such a function and its derivative, and denote them by special symbols. The unit-step function $u(t)$, shown in Figure 3.3(a), is defined as

$$u(t) = \begin{cases} 1 & \text{for } t > 0 \\ 0 & \text{for } t < 0 \\ \text{undefined} & \text{for } t = 0 \end{cases}$$

Figure 3.3 (a) The unit-step signal, $u(t)$; (b) the unit-impulse signal, $\delta(t)$; (c) the unit-ramp signal, $r(t)$

The unit-step signal has a value of one for $t > 0$ and has a value of zero for $t < 0$. The value $u(0)$, if required, can be assigned values such as 0, $\frac{1}{2}$, or 1 to suit a specific application. For example, the value $\frac{1}{2}$ is assigned in Fourier analysis.

The causal form of a signal $x(t)$, $x(t)$ is zero for $t < 0$, is obtained by multiplying it by the unit-step signal as $x(t)u(t)$. For example, $\cos(\frac{2\pi}{6}t)$ has nonzero values in the range $-\infty < t < \infty$, whereas the values of $\cos(\frac{2\pi}{6}t)u(t)$ are zero for $t < 0$ and $\cos(\frac{2\pi}{6}t)$ for $t > 0$. A time-shifted unit-step signal, for example $u(t - 2)$, is $u(t)$ shifted by two units to the right (changes from 0 to 1 at $t = 2$). Using scaled and shifted unit-step signals, a discontinuous signal, described differently over different intervals, can be specified, for easier mathematical analysis, by a single expression, valid for all t. For example, a signal that is identical to the first half period, beginning at $t = 0$, of the sine wave $\sin(t)$ and is zero otherwise can be expressed as $x(t) = \sin(t)(u(t) - u(t - \pi))$ or $x(t) = \sin(t)u(t) + \sin(t - \pi)u(t - \pi)$. The first expression can be interpreted as the sine wave multiplied by a pulse of unit height over the interval $0 < t < \pi$. The second expression can be interpreted as the sum of the causal form of the sine wave and its right-shifted version by π (a half period). The time scaled and shifted unit-step function $u(\pm at - t_0)$ is the same as $u(\pm t - (t_0/a))$, where $a \neq 0$ is a positive number.

3.2.2 Unit-impulse Signal

Consider a narrow unit-area rectangular pulse, $\delta_q(t)$, of width $2a$ and height $1/2a$ centered at $t = 0$ and the function $x(t) = 2 + e^{-t}$. The integral of their product, which is the local average of $x(t)$, is

$$\int_{-\infty}^{\infty} x(t)\delta_q(t)\, dt = \frac{1}{2a}\int_{-a}^{a}(2 + e^{-t})\, dt = 2 + \frac{e^a - e^{-a}}{2a}$$

The limiting value of the integral, as $a \to 0$, is

$$\lim_{a \to 0}\left(2 + \frac{e^a - e^{-a}}{2a}\right) = 2 + \lim_{a \to 0}\left(\frac{e^a + e^{-a}}{2}\right) = 3 = x(0)$$

In evaluating the limit, we used L'Hôpital's rule. As long as a is not equal to zero, the pulse is clearly defined by its width and height. The integral is an integral in the conventional sense. As $a \to 0$, the rectangular pulse, $\delta_q(t)$, degenerates into an impulse $\delta(t)$ and it is characterized only by its unit area at $t = 0$. Then, the integral becomes a definition

$$\lim_{a \to 0} \int_{-\infty}^{\infty} x(t)\delta_q(t)\, dt = \int_{-\infty}^{\infty} x(t)\delta(t)\, dt = x(0) \int_{0^-}^{0^+} \delta(t)\, dt = x(0)$$

The pulse, $\delta_q(t)$, and the signal $x(t) = 2 + e^{-t}$ are shown in Figure 3.4(a) with $a = 1$. Their product $x(t)\delta_q(t)$ is shown by the dotted line. The integral of the product is 3.1752 with four-digit precision. Figures 3.4(b), and (c) show the functions with $a = 0.4$, and $a = 0.1$, respectively. As the pulse width a is reduced, the variation in the amplitude of the function $x(t) = 2 + e^{-t}$ is also reduced and the integral of the product $\delta_q(t)x(t)$ (the local average of $x(t)$) approaches the value $x(0)$, as shown in Figure 3.4(d). The reason for associating the impulse in deriving the value of $x(0)$, rather than replacing t by 0 in $x(t)$, is to express $x(t)$ in terms of shifted and scaled impulses, as we shall see later.

The continuous unit-impulse signal $\delta(t)$, located at $t = 0$, is defined, in terms of an integral, as

$$\int_{-\infty}^{\infty} x(t)\delta(t)\, dt = x(0)$$

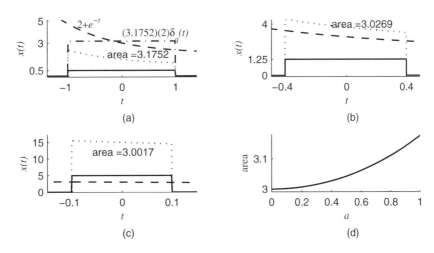

(a)

(b)

(c)

(d)

Figure 3.4 (a) The pulse $\delta_q(t)$ (solid line) with width $2a = 2$ and height $1/2a = 0.5$. The function $x(t) = 2 + e^{-t}$ (dashed line) and the product $\delta_q(t)x(t)$ (dotted line); (b) same as (a) with $a = 0.4$; (c) same as (a) with $a = 0.1$; (d) the area enclosed by the product $\delta_q(t)x(t)$ for various values of a

assuming that $x(t)$ is continuous at $t = 0$ (so that the value $x(0)$ is unique). The value of the function $x(t)$ at $t = 0$ has been sifted out or sampled by the defining operation. The impulse function is called a generalized function, since it is defined by the result of its operation (integration) on an ordinary function, rather than by its amplitude profile. A time-shifted unit-impulse signal $\delta(t - \tau)$, located at $t = \tau$, sifts out the value $x(\tau)$,

$$\int_{-\infty}^{\infty} x(t)\delta(t - \tau)\,dt = x(\tau),$$

assuming that $x(t)$ is continuous at $t = \tau$. As the amplitude profile of the impulse is undefined, the unit-impulse is characterized by its unit area concentrated at $t = 0$ (in general, whenever its argument becomes zero), called the strength of the impulse. The unit-impulse is represented by a small triangle (pointing upwards for a positive impulse and pointing downwards for a negative impulse), as shown in Figure 3.3(b). The power or energy of the impulse signal is undefined.

The area enclosed by a function over some finite duration is easy to visualize. For example, the distribution of mass along a line is defined by its density $\rho(x)$ and the mass between $x = 0$ and $x = 1$ is given by

$$\int_{0}^{1} \rho(x)dx$$

However, the symbol $\delta(t)$ stands for a function, whose shape and amplitude is such that its integral at the point $t = 0$ is unity. This is the limiting case of the density $\rho(x)$, when unit mass is concentrated at a single point $x = 0$. It is difficult to visualize such a function. But, it is easy to visualize a function of arbitrarily brief, but nonzero duration. For example, the impulse can be considered, for practical purposes, as a sufficiently narrow rectangular pulse of unit area. The width of the pulse Δt should be so short that the variation of any ordinary function $x(t)$, appearing in an expression involving an impulse, is negligible in Δt s. Therefore, to understand any operation involving the impulse, we start with a brief pulse, perform the operation, and take the limiting form as the width of the pulse approaches zero. As only its area is specified, it is possible to start with many functions of brief duration and apply the limiting process. The only condition is that its area must be unity throughout the limiting process. Some other functions, besides the rectangular pulse, that degenerate into the unit-impulse signal in the limit are shown in Figure 3.5. For practical purposes, any of these functions with a sufficiently brief duration is adequate. The point is that practical devices can produce a pulse of finite width only, whereas, in theory, we use zero-width pulses for the sake of mathematical convenience.

The product of an ordinary function $x(t)$, which is continuous at $t = \tau$, and $\delta(t - \tau)$ is given as $x(t)\delta(t - \tau) = x(\tau)\delta(t - \tau)$, since the impulse has unit area concentrated at $t = \tau$ and the value of $x(t)$ at that point is $x(\tau)$. That is, the product of an ordinary function with the unit-impulse is an impulse with its area or strength equal to the

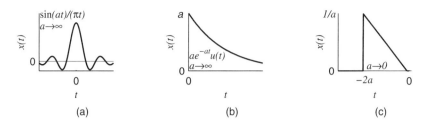

Figure 3.5 Some functions that degenerate into unit-impulse signal, $\delta(t)$, in the limit. (a) $\delta(t) = \lim_{a\to\infty} \frac{\sin(at)}{\pi t}$; (b) $\delta(t) = \lim_{a\to\infty} ae^{-at}u(t)$; (c) $\delta(t) = \lim_{a\to 0} \frac{t}{2a^2}(u(t) - u(t + 2a))$

value of the function at the location of the impulse. As the impulse is defined by an integral, any expression involving an impulse has to be eventually integrated to have a numerical value. An expression such as the product given above implies that the integral of the two sides are equal. For example,

$$\int_{-\infty}^{\infty} e^t \delta(t)dt = 1 \qquad \int_{-\infty}^{\infty} e^t \delta(t+2)dt = e^{-2} \qquad \int_{-2}^{2} e^t \delta(-t)dt = 1$$

$$\int_{2}^{4} e^t \delta(t)dt = 0 \qquad \int_{0^-}^{0^+} e^t \delta(t)dt = 1 \qquad \int_{-\infty}^{\infty} e^{t-3}\delta(t-3)dt = 1$$

In the fourth integral, the argument t of the impulse never becomes zero within the limits of the integral.

3.2.2.1 The Impulse Representation of Signals

A major application of the impulse is to decompose an arbitrary signal $x(t)$ into scaled and shifted impulses, so that the representation and analysis of $x(t)$ becomes easier. In the well-known rectangular rule of numerical integration, an arbitrary signal $x(t)$ is approximated by a series of rectangles. Each rectangle is of fixed width, say a, and height equal to a known value of $x(t)$ in that interval. The area of the rectangle is an approximation to that of $x(t)$ in that interval. The sum of areas of all such rectangles is an approximation of the *area* enclosed by the signal.

We can as well *represent* $x(t)$ approximately, in each interval of width a, by the area of the corresponding rectangle located at $t = t_0$ multiplied by a unit-area rectangular pulse, $\delta_q(t - t_0)$ of width a and height $1/a$, since the amplitude of the pulse $x(t_0)a\delta_q(t - t_0)$ is $x(t_0)$. For example, $x(t)$ can be represented by $(3.1752)(2)\delta_q(t)$ in Figure 3.4(a), shown by a dash–dot line. The sum of a succession of all such rectangles is an approximation to $x(t)$. As the width a is made smaller, the approximation becomes better. For example, $x(t)$ is represented by $(3.0269)(0.8)\delta_q(t)$ in Figure 3.4(b) and $(3.0017)(0.2)\delta_q(t)$ in Figure 3.4(c). Eventually, as a$\to$0, the pulse degenerates into impulse and the representation becomes exact.

Consider the product of a signal with a shifted impulse $x(t)\delta(t - \tau) = x(\tau)\delta(t - \tau)$. Integrating both sides with respect to τ, we get

$$\int_{-\infty}^{\infty} x(t)\delta(t - \tau)d\tau = x(t) \int_{-\infty}^{\infty} \delta(t - \tau)d\tau = x(t) = \int_{-\infty}^{\infty} x(\tau)\delta(t - \tau)d\tau$$

The integrand $x(\tau)\delta(t - \tau)d\tau$, which is one of the constituent impulses of $x(t)$, is a shifted impulse $\delta(t - \tau)$ located at $t = \tau$ with strength $x(\tau)d\tau$. The integration operation, with respect to τ, sums all these impulses to form $x(t)$. It should be emphasized that the integral, in this instance, represents a sum of a continuum of impulses (not an evaluation of an area). Therefore, the signal $x(t)$ is represented by the sum of scaled and shifted continuum of impulses with the strength of the impulse at any t being $x(t)dt$. The unit-impulse is the basis function and $x(t)dt$ is its coefficient. As the area enclosed by the integrand is nonzero only at the point $t = \tau$, the integral is effective only at that point. By varying the value of t, we can sift out all the values of $x(t)$.

Let a quasi-impulse, $\delta_q(t)$, is defined by a rectangular pulse with its base of width a, from $t = 0$ to $t = a$, and height $1/a$. Assume that the signal, $x(t) = e^{-1.2t}(u(t) - u(t - 1.5))$, is approximated by rectangles with width a and height equal to the value of $x(t)$ at the beginning of the corresponding rectangle. Figure 3.6 shows the approximation of $x(t)$ by rectangular pulses of width $a = 0.5$. We break up $x(t)$ so that it is expressed as a sum of sections of width $a = 0.5$.

$$x(t) = x0(t) + x1(t) + x2(t)$$
$$= e^{-1.2t}(u(t) - u(t - 0.5))$$
$$+ e^{-1.2t}(u(t - 0.5) - u(t - 1))$$
$$+ e^{-1.2t}(u(t - 1) - u(t - 1.5))$$

By replacing each section by a function that is constant with a value equal to that of $x(t)$ at the beginning of the section, we get

$$x(t) \approx xa(t) + xb(t) + xc(t)$$
$$= e^{-1.2(0)(0.5)}(u(t) - u(t - 0.5))$$
$$+ e^{-1.2(1)(0.5)}(u(t - 0.5) - u(t - 1))$$
$$+ e^{-1.2(2)(0.5)}(u(t - 1) - u(t - 1.5))$$

By multiplying and dividing by $a = 0.5$, we get

$$x(t) \approx xa(t) + xb(t) + xc(t)$$
$$= e^{-1.2(0)(0.5)} \left(\frac{u(t) - u(t - 0.5)}{0.5} \right) (0.5)$$

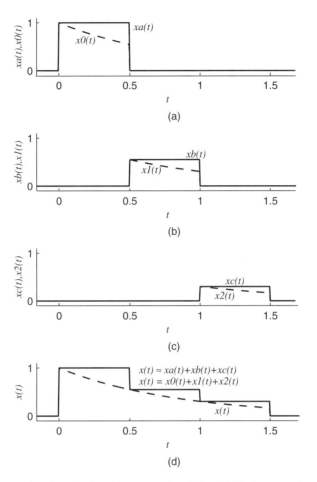

Figure 3.6 The approximation of a signal by a sum of scaled and shifted rectangular pulses. (a) $xa(t) = \delta_q(t)(0.5)$, $x0(t) = e^{-1.2t}(u(t) - u(t - 0.5))$; (b) $xb(t) = e^{-0.6}\delta_q(t - 0.5)(0.5)$, $x1(t) = e^{-1.2t}(u(t - 0.5) - u(t - 1))$; (c) $xc(t) = e^{-1.2}\delta_q(t - 1)(0.5)$, $x2(t) = e^{-1.2t}(u(t - 1) - u(t - 1.5))$; (d) $x(t) = x0(t) + x1(t) + x2(t) = e^{-1.2t}(u(t) - u(t - 1.5)) \approx \delta_q(t)(0.5) + e^{-0.6}\delta_q(t - 0.5)(0.5) + e^{-1.2}\delta_q(t - 1)(0.5)$

$$+ e^{-1.2(1)(0.5)} \left(\frac{u(t - 0.5) - u(t - 1)}{0.5} \right) (0.5)$$

$$+ e^{-1.2(2)(0.5)} \left(\frac{u(t - 1) - u(t - 1.5)}{0.5} \right) (0.5)$$

$$x(t) \approx xa(t) + xb(t) + xc(t)$$

$$= \delta_q(t)(0.5) + e^{-0.6}\delta_q(t - 0.5)(0.5) + e^{-1.2}\delta_q(t - 1)(0.5)$$

$$= \sum_{n=0}^{2} e^{-1.2(n)(0.5)}\delta_q(t - (n)(0.5))(0.5)$$

In general, we approximate an arbitrary $x(t)$ as

$$x(t) \approx \sum_{n=-\infty}^{\infty} x((n)(a))\delta_q(t - (n)(a))(a)$$

which reverts to the exact representation of $x(t)$

$$x(t) = \int_{-\infty}^{\infty} x(\tau)\delta(t - \tau)\mathrm{d}\tau,$$

as $a \to 0$ (a is replaced by the differential $\mathrm{d}\tau$ and $(n)(a)$ becomes the continuous variable τ).

3.2.2.2 The Unit-impulse as the Derivative of the Unit-step

A function, which is the derivative of the unit-step function, must have its integral equal to zero for $t < 0$ and one for $t > 0$. Therefore, such a function must be defined to have unit area at $t = 0$ and zero area elsewhere. Figure 3.7(a) shows the quasi-impulse $\delta_q(t)$ with width 1 and height $1/1 = 1$ (solid line) and Figure 3.7(b) shows its integral $u_q(t)$ (solid line), which is an approximation to the unit-step function. As the width of $\delta_q(t)$ is reduced and its height correspondingly increased, as shown in Figure 3.7(a) (dashed line with width 0.5 and dotted line with width 0.2), $\delta_q(t)$ is more like an impulse, and the corresponding integrals, shown in Figure 3.7(b) (dashed and dotted lines), become better approximations to the unit-step function. At any stage in the limiting process, $u_q(t)$ remains the integral of $\delta_q(t)$ and $\delta_q(t)$ remains the derivative (except at the corners) of $u_q(t)$ and is defined to be so even in the limit (for the sake of mathematical convenience) as the width of $\delta_q(t)$ tends to zero. $\delta_q(t)$ and $u_q(t)$ become, respectively, the unit-impulse and unit-step functions in the limit and

$$\frac{\mathrm{d}u(t)}{\mathrm{d}t} = \delta(t) \qquad \text{and} \qquad \int_{-\infty}^{t} \delta(\tau)\mathrm{d}\tau = u(t)$$

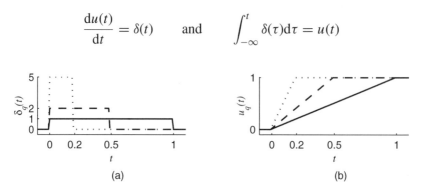

(a)   (b)

Figure 3.7 (a) The quasi-impulse $\delta_q(t)$ with width 1 and height $1/1 = 1$ (solid line), and with width 0.5 and height $1/0.5 = 2$ (dashed line), and with width 0.2 and height $1/0.2 = 5$ (dotted line); (b) their integrals $u_q(t)$, approaching the unit-step function as the width of the quasi-impulse tends to zero

For example, the voltage across a capacitor is proportional to the integral of the current through it. Therefore, a unit-impulse current passing through a capacitor of 1 F produces a unit-step voltage across the capacitor.

A signal $x(t)$, with step discontinuities, for example, at $t = t_1$ of height $(x(t_1^+) - x(t_1^-))$ and at $t = t_2$ of height $(x(t_2^+) - x(t_2^-))$, can be expressed as

$$x(t) = x_c(t) + (x(t_1^+) - x(t_1^-))u(t - t_1) + (x(t_2^+) - x(t_2^-))u(t - t_2)$$

where $x_c(t)$ is $x(t)$ with the discontinuities removed and $x(t_1^+)$ and $x(t_1^-)$ are, respectively, the right- and left-hand limits of $x(t)$ at $t = t_1$. The derivative of $x(t)$ is given by generalized function theory as

$$\frac{dx(t)}{dt} = \frac{dx_c(t)}{dt} + (x(t_1^+) - x(t_1^-))\delta(t - t_1) + (x(t_2^+) - x(t_2^-))\delta(t - t_2)$$

where $dx_c(t)/dt$ is the ordinary derivative of $x_c(t)$ at all t except at $t = t_1$ and $t = t_2$. Note that $dx_c(t)/dt$ may have step discontinuities. In the expression for $dx(t)/dt$, the impulse terms serve as *indicators* of step discontinuities in its integral, that is $x(t)$. Therefore, the use of impulses in this manner prevents the loss of step discontinuities in the integration operation and we get back $x(t)$ exactly by integrating its derivative. That is,

$$x(t) = x(t_0) + \int_{t_0}^{t} \frac{dx(t)}{dt} dt$$

For example, the derivative of the signal $x(t)$, shown in Figure 3.8(a) along with $x_c(t)$ (dashed line)

$$x(t) = u(-t - 1) + e^{-t}(u(t + 1) - u(t - 1)) + 2t(u(t - 1) - u(t - 2))$$

$$+ \cos\left(\frac{\pi}{2}t\right)u(t - 2)$$

$$= x_c(t) + 1.7183u(t + 1) + 1.6321u(t - 1) - 5u(t - 2),$$

Figure 3.8   (a) Signal $x(t)$ with step discontinuities; (b) its derivative

is

$$\frac{dx(t)}{dt} = 0 - e^{-t}(u(t+1) - u(t-1)) + 2(u(t-1) - u(t-2))$$

$$- \left(\frac{\pi}{2}\right) \sin\left(\frac{\pi}{2}t\right) u(t-2)$$

$$+ 1.7183\delta(t+1) + 1.6321\delta(t-1) - 5\delta(t-2),$$

shown in Figure 3.8(b).

3.2.2.3 The Scaling Property of the Impulse

The area enclosed by a time-scaled pulse $x(at)$ and that of its time-reversed version $x(-at)$ is equal to the area enclosed by $x(t)$ divided by $|a|$. Therefore, the scaling property of the impulse is given as

$$\delta(at) = \frac{1}{|a|}\delta(t) \qquad a \neq 0$$

With $a = -1$, $\delta(-t) = \delta(t)$ implying that the impulse is an even-symmetric signal. For example,

$$\delta(2t+1) = \delta\left(2\left(t+\frac{1}{2}\right)\right) = \frac{1}{2}\delta\left(t+\frac{1}{2}\right) \text{ and } \delta\left(\frac{1}{2}t-1\right) = \delta\left(\frac{1}{2}(t-2)\right)$$

$$= 2\delta(t-2)$$

3.2.3 Unit-ramp Signal

The unit-ramp signal, shown in Figure 3.3(c), is defined as

$$r(t) = \begin{cases} t & \text{for } t \geq 0 \\ 0 & \text{for } t < 0 \end{cases}$$

The unit-ramp signal increases linearly, with unit slope, for positive values of its argument and its value is zero for negative values of its argument.

The unit-impulse, unit-step, and unit-ramp signals are closely related. The unit-impulse signal $\delta(t)$ is equal to the derivative of the unit-step signal $du(t)/dt$, according to generalized function theory. The unit-step signal $u(t)$ is equal to $\int_{-\infty}^{t} \delta(\tau)d\tau$. The unit-step signal $u(t)$ is equal to $dr(t)/dt$, except at $t = 0$, where no unique derivative exists. The unit-ramp signal $r(t)$ is equal to $\int_{-\infty}^{t} u(\tau)d\tau$.

3.2.4 Sinusoids

3.2.4.1 The Polar Form of Sinusoids

The polar form specifies a sinusoid, in terms of its amplitude and phase, as

$$x(t) = A\cos(\omega t + \theta), \qquad -\infty < t < \infty$$

where A, ω, and θ are, respectively, the amplitude, the angular frequency, and the phase. The amplitude A is the distance of either peak of the waveform from the horizontal axis. Let the period of the sinusoid be T s. Then, as

$$\cos(\omega(t+T)+\theta) = \cos(\omega t + \omega T + \theta) = \cos(\omega t + \theta) = \cos(\omega t + \theta + 2\pi)$$

$T = 2\pi/\omega$. The cyclic frequency, denoted by f, of a sinusoid is the number of cycles per second and is equal to the reciprocal of the period, $f = 1/T = \omega/2\pi$ cycles per second (Hz). The angular frequency, the number of radians per second, of a sinusoid is 2π times its cyclic frequency, that is $\omega = 2\pi f$ radians per second. For example, consider the sinusoid $3\cos(\frac{\pi}{8}t + \frac{\pi}{3})$, with $A = 3$. The angular frequency is $\omega = \pi/8$ radians per second. The period is $T = \frac{2\pi}{\frac{\pi}{8}} = 16$ seconds. The cyclic frequency is $f = 1/T = 1/16$ Hz. The phase is $\theta = \pi/3$ radians. The phase can also be expressed in terms of seconds, as $\cos(\omega t + \theta) = \cos(\omega(t + \frac{\theta}{\omega}))$. The phase of $\pi/3$ radians corresponds to $8/3$ s. As it repeats a pattern over its period, the sinusoid remains the same by a shift of an integral number of its period. A phase-shifted sine wave can be expressed as a phase-shifted cosine wave, $A\sin(\omega t + \theta) = A\cos(\omega t + (\theta - \frac{\pi}{2}))$. The phase of the sinusoid

$$\sin\left(\frac{2\pi}{16}t + \frac{\pi}{3}\right) = \cos\left(\frac{2\pi}{16}t + \left(\frac{\pi}{3} - \frac{\pi}{2}\right)\right) = \cos\left(\frac{2\pi}{16}t - \frac{\pi}{6}\right)$$

is $-\pi/6$ radians. A phase-shifted cosine wave can be expressed as a phase-shifted sine wave, $A\cos(\omega t + \theta) = A\sin(\omega t + (\theta + \frac{\pi}{2}))$.

3.2.4.2 The Rectangular Form of Sinusoids

An arbitrary sinusoid is neither even- nor odd-symmetric. The even component of a sinusoid is the cosine waveform and the odd component is the sine waveform. That is, a sinusoid is a linear combination of cosine and sine waveforms of the same frequency as that of the sinusoid. Expressing a sinusoid in terms of its cosine and sine components is called its rectangular form and is given as

$$A\cos(\omega t + \theta) = A\cos(\theta)\cos(\omega t) - A\sin(\theta)\sin(\omega t) = C\cos(\omega t) + D\sin(\omega t),$$

where $C = A \cos \theta$ and $D = -A \sin \theta$. The inverse relation is $A = \sqrt{C^2 + D^2}$ and $\theta = \cos^{-1}(\frac{C}{A}) = \sin^{-1}(\frac{-D}{A})$. For example,

$$\cos\left(\frac{2\pi}{16}t + \frac{\pi}{3}\right) = \frac{1}{2}\cos\left(\frac{2\pi}{16}t\right) - \frac{\sqrt{3}}{2}\sin\left(\frac{2\pi}{16}t\right)$$

$$\frac{3}{\sqrt{2}}\cos\left(\frac{2\pi}{16}t\right) + \frac{3}{\sqrt{2}}\sin\left(\frac{2\pi}{16}t\right) = 3\cos\left(\frac{2\pi}{16}t - \frac{\pi}{4}\right)$$

3.2.4.3 The Sum of Sinusoids of the Same Frequency

The sum of sinusoids of arbitrary amplitudes and phases, but with the same frequency, is also a sinusoid of the same frequency. Let

$$x_1(t) = A_1 \cos(\omega t + \theta_1) \qquad \text{and} \qquad x_2(t) = A_2 \cos(\omega t + \theta_2)$$

Then, $x(t) = x_1(t) + x_2(t) = A \cos(\omega t + \theta)$, where

$$A = \sqrt{A_1^2 + A_2^2 + 2A_1 A_2 \cos(\theta_1 - \theta_2)}$$

$$\theta = \tan^{-1}\frac{A_1 \sin(\theta_1) + A_2 \sin(\theta_2)}{A_1 \cos(\theta_1) + A_2 \cos(\theta_2)}$$

Any number of sinusoids can be combined into a single sinusoid by repeatedly using the formulas. Note that the formula for the rectangular form of the sinusoid is a special case of the sum of two sinusoids, one sinusoid being the cosine and the other being the sine.

3.2.4.4 The Complex Sinusoids

The complex sinusoid is given as

$$x(t) = Ae^{j(\omega t + \theta)} = Ae^{j\theta}e^{j\omega t} \qquad -\infty < t < \infty$$

The term $e^{j\omega t}$ is the complex sinusoid with unit magnitude and zero phase. Its complex (amplitude) coefficient is $Ae^{j\theta}$. The amplitude and phase of the sinusoid is represented by the single complex number $Ae^{j\theta}$. By adding its complex conjugate, $Ae^{-j(\omega t + \theta)}$ and dividing by two, due to Euler's identity, we get

$$x(t) = \frac{A}{2}\left(e^{j(\omega t + \theta)} + e^{-j(\omega t + \theta)}\right) = A \cos(\omega t + \theta)$$

The use of two complex sinusoids to represent a single real sinusoid requires four real quantities instead of two. This redundancy in terms of storage and operations can be avoided.

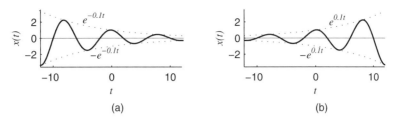

Figure 3.9 (a) Exponentially decreasing amplitude cosine wave, $x(t) = e^{-0.1t} \cos(\frac{2\pi}{8} t)$; (b) exponentially increasing amplitude cosine wave, $x(t) = e^{0.1t} \cos(\frac{2\pi}{8} t)$

3.2.4.5 Exponentially Varying Amplitude Sinusoids

An exponentially varying amplitude sinusoid, $Ae^{at} \cos(\omega t + \theta)$, is obtained by multiplying a sinusoid, $A \cos(\omega t + \theta)$, by a real exponential, e^{at}. The more familiar constant amplitude sinusoid results when $a = 0$. If ω is equal to zero, then we get a real exponential. Sinusoids, $x(t) = e^{-0.1t} \cos(\frac{2\pi}{8} t)$ and $x(t) = e^{0.1t} \cos(\frac{2\pi}{8} t)$, with exponentially varying amplitudes are shown, respectively, in Figures 3.9(a) and (b). The complex exponential representation of an exponentially varying amplitude sinusoid is given as

$$x(t) = \frac{A}{2} e^{at} \left(e^{j(\omega t + \theta)} + e^{-j(\omega t + \theta)} \right) = Ae^{at} \cos(\omega t + \theta)$$

Figure 3.1(b) shows exponentially varying amplitude complex sinusoid, $x(t) = e^{(-0.1 + j(3\frac{2\pi}{16}))t}$.

3.3 Signal Operations

In addition to the arithmetic operations, time shifting, time reversal, and time scaling operations are also commonly used in the analysis of continuous signals. The three operations described in this section are with respect to the independent variable, t.

3.3.1 Time Shifting

A signal $x(t)$ is time shifted by T s by replacing t by $t + T$. The value of $x(t)$ at $t = t_0$ occurs at $t = t_0 - T$ in $x(t + T)$. The rectangular pulse $x(t) = u(t - 1) - u(t - 3)$, shown in Figure 3.10 by a solid line, is a combination of two delayed unit-step signals. The right-shifted pulse $x(t - 1) = u(t - 2) - u(t - 4)$, shown in Figure 3.10 by a dashed line, is $x(t)$ shifted by 1 s to the right (delayed by one second, as the values of $x(t)$ occur 1 s late). For example, the first nonzero value occurs at $t = 2$ as $u(2 - 2) - u(2 - 4) = 1$. That is, the value of $x(t)$ at t_0 occurs in the shifted pulse 1 s later at $t_0 + 1$.

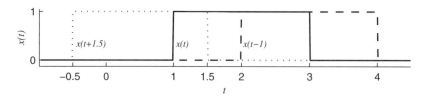

Figure 3.10   The rectangular pulse, $x(t) = u(t - 1) - u(t - 3)$, the right-shifted pulse, $x(t - 1)$, and the left-shifted pulse, $x(t + 1.5)$

The pulse $x(t + 1.5)$, shown in Figure 3.10 by a dotted line, is $x(t)$ shifted by 1.5 s to the left (advanced by 1.5 s, as the values of $x(t)$ occur 1.5 s early). For example, the first nonzero value occurs at $t = -0.5$ as $u(-0.5 + 0.5) - u(-0.5 - 1.5) = 1$. That is, the value of $x(t)$ at t_0 occurs in the shifted pulse 1.5 s earlier at $t_0 - 1.5$.

3.3.2 Time Reversal

Replacing the independent variable t in $x(t)$ by $-t$ results in the time-reversed or folded signal $x(-t)$. The value of $x(t)$ at $t = t_0$ occurs at $t = -t_0$ in $x(-t)$. The signal and its time-reversed version are mirror images of each other. The signal $x(t) = r(t + 1) - r(t - 1)$, shown in Figure 3.11 by a solid line, is a combination of two shifted unit-ramp signals. Consider the folded and shifted signal $x(-t - 1) = x(-(t + 1)) = r(-t) - r(-t - 2)$, shown in Figure 3.11 by a dashed line. This signal can be formed by first folding $x(t)$ to get $x(-t)$ and then shifting it to the left by 1 s (t is replaced by $(t + 1)$). This signal can also be formed by first shifting $x(t)$ to the right by 1 s to get $x(t - 1)$ and then folding it about the vertical axis at the origin (t is replaced by $-t$). The value of the signal $x(t)$ at t_0 occurs in the folded and shifted signal at $-t_0 - 1$. Consider the folded and shifted signal $x(-t + 3) = x(-(t - 3)) = r(-t + 4) - r(-t + 2)$, shown in Figure 3.11 by a dotted line. This signal can be formed by first folding $x(t)$ to get $x(-t)$ and then shifting it to the right by 3 s (t is replaced by $(t - 3)$). This signal can also be formed by first shifting $x(t)$ to the left by 3 s to get $x(t + 3)$ and then folding it about the vertical axis at the origin (t is replaced by $-t$). The value of $x(t)$ at t_0 occurs in the folded and shifted signal at $-t_0 + 3$.

Figure 3.11   The signal, $x(t) = r(t + 1) - r(t - 1)$, the shifted and folded signal, $x(-t - 1)$, and the shifted and folded signal, $x(-t + 3)$

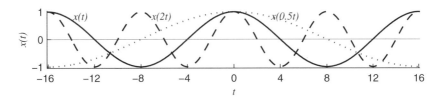

Figure 3.12 The cosine wave, $x(t) = \cos(\frac{\pi}{8}t)$, the compressed version $x(2t)$, and the expanded version $x(0.5t)$

3.3.3 Time Scaling

Replacing the independent variable t in $x(t)$ by at, $(a \neq 0)$, results in the time-scaled signal $x(at)$. With $|a| > 1$, we get a time-compressed version of $x(t)$. With $|a| < 1$, we get a time-expanded version. The value of $x(t)$ at $t = t_0$ occurs at $t = t_0/a$ in $x(at)$. The signal $x(t) = \cos(\frac{\pi}{8}t)$, shown in Figure 3.12 by a solid line, completes two cycles during 32 s. The time-compressed version with $a = 2$, $x(2t) = \cos(\frac{\pi}{8}(2t))$, shown in Figure 3.12 by a dashed line, completes four cycles during 32 s. The value of the signal $x(t)$ at t occurs at $t/2$ in $x(2t)$. For example, the negative peak at $t = 8$ in $x(t)$ occurs at $t = 4$ in $x(2t)$. The time-expanded version with $a = 0.5$, $x(0.5t) = \cos(\frac{\pi}{8}(0.5t))$, shown in Figure 3.12 by a dotted line, completes one cycle during 32 s. The value of the signal in $x(t)$ at t occurs at $t/0.5$ in $x(0.5t)$. For example, the negative peak at $t = 8$ in $x(t)$ occurs at $t = 16$ in $x(0.5t)$.

In general, the three operations described on a signal $x(t)$ can be expressed as $y(t) = x(at - b)$. The signal $y(t)$ can be generated by replacing t by $(at - b)$. However, it is instructive to consider it as the result of a sequence of two steps: (i) first shifting the signal $x(t)$ by b to get $x(t - b)$; and then (ii) time scaling (replace t by at) the shifted signal by a to get $x(at - b)$. An alternate sequence of two steps is to : (i) first time scale the signal $x(t)$ by a to get $x(at)$; and then (ii) shift (replace t by $t - b/a$) the time-scaled signal by b/a to get $x(a(t - b/a)) = x(at - b)$. Note that, time reversal operation is a part of the time scaling operation with a negative.

Let $x(t) = \cos(\frac{\pi}{8}t + \frac{\pi}{4})$, shown in Figure 3.13 by a solid line. It is required to find $x(-2t + 4)$. The shifted signal is $x(t + 4) = \cos(\frac{\pi}{8}(t + 4) + \frac{\pi}{4})$. Now scaling this signal by -2 yields the signal $x(-2t + 4) = \cos(\frac{\pi}{8}(-2t + 4) + \frac{\pi}{4}) = \cos(\frac{2\pi}{8}t - \frac{3\pi}{4})$, shown in Figure 3.13 by a dashed line. The value of the signal $x(t)$ at t_0 occurs at

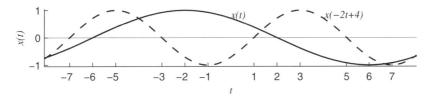

Figure 3.13 The sinusoid $x(t) = \cos(\frac{\pi}{8}t + \frac{\pi}{4})$ and $x(-2t + 4)$

$(-t_0 + 4)/2$ in $x(-2t + 4)$. We could have done the time scaling operation by -2 first to obtain $x(-2t) = \cos(\frac{\pi}{8}(-2t) + \frac{\pi}{4})$. Shifting this signal by $4/-2 = -2$ (replace t by $t - 2$), we get $x(-2t + 4) = \cos(\frac{\pi}{8}(-2t + 4) + \frac{\pi}{4})$.

3.4 Summary

- In this chapter, continuous signal classifications, basic signals, and signal operations have been presented.
- The representation and analysis of a signal depends on its type.
- As practical signals have arbitrary amplitude profile, these signals are usually decomposed and analyzed in terms of basic signals, such as the sinusoid or the impulse.
- In contrast to the characterization of the discrete impulse by its amplitude, the continuous impulse is characterized by its area.
- In addition to the arithmetic operations, time shifting, time reversal, and time scaling operations are also commonly used in the analysis of continuous signals.

Further Reading

1. Lathi, B. P., *Linear Systems and Signals*, Oxford University Press, New York, 2004.

Exercises

3.1 Is $x(t)$ an energy signal, a power signal, or neither? If it is an energy signal, find its energy. If it is a power signal, find its average power.

    3.1.1 $x(t) = 3$, $-1 < t < 1$ and $x(t) = 0$ otherwise.

    3.1.2 $x(t) = 2t$, $0 < t < 1$ and $x(t) = 0$ otherwise.

    3.1.3 $x(t) = 4e^{-0.2t}u(t)$.

    3.1.4 $x(t) = e^t$.

    3.1.5 $x(t) = Ce^{j(\frac{2\pi t}{T})}$.

    3.1.6 $x(t) = 2\cos\left(\frac{\pi t}{4} + \frac{\pi}{3}\right)$.

    3.1.7 $x(t) = u(t)$.

    3.1.8 $x(t) = t$.

  *3.1.9 $x(t) = 2\frac{1}{t}u(t - 1)$.

    3.1.10 $x(t) = 3e^{j(\frac{2\pi t}{6})}$.

    3.1.11 $x(t) = 3$.

    3.1.12 $x(t) = 3\sin\left(\frac{\pi t}{4} + \frac{\pi}{3}\right)$.

3.2 Is $x(t)$ even-symmetric, odd-symmetric, or neither? List the values of $x(t)$ at $t = -3, -2, -1, 0, 1, 2, 3$.

    3.2.1 $x(t) = 3\cos\left(\frac{2\pi}{6}t + \frac{\pi}{6}\right)$.

    3.2.2 $x(t) = 2\sin\left(\frac{2\pi}{6}t - \frac{\pi}{3}\right)$.

3.2.3 $x(t) = 4\cos\left(\frac{2\pi}{6}t\right)$.

3.2.4 $x(t) = 5$.

3.2.5 $x(t) = -2\sin\left(\frac{2\pi}{6}t\right)$.

3.2.6 $x(t) = t$.

3.2.7 $x(t) = \frac{\sin\left(\frac{\pi}{3}t\right)}{t}$.

3.2.8 $x(t) = \frac{\sin^2\left(\frac{\pi}{3}t\right)}{t}$.

3.2.9 $x(t) = e^{-t}$.

3.3 Find the even and odd components of $x(t)$. Verify that the integral of the odd component is zero. Verify that the integral of the even component and that of $x(t)$ are equal.

3.3.1 $x(t) = 2$, $-1 < t < 1$ and $x(t) = 0$ otherwise.

3.3.2 $x(t) = 3$, $-1 < t < 2$ and $x(t) = 0$ otherwise.

3.3.3 $x(t) = 2t$, $-1 < t < 1$ and $x(t) = 0$ otherwise.

*3.3.4 $x(t) = 3t$, $0 < t < 1$ and $x(t) = 0$ otherwise.

3.3.5 $x(t) = 2\cos\left(\frac{2\pi}{6}t - \frac{\pi}{3}\right)$.

3.3.6 $x(t) = e^{-0.2t}u(t)$.

3.3.7 $x(t) = u(t)$.

3.3.8 $x(t) = e^{j\left(\frac{2\pi}{6}t\right)}$.

3.3.9 $x(t) = t\,u(t)$.

3.3.10 $x(t) = \sin\left(\frac{2\pi}{6}t\right)$.

3.4 Evaluate the integral.

3.4.1 $\int_{-\infty}^{0} u(3t + 1)dt$.

3.4.2 $\int_{-\infty}^{-2} u\left(\frac{1}{3}t + 2\right)dt$.

3.4.3 $\int_{-\infty}^{5} u\left(\frac{1}{2}t - 4\right)dt$.

*3.4.4 $\int_{-11}^{\infty} u\left(-\frac{1}{2}t - 4\right)dt$.

3.5 Assume that the impulse is approximated by a rectangular pulse, centered at $t = 0$, of width $2a$ and height $1/2a$. Using this quasi-impulse, the signal $x(t)$ is sampled. What are the sample values of $x(t)$ at $t = 0$ with $a = 1$, $a = 0.1$, $a = 0.01$, $a = 0.001$, and $a = 0$?

3.5.1 $x(t) = 4e^{-3t}$.

3.5.2 $x(t) = 2\cos(t)$.

*3.5.3 $x(t) = 3\sin(t - \frac{\pi}{6})$.

3.5.4 $x(t) = \cos(t + \frac{\pi}{3})$.

3.5.5 $x(t) = \sin(t + \frac{\pi}{4})$.

3.6 Evaluate the integral.

3.6.1 $\int_{-1}^{\infty} \delta(t)e^t dt$.

3.6.2 $\int_{0}^{\infty} \delta(t + 1)e^t dt$.

*3.6.3 $\int_{0}^{\infty} \delta(t - 2)e^t dt$.

3.6.4 $\int_{-\infty}^{\infty} \delta(t+1)e^t dt$.

3.6.5 $\int_{-\infty}^{\infty} \delta(t+1)e^t u(t) dt$.

3.6.6 $\int_1^5 \delta(t+1)e^t dt$.

3.6.7 $\int_{-4}^{-1} \delta(t+2)e^t dt$.

3.6.8 $\int_0^{\infty} \delta(t-2)e^{(t-2)} dt$.

3.7   A quasi-impulse, $\delta_q(t)$, is defined by a rectangular pulse with its base of width a, from $t=0$ to $t=a$, and height $1/a$. Assume that the signal $x(t)$ is approximated by a series of rectangles with the height of each rectangle equal to the value of $x(t)$ at the beginning of the corresponding rectangle and width a. Express the signal $x(t)$ in terms of the quasi-impulse with $a=1$ and $a=0.5$.

3.7.1 $x(t) = e^t$, $0 \le t \le 5$ and $x(t) = 0$ otherwise.

*3.7.2 $x(t) = \cos(\frac{\pi}{6}t)$, $0 \le t \le 4$ and $x(t) = 0$ otherwise.

3.7.3 $x(t) = (t+3)$, $0 \le t \le 3$ and $x(t) = 0$ otherwise.

3.8   Find the derivative of the signal.

3.8.1 $\cos(\pi t)u(t)$.

3.8.2 $\sin(\pi t)u(t)$.

*3.8.3 $2e^{-3t}u(t)$.

3.9   Evaluate the integral.

3.9.1 $\int_{-\infty}^{\infty} \delta(3t+1) dt$.

3.9.2 $\int_{-\infty}^{\infty} \delta\left(\frac{1}{3}t+2\right) dt$.

3.9.3 $\int_{-\infty}^3 \delta\left(\frac{1}{2}t-2\right) dt$.

*3.9.4 $\int_{-\infty}^4 \delta\left(-\frac{1}{3}t+2\right) dt$.

3.9.5 $\int_{-\infty}^4 \delta\left(-\frac{1}{3}t-2\right) dt$.

3.10 Find the rectangular form of the sinusoid. Find the value of $t > 0$ where the first positive peak of the sinusoid occurs. Find the values of t at which the next two consecutive peaks, both negative and positive, occur.

3.10.1 $x(t) = -3\cos\left(\frac{2\pi}{8}t - \frac{\pi}{3}\right)$.

3.10.2 $x(t) = 2\sin\left(\frac{2\pi}{6}t + \frac{\pi}{6}\right)$.

*3.10.3 $x(t) = -5\sin\left(2\pi t + \frac{\pi}{4}\right)$.

3.10.4 $x(t) = 2\cos\left(2\pi t + \frac{\pi}{3}\right)$.

3.10.5 $x(t) = 4\cos\left(\frac{2\pi}{5}t - \frac{13\pi}{6}\right)$.

3.11 Find the polar form of the sinusoid. Find the values of $t > 0$ of the first three zeros of the sinusoid.

3.11.1 $x(t) = -\sqrt{3}\cos\left(\frac{2\pi}{6}t\right) - \sin\left(\frac{2\pi}{6}t\right)$.

3.11.2 $x(t) = \sqrt{2}\cos\left(\frac{2\pi}{6}t\right) - \sqrt{2}\sin\left(\frac{2\pi}{6}t\right)$.

3.11.3 $x(t) = -2\cos\left(\frac{2\pi}{6}t\right) + 2\sqrt{3}\sin\left(\frac{2\pi}{6}t\right)$.

*3.11.4 $x(t) = \cos\left(\frac{2\pi}{6}t\right) + \sin\left(\frac{2\pi}{6}t\right)$.

3.11.5 $x(t) = 3\cos\left(\frac{2\pi}{6}t\right) - \sqrt{3}\sin\left(\frac{2\pi}{6}t\right)$.

3.11.6 $x(t) = -2\sin\left(\frac{2\pi}{6}t\right)$.

3.12 Given the sinusoids $x_1(t) = A_1\cos(\omega t + \theta_1)$ and $x_2(t) = A_2\cos(\omega t + \theta_2)$, find the sinusoid $x(t) = x_1(t) - x_2(t) = A\cos(\omega t + \theta)$. First add a phase of π or $-\pi$ to the sinusoid $x_2(t)$ and then use the summation formulas given in the book. Find the sample values of the sinusoids $x_1(t)$ and $x_2(t)$ at $t = 0, 1, 2$ and verify that the sample values of $x_1(t) - x_2(t)$ are the same as those of $x(t)$.

3.12.1 $x_1(t) = -2\cos\left(\frac{2\pi}{6}t - \frac{\pi}{3}\right)$, $x_2(t) = 3\sin\left(\frac{2\pi}{6}t + \frac{\pi}{3}\right)$.

3.12.2 $x_1(t) = \sin\left(\frac{2\pi}{6}t + \frac{\pi}{4}\right)$, $x_2(t) = \cos\left(\frac{2\pi}{6}t + \frac{5\pi}{6}\right)$.

*3.12.3 $x_1(t) = 3\cos\left(\frac{2\pi}{6}t + \frac{\pi}{3}\right)$, $x_2(t) = 4\cos\left(\frac{2\pi}{6}t + \frac{\pi}{4}\right)$.

3.12.4 $x_1(t) = 2\cos\left(\frac{2\pi}{6}t + \frac{\pi}{6}\right)$, $x_2(t) = 5\cos\left(\frac{2\pi}{6}t + \frac{\pi}{3}\right)$.

3.13 Give the sample values of the exponentially varying amplitude sinusoid for $t = -1, 0, 1$.

3.13.1 $x(t) = e^{-t}\sin\left(\frac{2\pi}{6}t + \frac{\pi}{6}\right)$.

3.13.2 $x(t) = e^{2t}\cos\left(\frac{2\pi}{6}t - \frac{\pi}{3}\right)$.

3.13.3 $x(t) = e^{-2t}\cos(\pi t)$.

3.13.4 $x(t) = e^{2t}\sin\left(\frac{2\pi}{6}t\right)$.

3.14 The sinusoid $x(t)$ and the value k are specified. Find the value of $t > 0$ where the first positive peak of the sinusoid $x(t)$ occur. From the sinusoid $x(t + k)$, verify that its first positive peak, after $t > 0$, occurs as expected from the value of k.

3.14.1 $x(t) = 2\cos\left(\frac{2\pi}{6}t - \frac{\pi}{3}\right)$, $k = 2$.

*3.14.2 $x(t) = \sin\left(\frac{2\pi}{6}t + \frac{\pi}{6}\right)$, $k = -1$.

3.14.3 $x(t) = \sin\left(\frac{2\pi}{6}t - \frac{\pi}{4}\right)$, $k = 15$.

3.14.4 $x(t) = \cos\left(\frac{2\pi}{6}t + \frac{5\pi}{6}\right)$, $k = 12$.

3.14.5 $x(t) = \sin\left(\frac{2\pi}{6}t\right)$, $k = 1$.

3.15 The sinusoid $x(t)$ and the value k are specified. Find the value of $t > 0$ where the first positive peaks of the sinusoids $x(t)$ and $x(-t + k)$ occur.

3.15.1 $x(t) = 3\sin\left(\frac{2\pi}{6}t + \frac{\pi}{6}\right)$, $k = -1$.

*3.15.2 $x(t) = 2\cos\left(\frac{2\pi}{6}t - \frac{\pi}{4}\right)$, $k = 2$.

3.15.3 $x(t) = \sin\left(\frac{2\pi}{6}t - \frac{\pi}{3}\right)$, $k = -3$.

3.15.4 $x(t) = \sin\left(\frac{2\pi}{6}t + \frac{\pi}{3}\right)$, $k = -12$.

3.15.5 $x(t) = \cos\left(\frac{2\pi}{6}t + \frac{\pi}{6}\right)$, $k = 4$.

3.16 The sinusoid $x(t)$ and the values of a and k are specified. Find the value of $t > 0$ where the first positive peaks of the sinusoids $x(t)$ and $x(at + k)$ occur.

3.16.1　$x(t) = \cos\left(\frac{2\pi}{8}t + \frac{\pi}{3}\right),\ a = 2,\ k = 1.$

3.16.2　$x(t) = \sin\left(\frac{2\pi}{8}t + \frac{\pi}{6}\right),\ a = -\frac{1}{3},\ k = -2.$

3.16.3　$x(t) = \cos\left(\frac{2\pi}{8}t - \frac{\pi}{4}\right),\ a = \frac{3}{2},\ k = -1.$

*3.16.4　$x(t) = \sin\left(\frac{2\pi}{8}t - \frac{\pi}{3}\right),\ a = -\frac{2}{3},\ k = 2.$

3.16.5　$x(t) = \cos\left(\frac{2\pi}{8}t\right),\ a = 3,\ k = 1.$

3.17 The waveform $x(t)$ and the values k and a are specified. List the values at $t = -3, -2, -1, 0, 1, 2, 3$ of the waveforms $x(t)$, $x(t + k)$, and $x(at + k)$. Assume that the value of the function is its right-hand limit at any discontinuity.

3.17.1　$x(t) = e^{-0.1t}.\ a = 2,\ k = -1.$

3.17.2　$x(t) = e^{-0.2t}.\ a = \frac{1}{2},\ k = 2.$

3.17.3　$x(t) = e^{1.05t}.\ a = \frac{3}{2},\ k = 2.$

3.17.4　$x(t) = e^{1.2t}.\ a = \frac{1}{3},\ k = -3.$

*3.17.5　$x(t) = -2\sin\left(\frac{2\pi}{6}t + \frac{\pi}{3}\right)u(t).\ a = 2,\ k = 3.$

3.17.6　$x(t) = -2\sin\left(\frac{2\pi}{6}t + \frac{\pi}{6}\right)u(t).\ a = \frac{1}{2},\ k = -1.$

3.17.7　$x(t) = e^{-0.3t}\cos\left(\frac{2\pi}{6}t + \frac{\pi}{4}\right)u(t).\ a = -2,\ k = 1.$

3.17.8　$x(t) = e^{-0.4t}\cos\left(\frac{2\pi}{6}t - \frac{\pi}{6}\right)u(t).\ a = \frac{1}{3},\ k = 3.$

4

Time-domain Analysis of Discrete Systems

A system carries out some task in response to an input signal or produces an output signal that is an altered version of the input signal. For example, when we switch the power on to an electric motor, it produces mechanical power. A filter produces an output signal in which the various frequency components of the input signal are altered in a predefined way. A system is realized using physical components (hardware realization) or using a computer program (software realization) or a combination of both. In order to analyze a system, a mathematical model of the system has to be derived using the laws governing the behavior of its components and their interconnection. It is usually not possible to develop an accurate model of a system. Therefore, a model, with minimum mathematical complexity, is developed so that it is a sufficiently accurate representation of the actual system. Although systems can have multiple inputs and multiple outputs, we consider single-input and single-output systems only, for simplicity. Because the frequency-domain methods, described in later chapters, are easier for the analysis of higher-order systems, only first-order systems are considered in this chapter.

The difference equation model of a system is derived in Section 4.1. In Section 4.2, the various classifications of systems are described. The convolution–summation model of a system is developed in Section 4.3. In Section 4.4, the stability condition of a system is derived in terms of its impulse response. In Section 4.5, the basic components used in the implementation of discrete systems, implementation of a specific system, and the decomposition of higher-order systems are presented.

4.1 Difference Equation Model

The resistor–capacitor (RC) circuit, shown in Figure 4.1, is a lowpass filter, as the impedance of the capacitor is smaller at higher frequencies and larger at lower

A Practical Approach to Signals and Systems   D. Sundararajan
© 2008 John Wiley & Sons (Asia) Pte Ltd

Figure 4.1 An RC filter circuit.

frequencies. Therefore, the output voltage across the capacitor $y(t)$ is a filtered version of the input $x(t)$. The relationship between the current through the capacitor and the voltage across it is $i(t) = Cdy(t)/dt$. Then, due to Kirchhoff's voltage law, we get the differential equation (an equation that contains derivatives of functions) model of the circuit

$$RC\frac{dy(t)}{dt} + y(t) = x(t)$$

where R is in ohms and C is in farads. This model of the filter circuit can be approximated by a difference equation (an equation that contains differences of functions) by approximating the differential in the differential equation by a difference. One of the ways of this approximation is by replacing the term $dy(t)/dt$ by $[y(nT_s) - y((n-1)T_s)/T_s]$, where T_s is the sampling interval. The continuous variables $x(t)$ and $y(t)$ become $x(nT_s)$ and $y(nT_s)$, respectively. As usual, the sampling interval T_s in nT_s is suppressed and we get the difference equation as

$$RC\frac{y(n) - y(n-1)}{T_s} + y(n) = x(n)$$

Let $b_1 = T_s/(T_s + RC)$ and $a_0 = -RC/(T_s + RC)$. Then, we get the difference equation characterizing the circuit as

$$y(n) = b_1 x(n) - a_0 y(n-1) \qquad (4.1)$$

Let us assume that the input voltage is applied to the circuit at $n = 0$. Then, the output of the circuit at $n = 0$ is given by

$$y(0) = b_1 x(0) - a_0 y(0-1)$$

The voltage $y(-1)$ across the capacitor at $n = -1$, called the initial condition of the circuit, is required to find the output. The number of initial conditions required to find the output indicates the number of independent storage devices in the system. This number is also the order of the system. As only one value of initial condition is required, the model of the RC circuit is a first-order difference equation. Given the initial condition and the input, using this model, we can approximate the response of the circuit.

4.1.1 System Response

The response of a linear system is due to two independent causes, the input and the initial condition of the system at the time the input is applied. The response due to the initial condition alone is called the zero-input response, as the input is assumed to be zero. The response due to the input alone is called the zero-state response, as the initial condition or the state of the system is assumed to be zero. The complete response of a linear system is the sum of the zero-input and zero-state responses.

4.1.1.1 Zero-state Response

The difference equation characterizing a system has to be solved to get the system response. One way of solving a difference equation is by iteration. With the given initial condition $y(-1)$ and the inputs $x(0)$ and $x(-1)$, we can find the output $y(0)$ of a first-order difference equation. Then, in the next iteration, using $y(0)$, $x(1)$, and $x(0)$, we can compute $y(1)$. We repeat this process to get the desired number of outputs. Note that this method is suitable for programming in a digital computer. We can also deduce the closed-form solution by looking at the pattern of the expressions of the first few iterations. Let us solve Equation (4.1) by iteration. Assume that the initial condition is zero and the input signal is the unit-step, $u(n)$.

$$y(0) = b_1 x(0) + (-a_0) y(-1) = b_1$$
$$y(1) = b_1 x(1) + (-a_0) y(0) = b_1 (1 + (-a_0))$$

$$\vdots$$

$$y(n) = b_1 (1 + (-a_0) + (-a_0)^2 + \cdots + (-a_0)^n)$$
$$= b_1 \left(\frac{1 - (-a_0)^{(n+1)}}{1 - (-a_0)} \right), \quad (-a_0) \neq 1, \ n = 0, 1, 2, \ldots$$

4.1.1.2 Zero-input Response

Assume that the initial condition is $y(-1) = 3$. Since $x(n) = 0$ for all n, Equation (4.1) reduces to $y(n) = (-a_0) y(n-1)$, $y(-1) = 3$. Therefore,

$$y(0) = 3(-a_0), \quad y(1) = 3(-a_0)^2, \cdots, y(n) = 3(-a_0)^{(n+1)}$$

4.1.1.3 Complete Response

The complete response of the system is the sum of the zero-input and zero-state responses.

$$y(n) = b_1 \overbrace{\left(\frac{1 - (-a_0)^{(n+1)}}{1 - (-a_0)} \right)}^{\text{zero-state}} + \overbrace{3(-a_0)^{n+1}}^{\text{zero-input}}, \quad n = 0, 1, 2, \ldots$$

$$y(n) = b_1 \overbrace{\left(\frac{1}{1 - (-a_0)} \right)}^{\text{steady-state}} + \overbrace{b_1 \left(\frac{-(-a_0)^{(n+1)}}{1 - (-a_0)} \right) + 3(-a_0)^{(n+1)}}^{\text{transient}}$$

4.1.1.4 Transient and Steady-state Responses

The transient response of the system is

$$b_1 \left(\frac{-(-a_0)^{(n+1)}}{1 - (-a_0)} \right) + 3(-a_0)^{(n+1)}.$$

The steady-state response of the system,

$$b_1 \left(\frac{1}{1 - (-a_0)} \right)$$

is the response of the system after the transient response has decayed. The transient response of a stable system always decays with time. The form of the transient response depends solely on the characteristics of the system while that of the steady-state response solely depends on the input signal.

Figure 4.2 shows the various components of the response of the first-order system governed by the difference equation $y(n) = 0.1x(n) + 0.9y(n - 1)$ with the initial condition $y(-1) = 3$ and the input $x(n) = u(n)$, the unit-step signal. The zero-input response, shown in Figure 4.2(a), is $3(0.9)^{(n+1)}u(n)$. The zero-state response, shown in Figure 4.2(b), is $(1 - (0.9)^{(n+1)})u(n)$. The sum of the zero-input and zero-state responses is the complete response, shown in Figure 4.2(c), is $3(0.9)^{(n+1)} + 1 - (0.9)^{(n+1)} = (1 + 2(0.9)^{(n+1)})u(n)$. The transient response due to input alone, shown in Figure 4.2(d), is $-(0.9)^{(n+1)}u(n)$. The total transient response, shown in Figure 4.2(e), is $3(0.9)^{(n+1)} - (0.9)^{(n+1)} = 2(0.9)^{(n+1)}u(n)$. The steady-state response, shown in Figure 4.2(f), is $u(n)$. The sum of the transient and steady-state responses also forms the complete response.

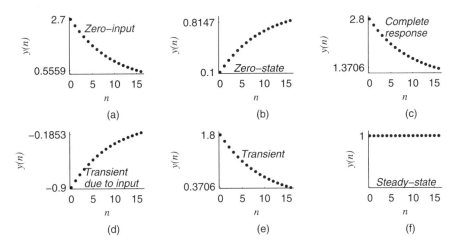

Figure 4.2   The response of a first-order system for unit-step input signal. (a) Zero-input response; (b) zero-state response; (c) complete response; (d) transient response due to input; (e) transient response; (f) steady-state response

4.1.1.5 Zero-input Response by Solving the Difference Equation

Consider the Nth order difference equation of a causal LTI discrete system relating the output $y(n)$ to the input $x(n)$

$$y(n) + a_{N-1}y(n-1) + a_{N-2}y(n-2) + \cdots + a_0 y(n-N)$$
$$= b_N x(n) + b_{N-1}x(n-1) + \cdots + b_0 x(n-N),$$

where N is the order of the system and the coefficients as and bs are real constants characterizing the system. If the input is zero, the difference equation reduces to

$$y(n) + a_{N-1}y(n-1) + a_{N-2}y(n-2) + \cdots + a_0 y(n-N) = 0$$

The solution to this equation gives the zero-input response of the system. This equation is a linear combination of $y(n)$ and its delayed versions equated to zero, for all values of n. Therefore, $y(n)$ and all its delayed versions must be of the same form. Only the exponential function has this property. Therefore, the solution is of the form $C\lambda^n$, where C and λ are to be found. Substituting $y(n) = C\lambda^n$, $y(n-1) = C\lambda^{n-1}$, etc., we get

$$(1 + a_{N-1}\lambda^{-1} + a_{N-2}\lambda^{-2} + \cdots + a_0\lambda^{-N})C\lambda^n = 0$$

Multiplying both sides by λ^N, we get

$$(\lambda^N + a_{N-1}\lambda^{N-1} + a_{N-2}\lambda^{N-2} + \cdots + a_0)C\lambda^n = 0$$

Assuming that the solution $C\lambda^n$ is nontrivial ($C \neq 0$),

$$(\lambda^N + a_{N-1}\lambda^{N-1} + a_{N-2}\lambda^{N-2} + \cdots + a_0) = 0 \qquad (4.2)$$

The characteristic polynomial on the left-hand side has N roots, $\lambda_1, \lambda_2, \ldots, \lambda_N$. Therefore, we get N solutions, $C_1\lambda_1^n, C_2\lambda_2^n, \ldots, C_N\lambda_N^n$. As the system is assumed to be linear and the solution has to satisfy N independent initial conditions of the system, the zero-input response of the system is given by

$$y(n) = C_1\lambda_1^n + C_2\lambda_2^n + \cdots + C_N\lambda_N^n$$

assuming all the roots of the characteristic polynomial are distinct. The constants can be found using the N independent initial conditions of the system. The zero-input response represents a behavior that is characteristic of the system. As the form of the zero-input response of any Nth order system is the same, it is the set of roots of the characteristic polynomial that distinguishes a specific system. Therefore, Equation (4.2) is called the characteristic equation of the system and the roots, $\lambda_1, \lambda_2, \ldots, \lambda_N$, are called the characteristic roots of the system. The corresponding exponentials, $\lambda_1^n, \lambda_2^n, \ldots, \lambda_N^n$, are called the characteristic modes of the system. The characteristic modes of a system are also influential in the determination of the zero-state response.

Example 4.1. Find the zero-input response of the system by solving its difference equation $y(n) = 0.1x(n) + 0.9y(n-1)$. The initial condition is $y(-1) = 3$.

Solution
The characteristic equation is $\lambda - 0.9 = 0$. The characteristic root of the system is $\lambda = 0.9$. The characteristic mode of the system is $(0.9)^n$. Therefore, the zero-input response is of the form

$$y(n) = C(0.9)^n$$

With $y(-1) = 3$ and letting $n = -1$, we get $C = 2.7$. Therefore, the zero-input response, as shown in Figure 4.2(a), is

$$y(n) = 2.7(0.9)^n u(n) \qquad \square$$

4.1.2 Impulse Response

The impulse response, $h(n)$, of a system is its response for a unit-impulse input signal with the initial conditions of the system zero. One way to find the impulse response of a system is by iteration. Another method is to find the zero-input response by solving the characteristic equation.

Example 4.2. Find the closed-form expression for the impulse response $h(n)$ of the system governed by the difference equation, with input $x(n)$ and output $y(n)$, $y(n) = 2x(n) + 3x(n-1) + \frac{1}{2}y(n-1)$ (i) by solving the difference equation and (ii) by iteration. Find the first four values of $h(n)$.

Solution
As the system is initially relaxed (initial conditions zero), we get from the difference equation $h(0) = 2$ and $h(1) = 4$ by iteration. As the value of the impulse signal is zero for $n > 0$, the response for $n > 0$ can be considered as zero-input response. The characteristic equation is

$$\left(\lambda - \frac{1}{2}\right) = 0$$

The zero-input response is of the form

$$h(n) = C \left(\frac{1}{2}\right)^n u(n-1)$$

As $u(n-1) = u(n) - \delta(n)$, the response is also given by

$$h(n) = C \left(\frac{1}{2}\right)^n u(n) - C\delta(n), \ n > 0$$

Letting $n = 1$, with $h(1) = 4$, we get $C = 8$. The impulse response is the sum of the response of the system at $n = 0$ and the zero-input response for $n > 0$. Therefore,

$$h(n) = 2\delta(n) + 8 \left(\frac{1}{2}\right)^n u(n) - 8\delta(n) = -6\delta(n) + 8 \left(\frac{1}{2}\right)^n u(n)$$

By iteration,

$$h(0) = 2$$

$$h(1) = \left(3 + 2\frac{1}{2}\right) = 4$$

$$h(2) = \left(\frac{1}{2}\right) 4$$

$$h(3) = \left(\frac{1}{2}\right)^2 4$$

$$\vdots$$

$$h(n) = \left(\frac{1}{2}\right)^{n-1} 4$$

$$h(n) = 2\delta(n) + \left(4\left(\frac{1}{2}\right)^{n-1}\right)u(n-1) = -6\delta(n) + 8\left(\frac{1}{2}\right)^n u(n)$$

The first four values of $h(n)$ are

$$\{h(0) = 2, h(1) = 4, h(2) = 2, h(3) = 1\} \qquad \square$$

In general, the impulse response of a first-order system governed by the differ-
ence equation $y(n) + a_0 y(n-1) = b_1 x(n) + b_0 x(n-1)$ is $h(n) = \frac{b_0}{a_0}\delta(n) + (b_1 - \frac{b_0}{a_0})(-a_0)^n u(n)$.

4.1.3 Characterization of Systems by their Responses to Impulse and Unit-step Signals

We can get information about the system behavior from the impulse and unit-step
responses. If the significant values of the impulse response are of longer duration, as
shown by filled circles in Figure 4.3(a), then the response of the system is sluggish.
The corresponding unit-step response is shown by filled circles in Figure 4.3(b). The
time taken for the unit-step response to rise from 10 to 90% of its final value is called
the rise time of the system. If the significant values of the impulse response are of
shorter duration, as shown by unfilled circles in Figure 4.3(a), then the response of the
system is faster, as shown by unfilled circles in Figure 4.3(b). A system with a shorter
impulse response has less memory and it is readily influenced by the recent values of
the input signal. Therefore, its response is fast. The faster is the rate of decay of the
impulse response, the faster the response approaches its steady-state value.

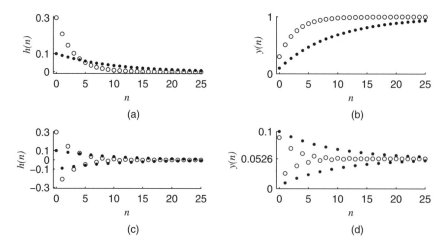

Figure 4.3 (a) Typical monotonically decreasing impulse responses; (b) the corresponding unit-step
responses; (c) typical alternating sequence impulse responses; (d) the corresponding unit-step responses.

The unit-step response, at n, is the sum of the first $n + 1$ terms of the impulse response, $y(n) = \sum_{m=0}^{n} h(m)$. As the final value tends to one in Figure 4.3(b) and as the unit-step signal, ultimately, acts like a dc signal, the monotonically decreasing impulse response indicates a system that passes low-frequency components of a signal well.

Figure 4.3(c) shows typical alternating sequence impulse responses. The corresponding unit-step responses are shown in Figure 4.3(d). In these cases also, the system response time is faster with a short-duration impulse response. However, note that the final value of the unit-step response approaches a very low value in Figure 4.3(d). This indicates a system that does not pass low-frequency components of a signal well.

4.2 Classification of Systems

4.2.1 Linear and Nonlinear Systems

Let the response of a system to signal $x_1(n)$ be $y_1(n)$ and the response to $x_2(n)$ be $y_2(n)$. Then, the system is linear if the response to the linear combination $ax_1(n) + bx_2(n)$ is $ay_1(n) + by_2(n)$, where a and b are arbitrary constants. A general proof is required to prove that a system is linear. However, one counterexample is enough to prove that a system is nonlinear. Nonlinear terms, such as $x^2(n)$ or $x(n)y(n-1)$ (terms involving the product of $x(n)$, $y(n)$, and their shifted versions) in the difference equation is an indication that the system is not linear. Any nonzero constant term is also an indication of a nonlinear system. The linearity condition implies that the total response of a linear system is the sum of zero-input and zero-state components. The linearity of a system with respect to zero-input and zero-state responses should be checked individually. In most cases, zero-state linearity implies zero-input linearity.

Example 4.3. Given the difference equation of a system, with input $x(n)$ and output $y(n)$, determine whether the system is linear. Verify the conclusion with the inputs $\{x_1(n), n = 0, 1, 2, 3\} = \{1, 4, 3, 2\}$, $\{x_2(n), n = 0, 1, 2, 3\} = \{2, 3, 4, 1\}$ and $x(n) = 2x_1(n) - 3x_2(n)$ by computing the first four values of the output. Assume that the initial condition $y(-1)$ is zero.

(a) $y(n) = x(n) + y(n-1) + 3$
(b) $y(n) = x(n) - (2n)y(n-1)$

Solution
(a) As the nonzero term indicates that the system is nonlinear, we try the counterexample method. By iteration, the first four output values of the system to the input signal $x_1(n)$ are

$$y_1(0) = x_1(0) + y_1(0-1) + 3 = 1 + 0 + 3 = 4$$

$$y_1(1) = x_1(1) + y_1(1-1) + 3 = 4 + 4 + 3 = 11$$

$$y_1(2) = x_1(2) + y_1(2 - 1) + 3 = 3 + 11 + 3 = 17$$

$$y_1(3) = x_1(3) + y_1(3 - 1) + 3 = 2 + 17 + 3 = 22$$

The output to $x_2(n)$ is $\{y_2(n), n = 0, 1, 2, 3\} = \{5, 11, 18, 22\}$. Now, $y(n) = \{2y_1(n) - 3y_2(n), n = 0, 1, 2, 3\} = \{-7, -11, -20, -22\}$.

The system response to the combined input $\{2x_1(n) - 3x_2(n), n = 0, 1, 2, 3\}$ $= \{-4, -1, -6, 1\}$ is $\{y(n), n = 0, 1, 2, 3\} = \{-1, 1, -2, 2\}$. As this output is different from that computed earlier, the system is nonlinear.

(b) The system output to $x_1(n)$ is $y_1(n) = x_1(n) - (2n) y_1(n - 1)$. The system output to $x_2(n)$ is $y_2(n) = x_2(n) - (2n) y_2(n - 1)$. Then,

$$ay_1(n) + by_2(n) = ax_1(n) - (2an) y_1(n - 1) + bx_2(n) - (2bn) y_2(n - 1)$$

The system output to $ax_1(n) + bx_2(n)$ is

$$ax_1(n) + bx_2(n) - (2n)(ay_1(n - 1) + by_2(n - 1))$$

As both the expressions for the output are the same, the system is linear. The output to $x_1(n)$ is $\{y_1(n), n = 0, 1, 2, 3\} = \{1, 2, -5, 32\}$. The output to $x_2(n)$ is $\{y_2(n), n = 0, 1, 2, 3\} = \{2, -1, 8, -47\}$. Now, $y(n) = \{2y_1(n) - 3y_2(n), n = 0, 1, 2, 3\} = \{-4, 7, -34, 205\}$.

The system response to the combined input $\{2x_1(n) - 3x_2(n), n = 0, 1, 2, 3\}$ $= \{-4, -1, -6, 1\}$ is $\{y(n), n = 0, 1, 2, 3\} = \{-4, 7, -34, 205\}$. This output is the same as that computed earlier. $\qquad\square$

4.2.2 Time-invariant and Time-varying Systems

The output of a time-invariant system to the input $x(n - m)$ must be $y(n - m)$ for all m, assuming that the output to the input $x(n)$ is $y(n)$ and the initial conditions are identical. A general proof is required to prove that a system is time-invariant. However, one counterexample is enough to prove that a system is time-variant. Terms, such as $x(2n)$ or $x(-n)$, with a nonzero and nonunity constant associated with the index n in the difference equation indicates a time-variant system. Any coefficient that is an explicit function of n in the difference equation also indicates a time-variant system.

Example 4.4. Given the difference equation of a system, with input $x(n)$ and output $y(n)$, determine whether the system is time-invariant. Verify the conclusion with the inputs $\{x(n), n = 0, 1, 2, 3\} = \{1, 4, 3, 2\}$ and $\{x(n - 2), n = 2, 3, 4, 5\} = \{1, 4, 3, 2\}$ by computing the first four values of the output. Assume that the initial condition $y(-1)$ is zero.

(a) $y(n) = nx(n)$
(b) $y(n) = 2x(n)$

Solution

(a) As the coefficient in the difference equation is the independent variable n, we try the counterexample method. The output of the system to $x(n)$ is $\{y(n), n = 0, 1, 2, 3\} = \{0, 4, 6, 6\}$. The output of the system to $x(n - 2)$ is $\{y(n), n = 2, 3, 4, 5\} = \{2, 12, 12, 10\}$. As the two outputs are different, the system is time-varying.

(b) The system output to $x(n)$ is $y(n) = 2x(n)$. By replacing n by $(n - 2)$, we get $y(n - 2) = 2x(n - 2)$. The system output to $x(n - 2)$ is $2x(n - 2)$. As the outputs are the same, the system is time-invariant. The output of the system to $x(n)$ is $\{y(n), n = 0, 1, 2, 3\} = \{2, 8, 6, 4\}$. The output of the system to $x(n - 2)$ is $\{y(n), n = 2, 3, 4, 5\} = \{2, 8, 6, 4\}$. □

Linear time-invariant (LTI) systems satisfy the linearity and time-invariant properties and are easier to analyze and design. Most practical systems, although not strictly linear and time-invariant, can be considered as LTI systems with acceptable error limits.

4.2.3 *Causal and Noncausal Systems*

Practical systems respond only to present and past input values, but not to future input values. These systems are called causal or nonanticipatory systems. If the present output $y(n)$ depends on the input $x(n + k)$ with $k > 0$, then the system is noncausal. This implies that the impulse response of a causal system $h(n)$ is zero for $n < 0$. Ideal systems, such as ideal filters, are noncausal. However, they are of interest because they set an upper bound for the system response. Practical systems approximate the ideal response, while being causal (that is physically realizable).

Example 4.5. Given the difference equation of a system, with input $x(n)$ and output $y(n)$, determine whether the system is causal. Find the impulse response.

(a) $y(n) = x(n + 2) + 2x(n) - 3x(n - 1)$
(b) $y(n) = 2x(n) - x(n - 1) + 3x(n - 4)$.

Solution

(a) As the output $y(n)$ is a function of the future input sample $x(n + 2)$, the system is noncausal. The impulse response of the system is obtained, by substituting $x(n) = \delta(n)$ in the input-output relation, as $y(n) = h(n) = \delta(n + 2) + 2\delta(n) - 3\delta(n - 1)$. That is, $h(-2) = 1$, $h(-1) = 0$, $h(0) = 2$, and $h(1) = -3$.

(b) The system is causal. The impulse response of the system is

$$\{h(0) = 2, h(1) = -1, h(2) = 0, h(3) = 0, h(4) = 3\}$$ □

4.2.4 Instantaneous and Dynamic Systems

With regard to system memory, systems are classified as instantaneous or dynamic. A system is instantaneous (no memory) if its output at an instant is a function of the input at that instant only. The system characterized by the difference equation $y(n) = 2x(n)$ is a system with no memory. An example is an electrical circuit consisting of resistors only. Any system with storage elements, such as inductors and capacitors, is a dynamic system, since the output at an instant of such systems is a function of past values of the input also. The discrete model of this type of systems will have terms, such as $x(n - 1)$ or $x(n - 2)$, that require memory units to implement. If the output depends only on a finite number of past input samples, then it is called a finite memory system. For example, $y(n) = x(n - 1) + x(n - 2)$ is the difference equation of a system with two memory units. Systems with capacitive or inductive elements are infinite memory systems, since their output is a function of entire past history of the input. Instantaneous systems are a special case of dynamic systems with zero memory.

4.2.5 Inverse Systems

A system is invertible if its input can be determined from its output. This implies that each input has a unique output. Systems with an input–output relationship such as $y(n) = x^2(n)$ are not invertible. If the impulse response of a system, made up of two systems connected in cascade, is $h(n) = \delta(n)$, then the two systems are the inverses of one another. For example, the inverse of the system with the input–output relationship $y(n) = 2x(n)$ is $x(n) = \frac{1}{2}y(n)$.

4.2.6 Continuous and Discrete Systems

In continuous systems, input, output, and all other signals are of continuous type and they are processed using devices such as resistors, inductors, and capacitors. In a discrete system, input, output, and all other signals are of discrete type and they are processed using discrete devices such as a digital computer. While most naturally occurring signals are of continuous type, they are usually analyzed and processed using discrete systems, as it is advantageous, by converting the continuous signals to discrete signals by sampling. These type of systems, in which both types of signals appear, are called hybrid systems.

4.3 Convolution–Summation Model

In the difference equation model of a system, we used some output and input values in formulating the model. In the convolution-summation model, the model is formulated in terms of all the input values applied to the system, assuming that the initial conditions are zero. The input signal is decomposed in terms of scaled and shifted unit-impulses. Therefore, with the knowledge of the response of the system to just the unit-impulse (called the impulse response), we find the response to each of the constituent impulses of an arbitrary input signal and sum the individual responses to find the total response. As the initial conditions are assumed to be zero, the response obtained using this model is the zero-state response.

Let us find the convolution of the impulse response $\{h(m), m = 0, 1, 2, 3\} = \{5, 0, 3, 2\}$ and the input $\{x(m), m = 0, 1, 2, 3\} = \{4, 1, 3, 2\}$ shown in Figure 4.4. The time-reversed impulse response, $\{h(0 - m), m = 3, 2, 1, 0\}$, is $\{2, 3, 0, 5\}$. There is only one nonzero product, $x(0)h(0) = 4 \times 5 = 20$, of $x(m)h(0 - m)$ with $m = 0$ and the convolution output is $y(0) = 20$. The product $x(0)h(0)$ is the response of the system at $n = 0$ to the present input sample $x(0)$. There is no contribution to the output at $n = 0$ due to input samples $x(1)$, $x(2)$, and $x(3)$ since the system is causal. The time-reversed impulse response is shifted to the right by one sample interval to get $h(1 - m) = h(-m + 1)) = h(-(m - 1))$. The convolution output $y(1)$ at $n = 1$ is the sum of products $x(m)h(1 - m)$, $m = 0, 1$. That is, $y(1) = x(0)h(1) + x(1)h(0) = 4 \times 0 + 1 \times 5 = 5$. The product $x(1)h(0)$ is the response of the system at $n = 1$ to the present input sample $x(1)$. The product $x(0)h(1)$ is the response of the system at $n = 1$ to the past input sample $x(0)$. Repeating the process, we find the remaining five output values. While $x(n)$ and $h(n)$ have four elements each, the output sequence $y(n)$ has seven elements. The duration of the convolution of two finite sequences of length N and M is $N + M - 1$ samples, as the overlap of nonzero portions can occur only over that length.

A more formal development of the convolution operation is as follows. An arbitrary signal can be decomposed, in terms of scaled and shifted impulses, as

$$x(n) = \sum_{m=-\infty}^{\infty} x(m)\delta(n - m)$$

The impulse response $h(n)$ of a LTI system is its response to an impulse $\delta(n)$ with the system initially relaxed (initial conditions zero). Due to the time-invariance property, a delayed impulse $\delta(n - m)$ will produce the response $h(n - m)$. Since a LTI

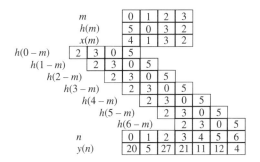

Figure 4.4   The linear convolution operation

system is linear, a scaled and shifted impulse $x(m)\delta(n-m)$ will produce the response $x(m)h(n-m)$. Therefore, using both the linearity and time-invariance properties, the system response $y(n)$ to an arbitrary signal $x(n)$ can be expressed as

$$y(n) = \sum_{m=-\infty}^{\infty} x(m)h(n-m) = x(n) * h(n)$$

The convolution-summation of the sequences $x(n)$ and $h(n)$ is denoted as $x(n) * h(n)$. For a causal system, as its impulse response $h(n)$ is zero for $n < 0$, the upper limit of the summation is n, instead of ∞, as $h(n-m) = 0$, $m > n$.

$$y(n) = \sum_{m=-\infty}^{\infty} x(m)h(n-m) = \sum_{m=-\infty}^{n} x(m)h(n-m)$$

If the signal $x(n)$ starts at any finite instant $n = n_0$, then the lower limit is equal to n_0. The effective range of the summation is easily determined by observing that if $x(m)$ or $h(n-m)$ or both are zero in a certain range, the product $x(m)h(n-m)$ is zero in that range.

Essentially, the convolution operation finds the sum of products of two sequences, each other's index running in opposite directions. To summarize, the output of a system is found by convolution with the repeated use of four operations (fold, shift, multiply, and add).

1. One of the two sequences to be convolved (say $h(m)$) is time-reversed, that is folded about the vertical axis at the origin to get $h(-m)$.
2. The time-reversed sequence, $h(-m)$, is shifted by n_0 sample intervals (right-shift for positive n_0 and left-shift for negative n_0), yielding $h(n_0 - m)$, to find the output at $n = n_0$.
3. The term by term products of the overlapping samples of the two sequences, $x(m)$, and $h(n_0 - m)$, are computed.
4. The sum of all the products is the output sample value at $n = n_0$.

Two finite sequences to be convolved overlap only partly at the beginning and the end of the convolution operation, as can be seen in Figure 4.4, and less arithmetic is required to find the convolution output in these cases. The convolution expression, requiring minimum arithmetic, for two finite sequences is given as follows. Let $x(n)$, $n = 0, 1, \ldots, N-1$ and $h(n)$, $n = 0, 1, \ldots, M-1$. Then,

$$y(n) = \sum_{m=\text{Max}(0,n-M+1)}^{\text{Min}(n,N-1)} x(m)h(n-m) = \sum_{m=\text{Max}(0,n-N+1)}^{\text{Min}(n,M-1)} h(m)x(n-m)$$

$$n = 0, 1, \ldots, N+M-2, \tag{4.3}$$

where Min and Max stand, respectively, for 'minimum of' and 'maximum of'. Along with the shift property of convolution presented shortly, this expression can be used to evaluate the convolution of two finite sequences starting from any n.

Example 4.6. Find the linear convolution of the sequences $\{x(n), n = 0, 1, 2\} = \{1, 2, 3\}$ and $\{h(n), n = 0, 1\} = \{2, -3\}$.

Solution
Using Equation (4.3), we get

$$y(0) = (1)(2) = 2$$
$$y(1) = (1)(-3) + (2)(2) = 1$$
$$y(2) = (2)(-3) + (3)(2) = 0$$
$$y(3) = (3)(-3) = -9$$

The values of the convolution of $x(n)$ and $h(n)$ are

$$\{y(0) = 2, \quad y(1) = 1, \quad y(2) = 0, \quad y(3) = -9\} \qquad \square$$

Example 4.7. Find the closed-form expression of the convolution of the sequences $x(n) = (0.6)^n u(n)$ and $h(n) = (0.5)^n u(n)$.

Solution

$$y(n) = \sum_{l=-\infty}^{\infty} x(l)h(n-l) = \sum_{l=0}^{n} (0.6)^l (0.5)^{n-l}, \ n \geq 0$$

$$= (0.5)^n \sum_{l=0}^{n} \left(\frac{0.6}{0.5}\right)^l = (0.5)^n \left(\frac{1 - \left(\frac{0.6}{0.5}\right)^{n+1}}{1 - \left(\frac{0.6}{0.5}\right)}\right)$$

$$= (6(0.6)^n - 5(0.5)^n)u(n)$$

The first four values of the convolution of $x(n)$ and $h(n)$ are

$$\{y(0) = 1, \quad y(1) = 1.1, \quad y(2) = 0.91, \quad y(3) = 0.671\} \qquad \square$$

4.3.1 Properties of Convolution–Summation

Convolution–summation is commutative, that is the order of the two sequences to be convolved is immaterial.

$$x(n) * h(n) = h(n) * x(n)$$

Convolution–summation is distributive. That is the convolution of a sequence with the sum of two sequences is the same as the sum of the individual convolution of the first sequence with the other two sequences.

$$x(n) * (h_1(n) + h_2(n)) = x(n) * h_1(n) + x(n) * h_2(n)$$

Convolution–summation is associative. That is the convolution of a sequence with the convolution of two sequences is the same as the convolution of the convolution of the first two sequences with the third sequence.

$$x(n) * (h_1(n) * h_2(n)) = (x(n) * h_1(n)) * h_2(n)$$

The shift property of convolution is that

$$\text{if} \quad x(n) * h(n) = y(n) \quad \text{then} \quad x(n - l) * h(n - m) = y(n - l - m)$$

The convolution of two shifted sequences is the convolution of the two original sequences shifted by the sum of the shifts of the individual sequences.

Convolution of a sequence $x(n)$ with the unit-impulse leaves the sequence unchanged, except for the translation of the origin of the sequence to the location of the impulse.

$$x(n) * \delta(n - k) = \sum_{m=-\infty}^{\infty} \delta(m - k)x(n - m) = x(n - k)$$

Example 4.8. Find the linear convolution of the sequences $\{x(n), n = 0, 1, 2\} = \{3, 2, 4\}$ and $h(n) = \delta(n + 3)$.

Solution
$$x(n) * \delta(n + 3) = \{x(n + 3), n = -3, -2, -1\} = \{3, 2, 4\} \qquad \square$$

Convolution of $x(n)$ with the unit-step is the running sum of $x(n)$.

$$x(n) * u(n) = \sum_{l=-\infty}^{n} x(l)$$

4.3.2 The Difference Equation and Convolution–Summation

The difference equation and the convolution–summation are two different mathematical models of a LTI system producing the same output for the same input. Therefore, these two models are related. Consider the first-order difference equation, with input

$x(n)$ and output $y(n)$.

$$y(n) = b_1 x(n) + (-a_0) y(n-1)$$

As the initial conditions are assumed to be zero for the convolution–summation model, $y(-1) = 0$. In order to derive the convolution–summation model, we have to express the past output term in terms of input samples.

$$y(0) = b_1 x(0)$$

$$y(1) = b_1 x(1) + (-a_0) y(0) = b_1 x(1) + (-a_0) b_1 x(0)$$

$$y(2) = b_1 x(2) + (-a_0) y(1) = b_1 x(2) + (-a_0) b_1 x(1) + (-a_0)^2 b_1 x(0)$$

$$\vdots$$

$$y(n) = b_1 x(n) + (-a_0) b_1 x(n-1) + \cdots + (-a_0)^n b_1 x(0)$$

Then, the impulse response, with $x(n) = \delta(n)$, is given as

$$h(0) = b_1, \ \ h(1) = (-a_0) b_1, \ \ h(2) = (-a_0)^2 b_1, \ \ldots, h(n) = (-a_0)^n b_1$$

The output $y(n)$, using $h(n)$, can be expressed as

$$y(n) = h(0) x(n) + h(1) x(n-1) + \cdots + h(n) x(0) = \sum_{m=0}^{n} h(m) x(n-m),$$

which is the convolution-summation. For any n, $h(0)$ determines the effect of the current input $x(n)$ on the output $y(n)$. In general, $h(m)$ determines the effect of the input $x(n-m)$, applied m iterations before, on the output $y(n)$. A system, whose impulse response is of finite duration, is called a finite impulse response system. A system, whose impulse response is of infinite duration, is called an infinite impulse response system. In the difference equation model of a system, a system is characterized by the coefficients, as and bs, of its difference equation. In the convolution–summation model of a system, the system is characterized by its impulse response $h(n)$.

4.3.3 Response to Complex Exponential Input

A complex exponential with frequency $j\omega_0$ is given as $x(n) = e^{j\omega_0 n}$, $-\infty < n < \infty$. Assuming a causal and stable system with impulse response $h(n)$, the output of the system is given by the convolution–summation as

$$y(n) = \sum_{m=0}^{\infty} h(m) e^{j\omega_0(n-m)} = e^{j\omega_0 n} \sum_{m=0}^{\infty} h(m) e^{-j\omega_0 m}$$

As the second summation is independent of n and letting

$$H(e^{j\omega_0}) = \sum_{m=0}^{\infty} h(m)e^{-j\omega_0 m}$$

we get,

$$y(n) = H(e^{j\omega_0})e^{j\omega_0 n} = H(e^{j\omega_0})x(n)$$

$H(e^{j\omega_0})$ is called the frequency response since it is a constant complex scale factor, indicating the amount of change in the amplitude and phase of an input complex exponential $e^{j\omega_0 n}$ with frequency $j\omega_0$ at the output. Since the impulse response is real-valued for practical systems, the scale factor for an exponential with frequency $-j\omega_0$ is $H^*(e^{j\omega_0})$, where the superscript $*$ indicates complex conjugation. The point is that the input–output relationship of a LTI system becomes a multiplication operation rather than the more complex convolution operation. As the complex exponential is the only signal that has this property, it is used predominantly as the basis for signal decomposition. Even if the exponent of the complex exponential input signal has a real part, $x(n) = e^{(\sigma + j\omega_0)n}$, the response of the system is still related to the input by the multiplication operation. A real sinusoidal input $A\cos(\omega_0 n + \theta)$ is also changed at the output by the same amount of amplitude and phase of the complex scale factor $H(e^{j\omega_0})$. That is, $A\cos(\omega_0 n + \theta)$ is changed to $(|H(e^{j\omega_0})|A)\cos[\omega_0 n + (\theta + \angle(H(e^{j\omega_0}))])$.

There was no transient component in the output expression $y(n)$, since the exponential signal was applied at $n = -\infty$. For finite values of n, any transient component in the output of a stable system must have died out. However, if we apply the exponential at any finite instant, say $n = 0$, there will be a transient component in the response, in addition to the steady-state component $H(e^{j\omega_0})e^{j\omega_0 n}u(n)$.

Example 4.9. Let the input signal to a stable system with impulse response $h(n) = b_1(-a_0)^n u(n)$ be $x(n) = e^{j\omega_0 n}u(n)$. Find the response of the system. Assume that $y(-1) = 0$.

Solution
Using the convolution–summation, we get

$$y(n) = \sum_{m=0}^{n} h(m)e^{j\omega_0(n-m)} = b_1 e^{j\omega_0 n} \sum_{m=0}^{n}(-a_0)^m e^{-j\omega_0 m}$$

$$= \left(\frac{b_1}{1-(-a_0)e^{-j\omega_0}}\right)\left(e^{j\omega_0 n} - (-a_0)^{(n+1)}e^{-j\omega_0}\right), \quad n = 0, 1, \ldots$$

The first term, the steady-state component

$$\left(\frac{b_1}{1 - (-a_0)e^{-j\omega_0}}\right) e^{j\omega_0 n},$$

is the same as the input complex exponential with a complex scale factor. The second term

$$\left(\frac{b_1}{1 - (-a_0)e^{-j\omega_0}}\right) (-(-a_0)^{(n+1)}e^{-j\omega_0}),$$

is the transient component that will die for sufficiently large values of n.   □

4.4 System Stability

One of the criteria for the stability of a system is that the system output is bounded if the input is bounded. A sequence $x(n)$ is bounded if $|x(n)| \leq M$ for all values of n, where M is a finite positive number. For example, the sequence $x(n) = (0.8)^n u(n)$ is bounded and $x(n) = (1.2)^n u(n)$ is unbounded. As convolution–summation is a sum of products, the sum is bounded if the input signal is bounded and the sum of the magnitude of the terms of the impulse response is finite. Let the sample values of the input signal $x(n)$ are bounded by the positive constant M. From the convolution–summation relation for a causal system with impulse response $h(n)$, we get

$$|y(n)| = |\sum_{m=0}^{\infty} h(m)x(n - m)|$$

$$\leq \sum_{m=0}^{\infty} |h(m)x(n - m)| = \sum_{m=0}^{\infty} |h(m)||x(n - m)|$$

$$|y(n)| \leq \sum_{m=0}^{\infty} |h(m)|M = M \sum_{m=0}^{\infty} |h(m)|$$

Therefore, if $\sum_{m=0}^{\infty} |h(m)|$ is bounded then $|y(n)|$ is bounded. Consequently, a necessary and sufficient stability condition is that the impulse response is absolutely summable,

$$\sum_{m=0}^{\infty} |h(m)| < \infty$$

As we used the convolution–summation to derive the stability condition, the stability condition ensures a bounded zero-state response. The stability of the zero-input response should be checked separately, and it is presented in Chapter 10.

Example 4.10. Is the system governed by the difference equation, with input $x(n)$ and output $y(n)$, stable?

(i) $y(n) = 9x(n) + 2y(n-1)$
(ii) $y(n) = 9x(n) + 0.8y(n-1)$

Solution
(i) The impulse response of the system is $h(n) = 9(2)^n u(n)$. As $h(n)$ is not absolutely summable, the system is unstable.
(ii) The impulse response of the system is $h(n) = 9(0.8)^n u(n)$. As $h(n)$ is absolutely summable, the system is stable. ◻

4.5 Realization of Discrete Systems

A discrete system can be realized in software or hardware or as a combination of both. In any case, the three basic components required in the realization of discrete systems are: (i) multiplier units; (ii) adder units; and (iii) delay units. A multiplier unit, shown in Figure 4.5(a), produces an output sequence $c\,x(n)$, in which each element is the product of the corresponding element in the input sequence $x(n)$ and the coefficient c. An adder unit, shown in Figure 4.5(b), produces an output sequence $x(n) + y(n)$, in which each element is the sum of the corresponding elements in the input sequences $x(n)$ and $y(n)$. By complementing the subtrahend and then adding it with the minuend, subtraction can be realized by an adder unit. A delay unit, shown in Figure 4.5(c), produces an output sequence $x(n-1)$, which is a delayed version of the input sequence $x(n)$ by one sampling interval.

By interconnecting the basic components, a discrete system is realized. Consider the realization, shown in Figure 4.6, of the system governed by the difference equation, with input $x(n)$ and output $y(n)$, $y(n) = 2x(n) + 0.8y(n-1)$. We get the delayed output term $y(n-1)$ by passing $y(n)$ through a delay unit. The product term $0.8y(n-1)$ is obtained by passing $y(n-1)$ through a multiplier unit with coefficient 0.8. The product term $2x(n)$ is obtained by passing $x(n)$ through a multiplier unit with coefficient 2. The adder unit combines the two partial results $2x(n)$ and $0.8y(n-1)$ to produce the output signal $y(n)$.

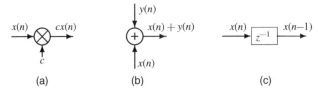

(a) (b) (c)

Figure 4.5 Basic components required in the realization of discrete systems. (a) multiplier unit; (b) adder unit; (c) delay unit

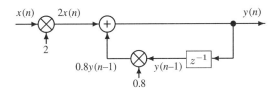

Figure 4.6 The realization of a discrete system

4.5.1 Decomposition of Higher-order Systems

To meet a given specification, a higher-order system is often required. Due to several advantages, a system is usually decomposed into first- and second-order systems connected in cascade or parallel. Figure 4.7(a) shows two systems with impulse responses $h_1(n)$ and $h_2(n)$ connected in parallel. The same input is applied to each system and the total response is the sum of the individual responses. The combined response of the two systems for the input $x(n)$ is $y(n) = x(n) * h_1(n) + x(n) * h_2(n)$. This expression, due to the distributive property of convolution, can be written as $y(n) = x(n) * (h_1(n) + h_2(n))$. That is, the parallel connection of the two systems is equivalent to a single system with impulse response $h(n) = h_1(n) + h_2(n)$, as shown in Figure 4.7(b).

Figure 4.8(a) shows two systems with impulse responses $h_1(n)$ and $h_2(n)$ connected in cascade. The output of one system is the input to the other. The response of the first system for the input $x(n)$ is $y_1(n) = x(n) * h_1(n)$. The response of the second system for the input $y_1(n) = x(n) * h_1(n)$ is $y(n) = (x(n) * h_1(n)) * h_2(n)$. This expression, due to the associative property of convolution, can be written as $y(n) = x(n) * (h_1(n) * h_2(n))$. That is, the cascade connection of the two systems is equivalent to a single system with impulse response $h(n) = h_1(n) * h_2(n)$, as shown in Figure 4.8(b). Due to the commutative property of convolution, the order of the systems in the cascade connection is immaterial, with respect to the input–output relationship.

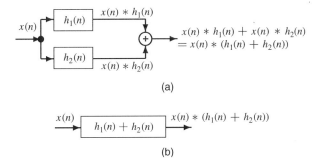

(a)

(b)

Figure 4.7 (a) Two systems connected in parallel; (b) a single system equivalent to the system in (a)

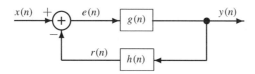

(a)

(b)

Figure 4.8   (a) Two systems connected in cascade; (b) a single system equivalent to the system in (a)

Figure 4.9   Two systems connected in a feedback configuration

4.5.2 Feedback Systems

Another configuration of systems, often used in control systems, is the feedback configuration shown in Figure 4.9. In feedback systems, a fraction of the output signal is fed
back and subtracted from the input signal to form the effective input signal. A feedback
signal $r(n)$ is produced by a system with impulse response $h(n)$ from the delayed samples of the output signal, $y(n-1)$, $y(n-2)$, etc. That is, $r(n) = \sum_{m=1}^{\infty} h(m)y(n-m)$.
This implies that $h(0) = 0$. The error signal $e(n)$ is the difference between the input
signal $x(n)$ and the feedback signal $r(n)$, $e(n) = x(n) - r(n)$. This error signal is the
input to a system with impulse response $g(n)$, which produces the output signal $y(n)$.
That is, $y(n) = \sum_{m=0}^{\infty} g(m)e(n-m)$.

4.6 Summary

- In this chapter, the time-domain analysis of LTI discrete systems has been presented.
- As discrete systems offer several advantages, they are mostly used instead of
  continuous systems. These systems can be designed to approximate continuous systems with a desired accuracy by selecting a sufficiently short sampling
  interval.
- The zero-input component of the response of a LTI system is its response due to the
  initial conditions alone with the input assumed to be zero. The zero-state component
  of the response of a LTI system is its response due to the input alone with the initial
  conditions assumed to be zero. The sum of the zero-input and zero-state responses
  is the complete response of the system.
- Two of the commonly used system models for time-domain analysis are the difference equation and convolution–summation models.

- The convolution–summation model gives the zero-state response of a LTI system. Both the zero-input and zero-state responses can be found using the difference equation model, either by solving the difference equation or by iteration.
- The impulse response of a system is its response to the unit-impulse input signal with zero initial conditions.
- The convolution–summation model is based on decomposing the input signal into a set of shifted and scaled impulses. The total response is found by summing the responses to all the constituent impulses of the input signal.
- The complete response of a system can also be considered as the sum of the transient component and the steady-state component. For a stable system, the transient component always decays with time. The steady-state component is the response after the transient response has decayed.
- A system is stable if its response is bounded for all bounded input signals. As the convolution–summation is a sum of products of the input and the impulse response, with the input bounded, the impulse response of a stable system must be absolutely summable for the convolution sum to be bounded.
- By interconnecting adder, multiplier, and delay units, any discrete system can be realized. A higher-order system is usually decomposed into a set of first- and second-order systems connected in cascade or parallel. A feedback system is obtained by feeding back some part of the output to the input.

Further Reading

1. Lathi, B. P., *Linear Systems and Signals*, Oxford University Press, New York, 2004.
2. Sundararajan, D., *Digital Signal Processing, Theory and Practice*, World Scientific, Singapore, 2003.
3. Sundararajan, D., *Discrete Fourier Transform, Theory, Algorithms, and Applications*, World Scientific, Singapore, 2001.

Exercises

4.1 Derive the closed-form expression for the impulse response $h(n)$, by iteration, of the system governed by the difference equation, with input $x(n)$ and output $y(n)$. List the values of the impulse response $h(n)$ at $n = 0, 1, 2, 3, 4, 5$.

4.1.1 $y(n) = x(n) + 2x(n-1) - 3y(n-1)$.

4.1.2 $y(n) = 2x(n) - 3x(n-1) + \frac{1}{2}y(n-1)$.

*4.1.3 $y(n) = 3x(n) - \frac{1}{3}y(n-1)$.

4.1.4 $y(n) = x(n) - 2x(n-1) + 2y(n-1)$.

4.1.5 $y(n) = 3x(n) - 4x(n-1) + y(n-1)$.

4.2 Find the closed-form expression for the impulse response $h(n)$ of the system by solving its difference equation, with input $x(n)$ and output $y(n)$. List the values of the impulse response $h(n)$ at $n = 0, 1, 2, 3, 4, 5$.

4.2.1 $y(n) = 3x(n) - x(n-1) + 2y(n-1)$.

4.2.2 $2y(n) = x(n) + x(n-1) - y(n-1)$.

4.2.3 $y(n) = 2x(n) + \frac{1}{4}y(n-1)$.

*4.2.4 $y(n) = 4x(n) + 3x(n-1) - y(n-1)$.

4.2.5 $y(n) = x(n) + x(n-1) - y(n-1)$.

4.3 Is the system governed by the given difference equation, with input $x(n)$ and output $y(n)$, linear? Let $\{x_1(n), n = 0, 1, 2, 3\} = \{1, 2, 3, 2\}$, $\{x_2(n), n = 0, 1, 2, 3\} = \{2, 3, 0, 4\}$ and $x(n) = 2x_1(n) - 3x_2(n)$. Assuming that the initial condition $y(-1)$ is zero, compute the first four output values and verify the conclusion.

4.3.1 $y(n) = 3x(n) - 2y(n-1) + 1$.

4.3.2 $y(n) = (x(n))^2 + y(n-1)$.

*4.3.3 $y(n) = x(n) - (n)y(n-1) + 2\cos(\frac{\pi}{2})$.

4.3.4 $y(n) = x(n) + x(n)y(n-1)$.

4.3.5 $y(n) = |x(n)|$.

4.3.6 $y(n) = (n)x(n) + y(n-1) - 3\cos(\pi)$.

4.4 Is the system governed by the given difference equation, with input $x(n)$ and output $y(n)$, time-invariant? Let $\{x(n), n = 0, 1, 2, 3, 4, 5, 6, 7, 8\} = \{2, 1, 3, 3, 4, 2, 5, 1, 3\}$. Assuming that the initial condition is zero, compute the first four output values and verify the conclusion to the input $\{x(n-2), n = 2, 3, 4, 5, 6, 7, 8, 9, 10\} = \{2, 1, 3, 3, 4, 2, 5, 1, 3\}$.

4.4.1 $y(n) = x(2n) + 2y(n-1)$.

4.4.2 $y(n) = 2x(n) - \sin(\frac{\pi}{2}n)y(n-1)$.

*4.4.3 $y(n) = (x(n))^2 - 2\cos(6\pi n)y(n-1)$.

4.4.4 $y(n) = x(n) + (n)y(n-1)$.

4.4.5 $y(n) = x(8-n)$.

4.5 Find the linear convolution of the sequences $x(n)$ and $h(n)$.

4.5.1 $\{x(n), n = 0, 1, 2\} = \{4, 2, 1\}$ and $\{h(n), n = 0, 1\} = \{-2, -3\}$.

4.5.2 $\{x(n), n = -2, -1, 0\} = \{2, -1, 4\}$ and $\{h(n), n = 3, 4, 5, 6\} = \{2, 1, 4, 3\}$.

*4.5.3 $\{x(n), n = -3, -2, -1, 0\} = \{2, 2, 1, 4\}$ and $\{h(n), n = 2, 3, 4, 5\} = \{3, 2, 3, 4\}$.

4.6 Find the closed-form expression for the convolution of the sequences $x(n)$ and $h(n)$. List the values of the convolution output at $n = 0, 1, 2, 3, 4, 5$.

4.6.1 $x(n) = u(n-1)$ and $h(n) = u(n-3)$.

4.6.2 $x(n) = (0.5)^n u(n-2)$ and $h(n) = (0.7)^n u(n-1)$.

4.6.3 $x(n) = (0.5)^{n-1} u(n-1)$ and $h(n) = (0.7)^{n-2} u(n-2)$.

4.6.4 $x(n) = (0.6)^n u(n)$ and $h(n) = x(n)$.

*4.6.5 $x(n) = (0.6)^n u(n-2)$ and $h(n) = u(n-1)$.

4.7 Find the linear convolution of the sequences $x(n)$ and $h(n)$.

4.7.1 $\{x(n), n = 1, 2, 3, 4\} = \{3, 2, 4, 1\}$ and $h(n) = \delta(n)$.

4.7.2 $\{x(n), n = -4, -3, -2\} = \{1, 3, 2\}$ and $h(n) = \delta(n-2)$.

4.7.3 $\{x(n), n = 3, 4, 5\} = \{5, 2, 3\}$ and $h(n) = \delta(n+3)$.

4.7.4 $x(n) = e^{j\frac{2\pi}{6}n}u(n)$ and $h(n) = \delta(n+4)$.

4.7.5 $x(n) = e^{j\frac{2\pi}{6}n}$ and $h(n) = \delta(n-6)$.

4.7.6 $x(n) = \cos(\frac{2\pi}{6}n)$ and $h(n) = \delta(n)$.

4.8   Verify the distributive and associative properties of convolution-summation
$$x(n) * (h_1(n) + h_2(n)) = x(n) * h_1(n) + x(n) * h_2(n)$$
and
$$x(n) * (h_1(n) * h_2(n)) = (x(n) * h_1(n)) * h_2(n)$$
where $\{h_1(n), n = 0, 1, 2, 3\} = \{1, 2, 3, 4\}$, $\{h_2(n), n = 0, 1, 2, 3\} = \{3, 2, 1, 5\}$, and $\{x(n), n = 0, 1, 2, 3\} = \{4, 4, 3, 2\}$.

4.9   Find the steady-state response of the system, with the impulse response
$$h(n) = -\frac{5}{3}\delta(n) + \frac{11}{3}(-0.6)^n u(n), \ n = 0, 1, 2, \dots,$$
to the input $x(n) = 3\sin(\frac{2\pi}{6}n - \frac{\pi}{6})u(n)$. Deduce the response to the input $e^{j\frac{2\pi}{6}n}$.

*4.10 Find the steady-state response of the system, with the impulse response
$$h(n) = -4\delta(n) + 7(0.5)^n u(n), \ n = 0, 1, 2, \dots,$$
to the input $x(n) = 2\cos(\frac{2\pi}{5}n + \frac{\pi}{4})u(n)$. Deduce the response to the input $e^{j\frac{2\pi}{5}n}$.

4.11   Derive the closed-form expression for the complete response (by finding the zero-state response using the convolution–summation and the zero-input response) of the system governed by the difference equation
$$y(n) = 2x(n) - x(n-1) + \frac{1}{3}y(n-1)$$
with the initial condition $y(-1) = 2$ and the input $x(n) = u(n)$, the unit-step function. List the values of the complete response $y(n)$ at $n = 0, 1, 2, 3, 4, 5$. Deduce the expressions for the transient and steady-state responses of the system.

4.12   Derive the closed-form expression for the complete response (by finding the zero-state response using the convolution–summation and the zero-input response) of the system governed by the difference equation
$$y(n) = x(n) - 2x(n-1) - \frac{1}{2}y(n-1)$$
with the initial condition $y(-1) = -3$ and the input $x(n) = (-1)^n u(n)$. List the values of the complete response $y(n)$ at $n = 0, 1, 2, 3, 4, 5$. Deduce the expressions for the transient and steady-state responses of the system.

*4.13 Derive the closed-form expression for the complete response (by finding the zero-state response using the convolution–summation and the zero-input response) of the system governed by the difference equation
$$y(n) = 3x(n) - 2x(n-1) + \frac{1}{4}y(n-1)$$
with the initial condition $y(-1) = 1$ and the input $x(n) = nu(n)$. List the values of the complete response $y(n)$ at $n = 0, 1, 2, 3, 4, 5$. Deduce the expressions for the transient and steady-state responses of the system.

4.14   Derive the closed-form expression for the complete response (by finding the zero-state response using the convolution–summation and the zero-input response) of the system governed by the difference equation

$$y(n) = x(n) + 3x(n-1) - \frac{3}{5}y(n-1)$$

with the initial condition $y(-1) = -2$ and the input $x(n) = (\frac{2}{5})^n u(n)$. List the values of the complete response $y(n)$ at $n = 0, 1, 2, 3, 4, 5$. Deduce the expressions for the transient and steady-state responses of the system.

4.15   Derive the closed-form expression for the complete response (by finding the zero-state response using the convolution–summation and the zero-input response) of the system governed by the difference equation

$$y(n) = 2x(n) - 4x(n-1) + \frac{1}{3}y(n-1)$$

with the initial condition $y(-1) = -3$ and the input $x(n) = 2\sin(\frac{2\pi}{6}n + \frac{\pi}{3})u(n)$. List the values of the complete response $y(n)$ at $n = 0, 1, 2, 3, 4, 5$. Deduce the expressions for the transient and steady-state responses of the system.

4.16   The impulse response of a LTI system is given. Is the system stable?

   4.16.1 $h(0) = 0$, $h(n) = \frac{(-1)^{n+1}}{n}$, $n = 1, 2, \ldots$.
   4.16.2 $h(0) = 0$, $h(n) = \frac{1}{n}$, $n = 1, 2, \ldots$.
   4.16.3 $h(0) = 0$, $h(n) = \frac{1}{n^2}$, $n = 1, 2, \ldots$.

4.17   Derive the closed-form expression of the impulse response $h(n)$ of the combined system consisting of systems governed by the given difference equations, with input $x(n)$ and output $y(n)$, if the systems are connected (i) in parallel and (ii) in cascade. List the first four values of the impulse response of the combined system.

   4.17.1

$$y_1(n) = 3x_1(n) + 2x_1(n-1) - \frac{1}{3}y_1(n-1) \quad \text{and}$$

$$y_2(n) = 2x_2(n) - 3x_2(n-1) - \frac{1}{4}y_2(n-1)$$

   *4.17.2

$$y_1(n) = x_1(n) - x_1(n-1) + \frac{1}{5}y_1(n-1) \quad \text{and} \quad y_2(n)$$

$$= x_2(n) + 2x_2(n-1) - \frac{3}{5}y_2(n-1)$$

   4.17.3

$$y_1(n) = 2x_1(n) + 2x_1(n-1) + \frac{2}{5}y_1(n-1) \quad \text{and} \quad y_2(n)$$

$$= 3x_2(n) - x_2(n-1) + \frac{5}{6}y_2(n-1)$$

5

Time-domain Analysis of Continuous Systems

While discrete systems, in general, are advantageous, we still need to study continuous systems. Continuous systems offer higher speed of operation. Even if we decide to use a discrete system, as the input and output signals are mostly continuous, we still need continuous systems for the processing of signals before and after the interface between the two types of system. The design of a discrete system can be made by first designing a continuous system and then using a suitable transformation of that design. As discrete systems usually approximate the continuous systems, comparing the exact analysis results of the continuous systems with that of the actual performance of the corresponding discrete system gives a measure of the approximation. For these reasons, the study of continuous systems is as much required as that of the discrete systems. In this chapter, we study two time-domain models of LTI continuous systems. We consider only first-order systems in this chapter as frequency-domain methods, described in later chapters, are easier for the analysis of higher-order systems. The analysis procedure remains essentially the same as that of discrete systems except that continuous systems are modeled using differential equation and convolution-integral methods, as the signals are of continuous type.

In Section 5.1, the various classifications of LTI continuous systems are described. In Sections 5.2 and 5.3, we develop the differential equation and convolution-integral models of a system, respectively. Using these models, in Section 5.4, the various components of the system response are derived. The important property of an exponential input signal to remain in the same form at the output of a stable LTI system is demonstrated. In Section 5.5, the stability of a system in terms of its impulse response is established. In Section 5.6, the basic components used in the implementation of continuous systems are presented and an implementation of a specific system is given. The decomposition of a higher-order system into a set of lower-order systems is also presented.

A Practical Approach to Signals and Systems   D. Sundararajan
© 2008 John Wiley & Sons (Asia) Pte Ltd

5.1 Classification of Systems

5.1.1 Linear and Nonlinear Systems

A system is linear if its response to a linear combination of input signals is the same linear combination of the individual responses to the inputs. Let the response of a system to signal $x_1(t)$ be $y_1(t)$ and the response to $x_2(t)$ be $y_2(t)$. Then, the system is linear if the response to a linear combination, $ax_1(t) + bx_2(t)$, is $ay_1(t) + by_2(t)$, where a and b are arbitrary constants. Nonlinear terms, such as $x^2(t)$ or $x(t)y(t)$ (the products involving $x(t)$, $y(t)$, and their derivatives), in the differential equation is an indication that the system is not linear. Any nonzero constant term is also an indication of a nonlinear system. The linearity condition implies that the total response of a linear system is the sum of zero-input and zero-state components. The zero-input component of the response of a system is its response due to the initial conditions alone with the input assumed to be zero. The zero-state component of the response of a system is its response due to the input alone with the initial conditions assumed to be zero. The linearity of a system with respect to zero-input and zero-state responses should be checked individually. In most cases, zero-state linearity implies zero-input linearity.

Example 5.1. Given the differential equation of a system, with output $y(t)$ and input $x(t)$, determine whether the system is linear. Assume that the initial condition $y(0)$ is zero.

(a) $y(t) = x(t) + t\frac{dy(t)}{dt}$

(b) $y(t) = x(t) + \left(\frac{dy(t)}{dt}\right)^2$

Solution

(a) Let $y_1(t)$ be the output to $x_1(t)$ and $y_2(t)$ be the output to $x_2(t)$. The system differential equation with $x_1(t)$ is $y_1(t) = x_1(t) + t\,dy_1(t)/dt$. The system differential equation with $x_2(t)$ is $y_2(t) = x_2(t) + t\,dy_2(t)/dt$. Then,

$$ay_1(t) + by_2(t) = ax_1(t) + at\frac{dy_1(t)}{dt} + bx_2(t) + bt\frac{dy_2(t)}{dt}$$

$$= ax_1(t) + bx_2(t) + t\frac{d}{dt}(ay_1(t) + by_2(t))$$

The system output to $x(t) = ax_1(t) + bx_2(t)$ is $y(t) = ay_1(t) + by_2(t)$ for a linear system. Substituting in the differential equation, we get

$$ay_1(t) + by_2(t) = ax_1(t) + bx_2(t) + t\frac{d}{dt}(ay_1(t) + by_2(t))$$

As both the differential equations are the same, the system is linear.

(b) The system differential equation with $x_1(t)$ is $y_1(t) = x_1(t) + (dy_1(t)/dt)^2$. The system differential equation with $x_2(t)$ is $y_2(t) = x_2(t) + (dy_2(t)/dt)^2$. Then,

$$ay_1(t) + by_2(t) = ax_1(t) + a\left(\frac{dy_1(t)}{dt}\right)^2 + bx_2(t) + b\left(\frac{dy_2(t)}{dt}\right)^2$$

The system output to $x(t) = ax_1(t) + bx_2(t)$ is $y(t) = ay_1(t) + by_2(t)$ for a linear system. Substituting in the differential equation, we get

$$ay_1(t) + by_2(t) = ax_1(t) + bx_2(t) + \left(a\frac{dy_1(t)}{dt} + b\frac{dy_2(t)}{dt}\right)^2$$

As the differential equations are different, the system is nonlinear. □

5.1.2 Time-invariant and Time-varying Systems

The output of a time-invariant system to the input $x(t - t_0)$ must be $y(t - t_0)$ for all t_0, assuming that the output to the input $x(t)$ is $y(t)$ and the initial conditions are identical. Terms, such as $x(2t)$ or $x(-t)$, with a nonzero and nonunit constant associated with the argument t in the differential equation indicates a time-variant system. Any coefficient that is an explicit function of t in the differential equation also indicates a time-variant system.

Example 5.2. Given the differential equation of a system, with output $y(t)$ and input $x(t)$, determine whether the system is time-invariant. Assume that the initial condition is zero.

(a) $y(t) = x(t) + t(dy(t)/dt)$
(b) $y(t) = x(t) + (dy(t)/dt)^2$

Solution
(a) By replacing t by $(t - a)$ in the differential equation, we get

$$y(t - a) = x(t - a) + (t - a)\frac{dy(t - a)}{dt}$$

The system output to $x(t - a)$ is $y(t - a)$ for a time-invariant system. Substituting in the differential equation, we get

$$y(t - a) = x(t - a) + t\frac{dy(t - a)}{dt}$$

As the differential equations are different, the system is time-varying.

(b) By replacing t by $(t - a)$ in the differential equation, we get

$$y(t - a) = x(t - a) + \left(\frac{dy(t - a)}{dt}\right)^2$$

The system output to $x(t - a)$ is $y(t - a)$ for a time-invariant system. Substituting in the differential equation, we get

$$y(t - a) = x(t - a) + \left(\frac{dy(t - a)}{dt}\right)^2$$

As both the differential equations are the same, the system is time-invariant.

$\square$

Linear time-invariant (LTI) systems satisfy the linearity and time-invariant properties and are easier to analyze and design. Most practical systems, although not strictly linear and time-invariant, can be considered as LTI systems with acceptable error limits.

5.1.3 Causal and Noncausal Systems

Practical systems respond only to present and past input values, but not to the future input values. These systems are called causal or nonanticipatory systems. This implies that the impulse response of a causal system $h(t)$ is zero for $t < 0$. If the present output $y(t)$ depends on the input $x(t + t_0)$ with $t_0 > 0$, then the system is noncausal. Ideal systems, such as ideal filters, are noncausal. However, they are of interest because they set an upper bound for the system response. Practical systems approximate the ideal response, while being causal (that is physically realizable).

Example 5.3. Given the differential equation of a system, with output $y(t)$ and input $x(t)$, determine whether the system is causal. Find the impulse response.

(a) $y(t) = x(t + 1) + 2x(t) - 3x(t - 1)$
(b) $y(t) = 2x(t) - x(t - 1) + 3x(t - 2)$.

Solution
(a) As the output $y(t)$ is a function of the future input sample $x(t + 1)$, the system is noncausal. The impulse response of the system is obtained, by substituting $x(t) = \delta(t)$ in the differential equation, as $y(t) = h(t) = \delta(t + 1) + 2\delta(t) - 3\delta(t - 1)$.
(b) The system is causal. The impulse response of the system is $y(t) = h(t) = 2\delta(t) - \delta(t - 1) + 3\delta(t - 2)$.

$\square$

5.1.4 Instantaneous and Dynamic Systems

With regard to system memory, systems are classified as instantaneous or dynamic. A system is instantaneous (no memory) if its output at an instant is a function of the input at that instant only. An example is an electrical circuit consisting of resistors only, with input–output relationship such as $v(t) = Ri(t)$. Any system with storage elements, such as inductors and capacitors, is called a dynamic system, since the output at an instant of such systems is also a function of past values of the input. If the output depends only on the input during T s of the immediate past, then it is called finite memory system. Systems with capacitive or inductive elements are infinite memory systems, since their output is a function of entire past history of the input. Instantaneous systems are a special case of the dynamic systems with zero memory.

5.1.5 Lumped-parameter and Distributed-parameter Systems

If the propagation time of a signal through a system is negligible, then that system is called a lumped-parameter system. For example, the current through a resistor in such a system is a function of time only, but not on the dimensions of the resistor. Such systems are modeled using ordinary differential equations. If the dimensions of a component are large compared with the wavelength of the highest frequency of interest, then the signal through that component is a function of time and the dimensions of the component. A system with that type of components is called a distributed-parameter system. Such systems, for example transmission lines, are modeled using partial differential equations.

5.1.6 Inverse Systems

A system is invertible if its input can be determined from its output. This implies that each input has a unique output. Systems with an input–output relationship such as $y(t) = x^2(t)$ are not invertible. If the impulse response of a system, made up of two systems connected in cascade, is $h(t) = \delta(t)$, then the two systems are the inverses of one another. For example, the inverse of the system with the input–output relationship $y(t) = 4x(t)$ is $x(t) = \frac{1}{4}y(t)$.

5.2 Differential Equation Model

Differential equations are used in one type of time-domain modeling of continuous systems. The input–output relationship of commonly used components of a system, such as inductors and capacitors, is governed by differential equations. Therefore, differential equations naturally arise in modeling systems. The interconnection of several elements leads to a model represented by higher-order differential equations. Consider the Nth-order differential equation of a causal LTI continuous system relating the output $y(t)$ to the input $x(t)$

$$\frac{d^N y(t)}{dt^N} + a_{N-1}\frac{d^{N-1} y(t)}{dt^{N-1}} + \cdots + a_1 \frac{dy(t)}{dt} + a_0 y(t)$$

$$= b_N \frac{d^N x(t)}{dt^N} + b_{N-1}\frac{d^{N-1} x(t)}{dt^{N-1}} + \cdots + b_1 \frac{dx(t)}{dt} + b_0 x(t)$$

where N is the order of the system and the coefficients as and bs are real constants characterizing the system. If the input is zero, the differential equation reduces to

$$\frac{d^N y(t)}{dt^N} + a_{N-1}\frac{d^{N-1} y(t)}{dt^{N-1}} + \cdots + a_1 \frac{dy(t)}{dt} + a_0 y(t) = 0$$

Denoting $d/dt = D$, we get

$$(D^N + a_{N-1} D^{N-1} + \cdots + a_1 D + a_0) y(t) = 0$$

The solution to this equation gives the zero-input response of the system. This equation is a linear combination of $y(t)$ and its N successive derivatives equated to zero, for all values of t. Therefore, $y(t)$ and all its N successive derivatives must be of the same form. Only the exponential function has this property. Therefore, the solution is of the form $Ce^{\lambda t}$, where C and λ are to be found. Substituting $y(t) = Ce^{\lambda t}$, $dy(t)/dt = C\lambda e^{\lambda t}$, etc., we get

$$(\lambda^N + a_{N-1}\lambda^{N-1} + \cdots + a_1\lambda + a_0)Ce^{\lambda t} = 0$$

Assuming that the solution is nontrivial $(C \neq 0)$,

$$(\lambda^N + a_{N-1}\lambda^{N-1} + \cdots + a_1\lambda + a_0) = 0 \tag{5.1}$$

The characteristic polynomial on the left-hand side has N roots, $\lambda_1, \lambda_2, \ldots, \lambda_N$. Therefore, we get N solutions, $C_1 e^{\lambda_1 t}, C_2 e^{\lambda_2 t}, \ldots, C_N e^{\lambda_N t}$. As the system is assumed to be linear and the solution has to satisfy N independent initial conditions of the system, the zero-input response of the system is given by

$$y(t) = C_1 e^{\lambda_1 t} + C_2 e^{\lambda_2 t} + \cdots + C_N e^{\lambda_N t},$$

assuming all the roots are distinct. The constants can be found using the N independent initial conditions of the system. The zero-input response represents a behavior that is characteristic of the system. As the form of the zero-input response of any Nth-order system is the same, it is the set of roots of the characteristic polynomial that distinguishes a specific system. Therefore, Equation (5.1) is called the characteristic equation of the system and the roots, $\lambda_1, \lambda_2, \ldots, \lambda_N$, are called the characteristic roots of the system. The corresponding exponentials, $e^{\lambda_1 t}$, $e^{\lambda_2 t}$, ..., $e^{\lambda_N t}$, are called the characteristic modes of the system. The characteristic modes of a system are also influential in the determination of the zero-state response.

5.3 Convolution-integral Model

We have shown, in Chapter 3, that an arbitrary signal $x(t)$ can be decomposed into scaled and shifted impulses as

$$x(t) = \int_{-\infty}^{\infty} x(\tau)\delta(t - \tau)d\tau$$

Let the response of a system to the impulse $\delta(t)$ be $h(t)$. Then, the system response of a LTI system to $x(\tau)\delta(t - \tau)d\tau$ is $x(\tau)h(t - \tau)d\tau$. The total response $y(t)$ of the system to the signal $x(t)$ is the sum of responses of all the constituent continuum of impulse components of $x(t)$,

$$y(t) = \int_{-\infty}^{\infty} x(\tau)h(t - \tau)d\tau = x(t) * h(t)$$

This relation is the convolution-integral of the signals $x(t)$ and $h(t)$ denoted as $x(t) * h(t)$. As the impulse response $h(t)$ of a causal system is zero for $t < 0$, the upper limit of the integral will be t in this case, instead of ∞, as $h(t - \tau) = 0$, $\tau > t$. If the signal $x(t)$ starts at the instant $t = t_0$ then the lower limit is equal to t_0. The convolution output is the integral of products of two signals, each other's argument running in opposite directions.

To summarize, the output of a system is found by convolution with the repeated use of four operations (fold, shift, multiply, and integrate).

1. One of the two signals to be convolved (say $h(\tau)$) is time-reversed, that is folded about the vertical axis at the origin to get $h(-\tau)$.
2. The time-reversed signal, $h(-\tau)$, is shifted by t_0 (right-shift for positive t_0 and left-shift for negative t_0), yielding $h(t_0 - \tau)$, to find the output at $t = t_0$.
3. The product of the two signals, $x(\tau)$ and $h(t_0 - \tau)$, is found.
4. The integral of the product is the output value at $t = t_0$.

Example 5.4. Find the convolution of the signals $x(t) = e^{-2t}u(t)$ and $h(t) = e^{-3t}u(t)$.

Solution

$$y(t) = \int_0^t e^{-2\tau}e^{-3(t-\tau)}d\tau = e^{-3t}\int_0^t e^{\tau}d\tau = (e^{-2t} - e^{-3t})u(t) \qquad \square$$

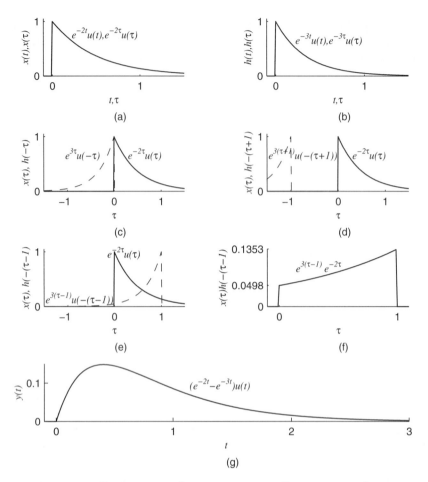

Figure 5.1 (a) $x(t) = e^{-2t}u(t)$, $x(\tau) = e^{-2\tau}u(\tau)$; (b) $h(t) = e^{-3t}u(t)$, $h(\tau) = e^{-3\tau}u(\tau)$; (c) $x(\tau) = e^{-2\tau}u(\tau)$ and the time-reversed signal $h(-\tau) = e^{3\tau}u(-\tau)$; (d) $x(\tau) = e^{-2\tau}u(\tau)$ and the time-reversed and advanced signal $h(-(\tau+1)) = e^{3(\tau+1)}u(-(\tau+1))$; (e) $x(\tau) = e^{-2\tau}u(\tau)$ and the time-reversed and delayed signal $h(-(\tau-1)) = e^{3(\tau-1)}u(-(\tau-1))$; (f) the product of $x(\tau)$ and $h(-(\tau-1))$, $e^{-2\tau}u(\tau)e^{3(\tau-1)}u(-(\tau-1))$; (g) the convolution output of $x(t) = e^{-2t}u(t)$ and $h(t) = e^{-3t}u(t)$, $y(t) = (e^{-2t} - e^{-3t})u(t)$

Figures 5.1(a) and (b) show the two signals to be convolved. These signals and the convolution output, shown in Figure 5.1(g), have the same independent variable t. However, the convolution-integral, for each value of t, is evaluated with respect to the dummy variable τ (a dummy variable exists only during the operation). Therefore, the two signals to be convolved are also shown with respect to τ in Figures 5.1(a) and (b). Figure 5.1(c) shows $x(\tau) = e^{-2\tau}u(\tau)$ and the time-reversed signal $h(-\tau) = e^{3\tau}u(-\tau)$. The convolution output at $t = 0$ is zero, since the area enclosed by the signal $e^{-2\tau}u(\tau)e^{3\tau}u(-\tau)$ is zero (there is no overlap of nonzero portions of the signals). Figure 5.1(d) shows $x(\tau) = e^{-2\tau}u(\tau)$ and the time-reversed and advanced signal $h(-(\tau + 1)) = e^{3(\tau+1)}u(-(\tau + 1))$. The convolution output at $t = -1$ is zero, since there is no overlap of nonzero portions of the signals. Figure 5.1(e) shows $x(\tau) = e^{-2\tau}u(\tau)$ and the time-reversed and delayed signal $h(-(\tau - 1)) = e^{3(\tau-1)}u(-(\tau - 1))$. The nonzero portions of the two signals overlap in the interval from $\tau = 0$ and $\tau = 1$. The product of the signals, $e^{-2\tau}u(\tau)e^{3(\tau-1)}u(-(\tau - 1))$, in this interval is shown in Figure 5.1(f). The area enclosed by this signal is the convolution output at $t = 1$. The convolution output of $x(t) = e^{-2t}u(t)$ and $h(t) = e^{-3t}u(t)$ is $y(t) = (e^{-2t} - e^{-3t})u(t)$, and is shown in Figure 5.1(g).

5.3.1 Properties of the Convolution-integral

The convolution-integral is commutative, that is the order of the two signals to be convolved is immaterial.

$$x(t) * h(t) = h(t) * x(t)$$

The convolution-integral is distributive. That is the convolution of a signal with the sum of two signals is the same as the sum of the individual convolutions of the first signal with the other two signals.

$$x(t) * (h_1(t) + h_2(t)) = x(t) * h_1(t) + x(t) * h_2(t)$$

The convolution-integral is associative. That is the convolution of a signal with the convolution of two signals is the same as the convolution of the convolution of the first two signals with the third signal.

$$x(t) * (h_1(t) * h_2(t)) = (x(t) * h_1(t)) * h_2(t)$$

The shift property of convolution is that

$$\text{if} \quad x(t) * h(t) = y(t) \quad \text{then} \quad x(t - t_1) * h(t - t_2) = y(t - t_1 - t_2)$$

The convolution of two shifted signals is the convolution of the two original signals shifted by the sum of the shifts of the individual signals.

The duration of the convolution of two finite length signals of duration T_1 and T_2 is $T_1 + T_2$, as the overlap of nonzero portions can occur only over that length.

Convolution of a signal $x(t)$ with the unit-impulse leaves the signal unchanged, except for the translation of the origin of the signal to the location of the impulse.

$$x(t) * \delta(t - t_1) = \int_{-\infty}^{\infty} \delta(\tau - t_1)x(t - \tau)d\tau = x(t - t_1)$$

Convolution of $x(t)$ with the unit-step is the running integral of $x(t)$.

$$x(t) * u(t) = \int_{-\infty}^{t} x(\tau)d\tau$$

5.4 System Response

As the amplitude profile of practical signals is usually arbitrary, the output of a system to such signals is found by decomposing the input signals into mathematically well-defined impulse or sinusoidal (in general, exponential) signals. While we are interested in the response of a system to a specific input signal, we use the impulse and the sinusoidal signals as intermediaries. In the convolution-integral model of a system, the impulse signal is used as an intermediary. While these intermediary signals are mathematical idealizations, they can be approximated to a required accuracy for practical purposes. Therefore, it is important to find the response of systems to these signals. In addition, system characteristics, such as rise time, time constant, and frequency selectivity, can be obtained from these responses.

5.4.1 Impulse Response

The impulse response, $h(t)$, of a system is its response to the unit-impulse input signal with the initial conditions of the system zero.

Example 5.5. Find the closed-form expression for the impulse response of the system governed by the differential equation, with output $y(t)$ and input $x(t)$,

$$\frac{dy(t)}{dt} + a_0 y(t) = b_1 \frac{dx(t)}{dt} + b_0 x(t)$$

Solution
The input signal $x(t) = \delta(t)$ is effective only at the instant $t = 0$ and establishes nonzero initial conditions in the system, by storing energy in system components such as capacitor, at the instant immediately after $t = 0$. Therefore, for $t > 0$, this problem can be considered as finding the zero-input response of the system with the initial condition $y(0^+)$. The symbol $y(0^+)$ indicates the value of $y(t)$ at the instant immediately after $t = 0$ and $y(0^-)$ indicates the value of $y(t)$ at the instant immediately before $t = 0$. Therefore, we have to find the initial condition $y(0^+)$ first and then the response to $\delta(t)$. The response to the input $b_1(d\delta(t)/dt) + b_0\delta(t)$ is found using the linearity property of the system. The value $y(0^+)$ is obtained by integrating the differential equation

$$\frac{dy(t)}{dt} + a_0 y(t) = \delta(t)$$

from $t = 0^-$ to $t = 0^+$.

$$\int_{0^-}^{0^+} \frac{dy(t)}{dt} dt + \int_{0^-}^{0^+} a_0 y(t) dt = \int_{0^-}^{0^+} \delta(t) dt$$

The right-hand side is equal to one. The first term on the left-hand side reduces to $y(0^+)$ as $y(0^-) = 0$. Remember that the impulse response is defined as the response of a system to the unit-impulse input with the initial conditions zero. An impulse on the right-hand side implies an impulse on the left-hand side. This impulse must occur in the highest derivative, $dy(t)/dt$, of $y(t)$ since an impulse in $y(t)$ requires the first term to contain the derivative of the impulse and the input does not contain any such function. Therefore, the second term reduces to zero, since the function $y(t)$ is known to be finite (a step function as it the integral of the first term) in the infinitesimal interval of integration. Therefore, the equation reduces to $y(0^+) = 1$. In general, the integrals of all the lower order derivative terms of an Nth-order differential equation evaluate to zero at $t = 0^+$ and the only nonzero initial condition is

$$\left. \frac{d^{N-1} y(t)}{dt^{N-1}} \right|_{t=0^+} = 1$$

For a first-order system, the zero-input response is of the form $Ce^{-a_0 t}$. With the initial condition $y(0^+) = 1$, we get the zero-input response as $e^{-a_0 t} u(t)$. For the input $b_0 \delta(t)$, the response is $b_0 e^{-a_0 t} u(t)$. For the input $b_1 (d\delta(t)/dt)$, by differentiating $b_1 e^{-a_0 t} u(t)$, we get the response as $b_1 \delta(t) - b_1 a_0 e^{-a_0 t} u(t)$. Note that, for linear systems, if $y(t)$ is the output to $x(t)$, then $dy(t)/dt$ is the output to $dx(t)/dt$. Therefore, the impulse response of the system is

$$h(t) = b_0 e^{-a_0 t} u(t) + b_1 \delta(t) - b_1 a_0 e^{-a_0 t} u(t) = b_1 \delta(t) + (b_0 - b_1 a_0) e^{-a_0 t} u(t) \quad \square$$

5.4.2 Response to Unit-step Input

Example 5.6. Find the complete response of the system characterized by the differential equation

$$\frac{dy(t)}{dt} + 4y(t) = 3x(t)$$

with $x(t) = u(t)$, the unit-step input signal. Assume that $y(0^-) = 2$.

Solution
Zero-input response. The characteristic equation of the system is $\lambda + 4 = 0$. The zero-input response is of the form Ce^{-4t}. Using the given initial condition, we get $C = 2$ and the zero-input response is $2e^{-4t}$.

Zero-state response. The impulse response of the system is $h(t) = 3e^{-4t}u(t)$. Using the convolution-integral, we get the zero-state response as

$$y(t) = \int_0^t u(t - \tau)3e^{-4\tau}d\tau = 3\int_0^t e^{-4\tau}d\tau = \frac{3}{4}(1 - e^{-4t})u(t)$$

As the unit-step signal is the integral of the unit-impulse, the unit-step response is the integral of the of the of the unit-impulse response. The unit-impulse response $h(t)$ is the derivative of the unit-step response $y(t)$.

Complete response. The complete response of the system is the sum of the zero-input and zero-state responses.

$$y(t) = \overbrace{\frac{3}{4} - \frac{3}{4}e^{-4t}}^{\text{zero-state}} + \overbrace{2e^{-4t}}^{\text{zero-input}} \qquad t \geq 0$$

$$y(t) = \overbrace{\frac{3}{4}}^{\text{steady-state}} \overbrace{-\frac{3}{4}e^{-4t} + 2e^{-4t}}^{\text{transient}} = \frac{3}{4} + \frac{5}{4}e^{-4t}$$

Transient and steady-state responses. The transient response of the system is $(5/4)e^{-4t}$. The steady-state response of the system, $3/4$, is the response of the system after the transient response has decayed. The transient response of a stable system always decays with time. The form of the transient response depends solely on the characteristics of the system while the form of the steady-state response depends solely on the input signal. The various components of the response are shown in Figure 5.2. □

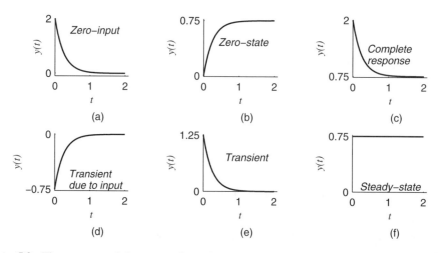

Figure 5.2 The response of the system for unit-step signal: (a) zero-input response; (b) zero-state response; (c) complete response; (d) transient response due to input; (e) transient response; (f) steady-state response

5.4.3 Characterization of Systems by their Responses to Impulse and Unit-step Signals

We can get information about the system behavior from the impulse and unit-step responses. If the significant portions of the impulse response is of longer duration, as shown by solid line in Figure 5.3(a), then the response of the system is sluggish. The corresponding unit-step response is shown by solid line in Figure 5.3(b). The time taken for the unit-step response to rise from 10 to 90% of its final value is called the rise time of the system. If the significant portions of the impulse response is of shorter duration, as shown by dashed line in Figure 5.3(a), then the response of the system is faster, as shown by dashed line in Figure 5.3(b). A system with a shorter impulse response has less memory and it is readily influenced by the recent values of the input signal. Therefore, its response is fast. The faster is the rate of decay of the impulse response, the faster the response approaches its steady-state value.

The unit-step response is the integral of the unit-impulse response, $y(t) = \int_0^t h(\tau)d\tau$. As the final value tends to one in Figure 5.3(b) and as the unit-step signal, ultimately, acts like a dc signal, the monotonically decreasing impulse response indicates a system that passes low-frequency components of a signal well.

Figure 5.3(c) shows the impulse response $\delta(t) - e^{-t}u(t)$. The corresponding unit-step response is shown in Figure 5.3(d). Note that the final value of the unit-step response approaches a very low value in Figure 5.3(d). This indicates a system that does not pass low-frequency components of a signal well.

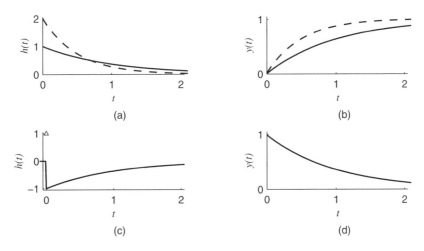

Figure 5.3   (a) Typical monotonically decreasing impulse responses; (b) the corresponding unit-step responses; (c) an impulse response that is a combination of an impulse and an exponential; (d) the corresponding unit-step response

5.4.4 Response to Complex Exponential Input

A complex exponential with frequency $j\omega_0$ is given as $x(t) = e^{j\omega_0 t}$, $-\infty < t < \infty$. Assuming a causal and stable system with impulse response $h(t)$, the output is given by the convolution-integral as

$$y(t) = \int_0^\infty h(\tau)e^{j\omega_0(t-\tau)}d\tau = e^{j\omega_0 t}\int_0^\infty h(\tau)e^{-j\omega_0\tau}d\tau$$

Note that the second integral is independent of t. Let

$$H(j\omega_0) = \int_0^\infty h(\tau)e^{-j\omega_0\tau}d\tau$$

Then,

$$y(t) = H(j\omega_0)e^{j\omega_0 t} = H(j\omega_0)x(t)$$

$H(j\omega_0)$ is called the frequency response since it is a constant complex scale factor, indicating the amount of change in the amplitude and phase of an input complex exponential $e^{j\omega_0 t}$ with frequency $j\omega_0$ at the output. The point is that the input–output relationship of a LTI system becomes a multiplication operation rather than the more complex convolution operation. As the complex exponential is the only signal that has this property, it is used predominantly as the basis for signal decomposition. Even if the exponent of the exponential input signal has a real part, $x(t) = e^{(\sigma+\omega_0)t} = e^{s_0 t}$, the response of the system is still related to the input by the multiplication operation. A real sinusoid input $A\cos(\omega_0 t + \theta)$ is also changed at the output by the same amount of amplitude and phase of the complex scale factor $H(j\omega_0)$. That is, $A\cos(\omega_0 t + \theta)$ is changed to $(|H(j\omega_0)|A)\cos(\omega_0 t + (\theta + \angle(H(j\omega_0))))$.

There was no transient component in the output expression $y(t)$, since the exponential signal was applied at $t = -\infty$. For finite values of t, any transient component in the output of a stable system must have died out. However, if we apply the exponential at any finite instant, say $t = 0$, there will be transient component, in addition to the steady-state component $H(j\omega_0)e^{j\omega_0 t}u(t)$.

Example 5.7. Let the input signal to a stable system with impulse response $h(t) = e^{-t}u(t)$ be $x(t) = e^{s_0 t}u(t)$. Find the response $y(t)$ of the system. Assume that $y(0^-) = 0$.

Solution
Using the convolution-integral, we get

$$y(t) = \int_0^t e^{-\tau}e^{s_0(t-\tau)}d\tau = e^{s_0 t}\int_0^t e^{-\tau(1+s_0)}d\tau$$

$$= \left(\frac{1}{s_0 + 1}\right)(e^{s_0 t} - e^{-t})u(t), \quad s_0 \neq -1$$

The steady-state component, $(1/(s_0 + 1))(e^{s_0 t})u(t)$, is the same as the input complex exponential with a complex scale factor. The second term, $(-e^{-t}/(s_0 + 1))u(t)$, is the transient component that will die for sufficiently large values of t.   □

5.5 System Stability

One of the criteria for the stability of a system is that the system output is bounded if the input is bounded. A signal $x(t)$ is bounded if $|x(t)| \leq P$ for all values of t, where P is a finite positive number. For example, the signal $x(t) = e^{-0.8t}u(t)$ is bounded and $x(t) = e^{0.8t}u(t)$ is unbounded. As convolution-integral is an integral of products, its value is bounded if the input signal is bounded and the value of the integral of the magnitude of the impulse response is bounded. Let the input signal $x(t)$ be bounded by the positive constant P. From the convolution-integral relation for a causal system with impulse response $h(t)$, we get

$$|y(t)| = |\int_0^\infty h(\tau)x(t - \tau)d\tau|$$

$$\leq \int_0^\infty |h(\tau)x(t - \tau)d\tau| = \int_0^\infty |h(\tau)||x(t - \tau)|d\tau$$

$$|y(t)| \leq \int_0^\infty |h(\tau)|Pd\tau = P\int_0^\infty |h(\tau)|d\tau$$

Therefore, if $\int_0^\infty |h(\tau)|d\tau$ is bounded then $|y(t)|$ is bounded. Consequently, a necessary and sufficient stability condition is that the impulse response is absolutely integrable,

$$\int_0^\infty |h(\tau)|d\tau < \infty$$

As we used the convolution-integral to derive the stability condition, the stability condition ensures a bounded zero-state response. The stability of the zero-input response should be checked separately and it is presented in Chapter 11.

Example 5.8.   Find the condition so that the causal LTI system governed by the differential equation, with output $y(t)$ and input $x(t)$,

$$\frac{dy(t)}{dt} + a_0 y(t) = b_0 x(t)$$

is stable.

Solution

As the impulse response of this system, $h(t) = b_0 e^{-a_0 t} u(t)$, is an exponential signal, the condition $a_0 > 0$ ensures that $h(t)$ is absolutely integrable.   ☐

5.6 Realization of Continuous Systems

The three basic components required in the realization of continuous systems are: (i) multiplier unit; (ii) adder unit; and (iii) integrator unit. A multiplier unit, shown in Figure 5.4(a), produces an output signal $c\,x(t)$, which is the product of the input signal $x(t)$ with the coefficient c. An adder unit, shown in Figure 5.4(b), produces an output signal $x(t) + y(t)$, which is the sum of the input signals $x(t)$ and $y(t)$. By changing the sign of the subtrahend and then adding it with the minuend, the subtraction operation can be realized by an adder unit. An integrator unit, shown in Figure 5.4(c), produces an output $\int_{-\infty}^{t} x(\tau)\,d\tau$ for an input $x(t)$. The output is the integral of the input.

The realization of a continuous system is an interconnection of the basic components. Consider the realization, shown in Figure 5.5, of a first-order system governed by the differential equation, with output $y(t)$ and input $x(t)$,

$$\frac{dy(t)}{dt} + 3y(t) = 2x(t)$$

A multiplier unit with coefficient -3 and input $y(t)$ produces $-3y(t)$. A multiplier unit with coefficient 2 and input $x(t)$ produces $2x(t)$. The adder unit combines the two partial results to produce the signal $-3y(t) + 2x(t)$, which is equal to $dy(t)/dt$. By passing this signal through an integrator unit, we get $y(t)$.

5.6.1 Decomposition of Higher-order Systems

To meet a given specification, an higher-order system is often required. Due to several advantages, a system is usually decomposed into first- and second-order systems connected in cascade or parallel. Consider two systems with impulse responses $h_1(t)$ and $h_2(t)$ connected in parallel, shown in Figure 5.6. The same input is applied to each system and the total response is the sum of the individual responses. The combined

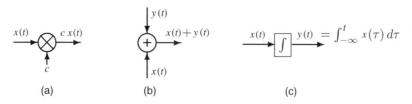

(a)   (b)   (c)

Figure 5.4 Basic components required in the realization of continuous systems: (a) multiplier unit; (b) adder unit; (c) integrator unit

Figure 5.5   A continuous system realization

response of the two systems for the input $x(t)$ is $y(t) = x(t) * h_1(t) + x(t) * h_2(t)$. This expression, due to the distributive property of convolution, can be written as $y(t) = x(t) * (h_1(t) + h_2(t))$. That is, the parallel connection of the two systems is equivalent to a single system with impulse response $h(t) = h_1(t) + h_2(t)$, as shown in Figure 5.6.

Consider two systems with impulse responses $h_1(t)$ and $h_2(t)$ connected in cascade, shown in Figure 5.7. The output of one system is the input to the other. The response of the first system for the input $x(t)$ is $y_1(t) = x(t) * h_1(t)$. The response of the second system for the input $y_1(t) = x(t) * h_1(t)$ is $y(t) = (x(t) * h_1(t)) * h_2(t)$. This expression, due to the associative property of convolution, can be written as $y(t) = x(t) * (h_1(t) * h_2(t))$. That is, the cascade connection of the two systems is equivalent to a single system with impulse response $h(t) = h_1(t) * h_2(t)$, as shown in Figure 5.7. Due to the commutative property of convolution, the order of the systems in the cascade connection is immaterial, with respect to the input–output relationship.

5.6.2 Feedback Systems

Another configuration of systems, often used in control systems, is the feedback configuration shown in Figure 5.8. In feedback systems, a fraction of the output signal is fed back and subtracted from the input signal to form the effective input signal. A

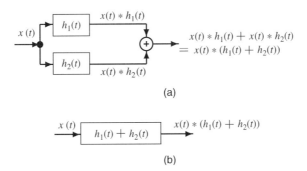

(a)

(b)

Figure 5.6   (a) Two systems connected in parallel; (b) a single system equivalent to the system in (a)

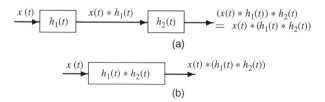

Figure 5.7   (a) Two systems connected in cascade; (b) a single system equivalent to the system in (a)

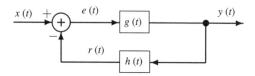

Figure 5.8   Two systems connected in a feedback configuration

feedback signal $r(t)$ is produced by a causal system with impulse response $h(t)$ from the the output signal, $y(t)$. That is, $r(t) = \int_0^\infty h(\tau)y(t - \tau)d\tau$. The error signal $e(t)$ is difference between the input signal $x(t)$ and the feedback signal $r(t)$, $e(t) = x(t) - r(t)$. This error signal is the input to a causal system with impulse response $g(t)$, which produces the output signal $y(t)$. That is, $y(t) = \int_0^\infty g(\tau)e(t - \tau)d\tau$.

5.7 Summary

- In this chapter, the time-domain analysis of LTI continuous systems has been presented.
- The zero-input component of the response of a LTI system is its response due to the initial conditions alone with the input assumed to be zero. The zero-state component of the response of a LTI system is its response due to the input alone with the initial conditions assumed to be zero. The sum of the zero-input and zero-state responses is the complete response of the system.
- Two of the commonly used system models for time-domain analysis are the differential equation and convolution-integral models.
- The convolution-integral model gives the zero-state response of a LTI system. Both the zero-input and zero-state responses can be found by solving the differential equation.
- The impulse response of a system is its response to the unit-impulse input signal with the initial conditions zero.
- The convolution-integral model is based on decomposing the input signal into continuum of shifted and scaled impulses. The total response is found by the integral of the responses to all the constituent impulses of the input signal.

- The complete response of a system can also be considered as the sum of transient and steady-state components. For a stable system, the transient component decays with time. The steady-state component is the response after the transient response has decayed.
- A system is stable if its response is bounded for all bounded input signals. As the convolution-integral is an integral of the product of input and impulse responses, with the input bounded, the impulse response of a stable system must be absolutely integrable for the value of the convolution-integral to be bounded.
- By interconnecting adder, multiplier, and integrator units, any continuous system can be realized. A higher-order system is usually decomposed into a set of first- and second-order systems connected in cascade or parallel. A feedback system is obtained by feeding back some part of the output to the input.

Further Reading

1. Lathi, B. P., *Linear Systems and Signals*, Oxford University Press, New York, 2004.

Exercises

5.1   Is the system governed by the given differential equation, with output $y(t)$ and input $x(t)$, linear?

5.1.1 $\frac{dy(t)}{dt} + 2y(t) + 2 = x(t)$.

5.1.2 $\left(\frac{dy(t)}{dt}\right)^2 + y(t) = \frac{dx(t)}{dt} + x(t)$.

5.1.3 $\frac{dy(t)}{dt} + ty(t) = 3\frac{dx(t)}{dt} + 2x(t)$.

5.1.4 $\frac{dy(3t)}{dt} + y(3t) = x(t)$.

5.1.5 $\frac{dy(t)}{dt} + y(t) + \sin(\pi) = x(t)$.

5.1.6 $\frac{dy(t)}{dt} + y(t) + \cos(\pi) = x(t)$.

5.1.7 $\frac{dy(t)}{dt} + y(t) = x(t)\frac{dx(t)}{dt}$.

5.1.8 $\frac{dy(t)}{dt} + e^{y(t)} = x(t)$.

*5.1.9 $\frac{dy(t)}{dt} = |x(t)|$.

5.2   Is the system governed by the given differential equation, with output $y(t)$ and input $x(t)$, time-invariant?

5.2.1 $\frac{dy(t)}{dt} + y(2t) = x(t)$.

5.2.2 $\frac{dy(t)}{dt} + \cos\left(\frac{\pi}{2}t\right)y(t) = x(t)$.

5.2.3 $\frac{dy(t)}{dt} + y(t) = tx(t)$.

*5.2.4 $y(t) = x(t - 5)$.

5.2.5 $\frac{dy(t)}{dt} + ty(t) = x(t)$.

5.2.6 $\frac{dy(t)}{dt} + y(t) = x(t)\frac{dx(t)}{dt}$.

5.2.7 $\frac{dy(t)}{dt} + e^{y(t)} = x(t)$.

5.2.8 $\frac{dy(t)}{dt} + y(-t) = x(t)$.

5.3   Find the closed-form expression for the convolution of the signals $x(t)$ and $h(t)$. List the values of the convolution output at $t = 0, 1, 2, 3, 4, 5$.

    5.3.1 $x(t) = 3u(t + 1)$ and $h(t) = 2u(t - 3)$.

    5.3.2 $x(t) = 2e^{-2t}u(t - 1)$ and $h(t) = 2e^{-2t}u(t + 3)$.

    5.3.3 $x(t) = 4u(t - 1)$ and $h(t) = 2e^{-2(t+2)}u(t + 2)$.

    *5.3.4 $x(t) = (u(t) - u(t - 3))$ and $h(t) = (u(t) - u(t - 3))$.

5.4   Find the convolution of the signals $x(t)$ and $h(t)$.

    5.4.1 $x(t) = e^{j\frac{2\pi}{6}t}u(t)$ and $h(t) = \delta(t + 4)$.

    5.4.2 $x(t) = e^{j\frac{2\pi}{6}t}$ and $h(t) = \delta(t + 12)$.

    5.4.3 $x(t) = \cos\left(\frac{2\pi}{6}t\right)$ and $h(t) = \delta(t)$.

5.5   Verify the distributive property of convolution integral, $x(t) * (h_1(t) + h_2(t)) = x(t) * h_1(t) + x(t) * h_2(t)$.

    5.5.1 $h_1(t) = 2e^{-2t}u(t)$, $h_2(t) = 3e^{-2t}u(t)$, $x(t) = u(t)$.

    5.5.2 $h_1(t) = 3e^{-3t}u(t)$, $h_2(t) = 5e^{-3t}u(t)$, $x(t) = e^{-t}u(t)$.

5.6   Verify the associative property of convolution integral, $x(t) * (h_1(t) * h_2(t)) = (x(t) * h_1(t)) * h_2(t)$.

    5.6.1 $h_1(t) = e^{-2t}u(t)$, $h_2(t) = e^{-3t}u(t)$, $x(t) = u(t)$.

    5.6.2 $h_1(t) = e^{-2t}u(t)$, $h_2(t) = e^{-3t}u(t)$, $x(t) = e^{-t}u(t)$.

5.7   Find the closed-form expression for the impulse response $h(t)$ of the system characterized by the differential equation, with output $y(t)$ and input $x(t)$. Deduce the closed-form expression for the unit-step response $y(t)$ of the system.

    5.7.1 $\frac{dy(t)}{dt} + 2y(t) = -\frac{dx(t)}{dt} + x(t)$.

    *5.7.2 $\frac{dy(t)}{dt} - y(t) = 2\frac{dx(t)}{dt} + 3x(t)$.

    5.7.3 $\frac{dy(t)}{dt} + 3y(t) = 2x(t)$.

    5.7.4 $\frac{dy(t)}{dt} + 4y(t) = -2\frac{dx(t)}{dt} + x(t)$.

    5.7.5 $\frac{dy(t)}{dt} + 2y(t) = 4x(t)$.

5.8   Derive the closed-form expression for the complete response (by finding the zero-state response by convolution and the zero-input response) of the system governed by the differential equation

$$\frac{dy(t)}{dt} + y(t) = 3\frac{dx(t)}{dt} + 2x(t)$$

with the initial condition $y(0^-) = 2$ and the input $x(t) = u(t)$, the unit-step function. Deduce the expressions for the transient and steady-state responses of the system.

5.9   Derive the closed-form expression for the complete response (by finding the zero-state response by convolution and the zero-input response) of the system governed by the differential equation

$$\frac{dy(t)}{dt} - 2y(t) = -3\frac{dx(t)}{dt} + x(t)$$

with the initial condition $y(0^-) = 1$ and the input $x(t) = tu(t)$, the unit-ramp function. Deduce the expressions for the transient and steady-state responses of the system.

5.10 Derive the closed-form expression for the complete response (by finding the zero-state response by convolution and the zero-input response) of the system governed by the differential equation

$$\frac{dy(t)}{dt} + 4y(t) = 2\frac{dx(t)}{dt} - 3x(t)$$

with the initial condition $y(0^-) = 2$ and the input $x(t) = e^{-3t}u(t)$. Deduce the expressions for the transient and steady-state responses of the system.

*5.11 Derive the closed-form expression for the complete response (by finding the zero-state response by convolution and the zero-input response) of the system governed by the differential equation

$$\frac{dy(t)}{dt} + y(t) = -\frac{dx(t)}{dt} + x(t)$$

with the initial condition $y(0^-) = 3$ and the input $x(t) = 2\cos(t)u(t)$. Deduce the expressions for the transient and steady-state responses of the system.

5.12 Derive the closed-form expression for the complete response (by finding the zero-state response by convolution and the zero-input response) of the system governed by the differential equation

$$\frac{dy(t)}{dt} + 5y(t) = 3\frac{dx(t)}{dt} - x(t)$$

with the initial condition $y(0^-) = -2$ and the input $x(t) = \sin(t)u(t)$. Deduce the expressions for the transient and steady-state responses of the system.

5.13 Find the steady-state response of the system, with the impulse response

$$h(t) = 3\delta(t) - 2e^{-2t}u(t),$$

to the input $x(t) = 3\cos(\frac{2\pi}{8}t + \frac{\pi}{3})u(t)$. Deduce the response for the input $e^{j\frac{2\pi}{8}t}$.

*5.14 Find the steady-state response of the system, with the impulse response

$$h(t) = 2\delta(t) - 4e^{-t}u(t),$$

to the input $x(t) = 2\sin(\frac{2\pi}{6}t - \frac{\pi}{6})u(t)$. Deduce the response for the input $e^{j\frac{2\pi}{6}t}$.

5.15 The impulse response of a LTI system is given. Use the bounded input bounded output test to find whether the system is stable?

5.15.1 $h(t) = e^{-2t}u(t)$.

5.15.2 $h(t) = u(t)$.

5.15.3 $h(t) = \frac{\sin(2t)}{t}u(t)$.

5.15.4 $h(t) = \left(\frac{\sin(2t)}{t}\right)^2 u(t)$.

5.15.5 $h(t) = -e^{3t}u(t)$.

5.16 Derive the closed-form expression of the impulse response $h(t)$ of the combined system consisting of systems with impulse responses $h_1(t)$ and $h_2(t)$, if the systems are connected in: (i) parallel; and (ii) cascade.

5.16.1 $h_1(t) = e^{-2t}u(t)$ and $h_2(t) = e^{-5t}u(t)$.

5.16.2 $h_1(t) = \delta(t) + e^{-3t}u(t)$ and $h_2(t) = \delta(t) - e^{-2t}u(t)$.

5.16.3 $h_1(t) = 2\delta(t) - e^{-4t}u(t)$ and $h_2(t) = e^{-3t}u(t)$.

6

The Discrete Fourier Transform

In this chapter, the most often used tools for the transformation of signals from the time-
to the frequency-domain and back again, the DFT and the IDFT, are presented. The
frequency-domain representation of signals and systems is introduced in Section 6.1.
In Section 6.2, a brief review of Fourier analysis is presented. The DFT and the IDFT
are derived in Section 6.3. The properties of the DFT are presented in Section 6.4.
Some applications of the DFT are presented in Section 6.5.

6.1 The Time-domain and the Frequency-domain

The independent variable, in the time-domain representation of signals and systems,
is time. In this domain, we analyze arbitrary signals in terms of scaled and shifted
impulses. A system is characterized in terms of its impulse response (Chapters 4
and 5). We still look for simple signals that provide more efficient signal and system
analysis. This leads us to an alternate representation of signals and systems, called the
frequency-domain representation. In this representation (which can be considered as
the transformation of the independent variable), the variation of a signal with respect
to the frequency of its constituent sinusoids is used in its characterization. At each
frequency, the amplitude and phase or, equivalently, the amplitudes of the cosine
and sine components of the sinusoid, are used for representing a signal. Systems
are characterized in terms of their responses to sinusoids. Both the time-domain and
frequency-domain representations completely specify a signal or a system. In the
frequency-domain, the independent variable is frequency, thereby explicitly specifying
the frequency components of a signal. While there are other basic signals, the sinusoid
is mostly used for signal and LTI system analysis because it provides ease of signal
decomposition, simpler system analysis, and more insight into the signal and system
characteristics. Except for the fact that the independent variable is frequency, the
system analysis is very similar to that used in the time-domain. That is, we decompose

A Practical Approach to Signals and Systems   D. Sundararajan
© 2008 John Wiley & Sons (Asia) Pte Ltd

an input signal in terms of sinusoids, find the response of the system to each sinusoid, and, using the linearity and time-invariant properties of LTI systems, sum up all the responses to find the complete response of the system. The big advantage of the sinusoids is that the steady-state output of a stable LTI system for a sinusoidal input is of the same form. Therefore, the output of a system can be found using the much simpler multiplication operation compared with the convolution operation required using the impulse signal.

A set of complex exponentials or sinusoids is used as the basis signals in the principal transforms used in signal and LTI system analysis. While sinusoidal waveforms are easy to visualize, the complex exponential, which is a functionally equivalent mathematical representation of a sinusoid, is often used in signal and system analysis, due to its compact form and ease of manipulation. In Fourier analysis, sinusoids with constant amplitudes (or exponentials with pure imaginary exponents) are used as basis signals. Sinusoids with exponentially varying amplitudes (or exponentials with complex exponents) are used in Laplace and z-transforms. Each transform is more suitable for the analysis of certain classes of signals and systems.

6.2 Fourier Analysis

The theory of Fourier analysis is that any periodic signal satisfying certain conditions, which are met by most signals of practical interest, can be represented uniquely as the sum of a constant value and an infinite number of harmonics. Harmonically related sinusoids, called harmonics, are a set of sinusoids consisting a fundamental harmonic with frequency f and other harmonics having frequencies those are integral multiples of f. The sum of a set of harmonically related sinusoids is not a sinusoid, but is a periodic waveform with period the same as that of the fundamental. Given a waveform, finding the amplitude of its constituent sinusoids is called the Fourier analysis. To sum up a set of sinusoids to synthesize an arbitrary waveform is called the Fourier synthesis. Consider the discrete periodic waveform, $x(n) = 2 + 3\sin(\frac{2\pi}{4}n) + \cos(2\frac{2\pi}{4}n)$, with period 4 samples, shown in Figure 6.1(a). The independent variable n (actually nT_s, where T_s is the sampling interval) is time and the dependent variable is amplitude. Figure 6.1(b) shows the frequency-domain representation of the waveform in (a). It shows the complex amplitude, multiplied by 4, of its constituent complex exponentials. To find the real sinusoids, shown in Figure 6.1(c), those constitute the signal, we add up the complex exponentials.

$$x(n) = \frac{1}{4}\left(8e^{j0\frac{2\pi}{4}n} - j6e^{j\frac{2\pi}{4}n} + 4e^{j2\frac{2\pi}{4}n} + j6e^{j3\frac{2\pi}{4}n}\right)$$

$$= 2 + 3\sin\left(\frac{2\pi}{4}n\right) + \cos\left(2\frac{2\pi}{4}n\right)$$

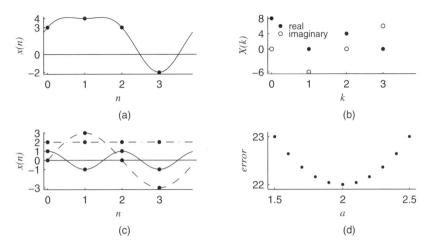

Figure 6.1 (a) A periodic waveform, $x(n) = 2 + 3\sin(\frac{2\pi}{4}n) + \cos(2\frac{2\pi}{4}n)$, with period 4 samples; (b) its frequency-domain representation; (c) the frequency components of the waveform in (a); (d) the square error in approximating the waveform in (a) using only the dc component with different amplitudes

As can be seen from this example, Fourier analysis represents a signal as a linear combination of sinusoids or, equivalently, complex exponentials with pure imaginary exponents.

The Fourier reconstruction of a waveform is with respect to the least-squares error criterion. That is, the mean value for power signals or the total value for the energy signals of the integral or sum of the squared magnitude of the error between the given waveform and the corresponding Fourier reconstructed waveform is guaranteed to be the minimum if part of the constituent sinusoids of a waveform is used in the reconstruction and will be zero if all the constituent sinusoids are used. The reason this criterion, based on signal energy or power, is used rather than a minimum uniform deviation criterion is that: (i) it is acceptable for most applications; and (ii) it leads to closed-form formulas for the analytical determination of Fourier coefficients.

Let $x_a(n)$ be an approximation to a given waveform $x(n)$ of period N, using fewer harmonics than required. The square error between $x(n)$ and $x_a(n)$ is defined as

$$\text{error} = \sum_{n=0}^{N-1} |x(n) - x_a(n)|^2$$

For a given number of harmonics, there is no better approximation for the signal than that provided by the Fourier approximation when the least-squares error criterion is applied. Assume that, we are constrained to use only the dc component to approximate the waveform in Figure 6.1(a). Let the optimal value of the dc component be a. To minimize the square error,

$$(3 - a)^2 + (4 - a)^2 + (3 - a)^2 + (-2 - a)^2$$

must be minimum. Differentiating this expression with respect to a and equating it to zero, we get

$$2(3 - a)(-1) + 2(4 - a)(-1) + 2(3 - a)(-1) + 2(-2 - a)(-1) = 0$$

Solving this equation, we get $a = 2$ as given by the Fourier analysis. The square error, for various values of a, is shown in Figure 6.1(d).

6.2.1 Versions of Fourier Analysis

Fourier analysis has four different versions, each of them using a set of constant-amplitude sinusoids, differing in some respect, as the basis signals. Continuous periodic signals are analyzed using an infinite number of harmonically related continuous sinusoids in FS, described in Chapter 7. Discrete aperiodic signals are analyzed using a continuum of discrete sinusoids over a finite frequency range in the DTFT, presented in Chapter 8. Continuous aperiodic signals are analyzed using a continuum of continuous sinusoids over an infinite frequency range in the FT, described in Chapter 9. The topic of the present chapter is the DFT, which analyzes the periodic extension of a finite duration discrete signal using a finite number of harmonically related discrete sinusoids. The DFT, because of its finite and discrete nature, is the simplest of the four different versions of Fourier analysis to visualize the analysis and synthesis of waveforms. Problems in understanding the concepts in other versions of Fourier analysis may be resolved by considering an equivalent DFT version.

6.3 The Discrete Fourier Transform

6.3.1 The Approximation of Arbitrary Waveforms with a Finite Number of Samples

We need a minimum of $2k + 1$ samples to represent a sinusoid uniquely, which completes k cycles in a period, as presented in Chapter 2. To approximate a periodic waveform in terms of dc, we need a minimum of one sample in a period. If we use the fundamental or first harmonic, which has the same period as that of the waveform to be analyzed, we need a minimum of three samples $(2k + 1 = 2(1) + 1 = 3)$ in a period, since the first harmonic completes one cycle. In the frequency-domain, we need one value to represent the dc and two values (the amplitude and the phase or the amplitudes of its cosine and sine components) to specify the first harmonic. That is, three samples are required in the both the time- and frequency-domains. With

N independent values in one domain, we can generate only N independent values in the other domain. Therefore, we need $2k + 1$ samples in both the time- and frequency-domains to represent a waveform with the dc and the first k harmonically related sinusoids.

In general, an infinite number of sinusoids is required to represent a waveform exactly. The concept of using a finite number of sinusoids is based on the fact that the waveforms encountered in practice can be approximated by a finite number of sinusoids with a finite but arbitrarily small tolerance, since, beyond some range, the spectral values become negligible. That is, all practical signals can be considered as band-limited. If the magnitude of the frequency components of a signal is identically zero outside some finite frequency range, then the signal is called band-limited. In addition, the waveforms are generally aperiodic. In order to make it finite duration, we have to truncate some part of it. Then, a periodic extension of the waveform is represented by discrete sinusoids. The truncation is acceptable because waveforms, in practice, have negligible values beyond a certain range. That is, all practical signals can be considered as time-limited. If the amplitude of a signal is identically zero outside some finite time interval, then the signal is called time-limited. Therefore, we can represent any waveform, encountered in practice, by a finite number of samples in both the time- and frequency-domains with adequate accuracy. This representation, using a finite number of samples in both the domains, is the feature of the DFT version of Fourier analysis. That is to make the essential information, characterizing a waveform, available in any one period, in both the domains, with sufficient accuracy. The point is that, while the representation of a waveform can be made adequate, the discrete and finite nature of the DFT makes it inherently suitable for numerical analysis. And, finally, the fact that Fourier analysis plays a central part in signal and system analysis, and fast algorithms are available for computing the DFT makes the DFT the heart of practical signal and system analysis.

6.3.2 The DFT and the IDFT

In the DFT, a set of N samples represents a waveform in both the time- and frequency-domains, whether the waveform is periodic or aperiodic and continuous or discrete. It is understood that the number of samples is adequate to represent the waveform with sufficient accuracy. The set of N samples is periodically extended and N harmonically related complex exponentials are used to represent the waveform. That is, for a real-valued signal with N samples, we are using real sinusoids with frequency indices $0, 1, 2, \ldots, N/2$ only. Frequency index zero represents the dc and $N/2$ represents a cosine waveform, assuming N is even.

The frequency components of a waveform are separated using the orthogonality property of the exponentials. For two complex exponentials $e^{j\frac{2\pi}{N}ln}$ and $e^{j\frac{2\pi}{N}kn}$ over a period of N samples, the orthogonality property is defined as

$$\sum_{n=0}^{N-1} e^{j\frac{2\pi}{N}(l-k)n} = \begin{cases} N & \text{for } l = k \\ 0 & \text{for } l \neq k \end{cases}$$

where $l, k = 0, 1, \ldots, N - 1$. If $l = k$, the summation is equal to N as $e^{j\frac{2\pi}{N}(l-k)n} = e^0 = 1$. Otherwise, by using the closed-form expression for the sum of a geometric progression, we get

$$\sum_{n=0}^{N-1} e^{j\frac{2\pi}{N}(l-k)n} = \frac{1 - e^{j2\pi(l-k)}}{1 - e^{j\frac{2\pi(l-k)}{N}}} = 0, \text{ for } l \neq k$$

That is, in order to find the coefficient, with a scale factor N, of a complex exponential, we multiply the samples of a signal with the corresponding samples of the complex conjugate of the complex exponential. Using each complex exponential in turn, we get the frequency coefficients of all the components of a signal as

$$X(k) = \sum_{n=0}^{N-1} x(n)W_N^{nk}, \quad k = 0, 1, \ldots, N - 1 \tag{6.1}$$

where $W_N = e^{-j\frac{2\pi}{N}}$. This is the DFT equation analyzing a waveform with harmonically related discrete complex sinusoids. $X(k)$ is the coefficient, scaled by N, of the complex sinusoid $e^{j\frac{2\pi}{N}kn}$ with a specific frequency index k (frequency $(2\pi/N)k$ radians per sample). The summation of the sample values of the N complex sinusoids multiplied by their respective frequency coefficients $X(k)$ is the IDFT operation. The N-point IDFT of the frequency coefficients $X(k)$ is defined as

$$x(n) = \frac{1}{N}\sum_{k=0}^{N-1} X(k)W_N^{-nk}, \quad n = 0, 1, \ldots, N - 1 \tag{6.2}$$

The sum of the sample values is divided by N in Equation (6.2) as the coefficients $X(k)$ have been scaled by the factor N in the DFT computation.

The DFT equation can also be written using matrices. With $N = 4$, the DFT is given by

$$\begin{bmatrix} X(0) \\ X(1) \\ X(2) \\ X(3) \end{bmatrix} = \begin{bmatrix} W_4^0 & W_4^0 & W_4^0 & W_4^0 \\ W_4^0 & W_4^1 & W_4^2 & W_4^3 \\ W_4^0 & W_4^2 & W_4^4 & W_4^6 \\ W_4^0 & W_4^3 & W_4^6 & W_4^9 \end{bmatrix} \begin{bmatrix} x(0) \\ x(1) \\ x(2) \\ x(3) \end{bmatrix}$$

If we use the definition of the DFT, we need N complex multiplications and $(N-1)$ complex additions for computing each of the N coefficients. The computational complexity of computing all the N coefficients is of the order of N^2. Fast algorithms reduce this computational complexity to the order of $N \log_2 N$. Because of these algorithms, the use of the DFT is more efficient in most applications compared with alternate methods.

Let us compute the DFT of $\{x(0) = 3, x(1) = 4, x(2) = 3, x(3) = -2\}$. The DFT of this set of data is computed as

$$
\begin{bmatrix} X(0) \\ X(1) \\ X(2) \\ X(3) \end{bmatrix} = \begin{bmatrix} 1 & 1 & 1 & 1 \\ 1 & -j & -1 & j \\ 1 & -1 & 1 & -1 \\ 1 & j & -1 & -j \end{bmatrix} \begin{bmatrix} 3 \\ 4 \\ 3 \\ -2 \end{bmatrix} = \begin{bmatrix} 8 \\ -j6 \\ 4 \\ j6 \end{bmatrix}
$$

The DFT spectrum is $\{X(0) = 8, X(1) = -j6, X(2) = 4, X(3) = j6\}$, as shown in Figure 6.1(b). Now, let us compute the sample values of the waveform from its DFT coefficients using the IDFT.

$$
\begin{bmatrix} x(0) \\ x(1) \\ x(2) \\ x(3) \end{bmatrix} = \frac{1}{4} \begin{bmatrix} 1 & 1 & 1 & 1 \\ 1 & j & -1 & -j \\ 1 & -1 & 1 & -1 \\ 1 & -j & -1 & j \end{bmatrix} \begin{bmatrix} 8 \\ -j6 \\ 4 \\ j6 \end{bmatrix} = \begin{bmatrix} 3 \\ 4 \\ 3 \\ -2 \end{bmatrix}
$$

We get back the time-domain sample values, confirming that the DFT and IDFT form a transform pair. What one operation does the other undoes.

6.3.3 DFT of Some Basic Signals

While the primary purpose of the DFT is to approximate the spectra of arbitrary signals using numerical procedures, it is useful, for understanding, to derive the DFT of some simple signals analytically. The DFT of the impulse signal $x(n) = 2\delta(n)$ is simply $X(k) = 2$. As the impulse signal is nonzero only at $n = 0$, the DFT equation reduces to $x(0)$ for any value of k. A signal and its DFT form a transform pair and is denoted as $x(n) \Longleftrightarrow X(k)$. For the specific example, the transform pair is denoted as $2\delta(n) \Longleftrightarrow 2$. The DFT, with $N = 16$, is shown in Figure 6.2(a). A plot of the complex coefficients $X(k)$ of the constituent complex sinusoids of a signal $x(n)$ versus k is called the complex spectrum of $x(n)$. The spectral value of two for all the frequency components imply that the impulse signal, with a value of two, is the sum of all the exponentials $\frac{2}{16} e^{j\frac{2\pi}{16}kn}$, $k = 0, 1, \ldots, 15$. In terms of real sinusoids, this impulse signal is the sum of dc component $\frac{2}{16}$, cosine waves $\frac{2}{8} \cos(\frac{2\pi}{16} kn)$, $k = 1, 2, \ldots, 7$, and $\frac{2}{16} \cos(\pi n)$.

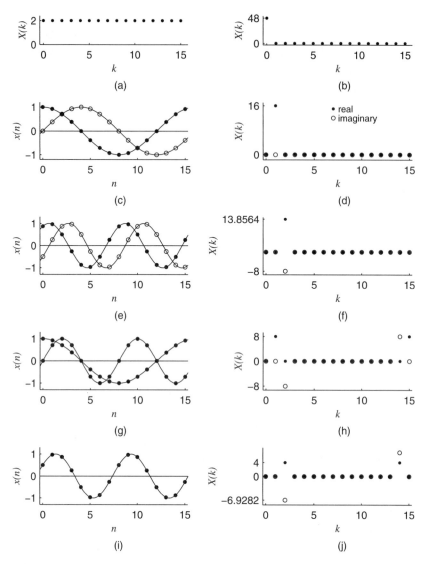

Figure 6.2 (a) The spectrum of the impulse $x(n) = 2\delta(n)$ with $N = 16$; (b) the spectrum of the dc signal $x(n) = 3$ with $N = 16$; (c) the complex sinusoid $x(n) = e^{j\frac{2\pi}{16}n}$; (d) its spectrum; (e) the complex sinusoid $x(n) = e^{j(2\frac{2\pi}{16}n - \frac{\pi}{6})}$ and (f) its spectrum; (g) the sinusoids $x(n) = \cos(\frac{2\pi}{16}n)$ and $x(n) = \sin(2\frac{2\pi}{16}n)$, and (h) their spectra; (i) the sinusoid $x(n) = \cos(2\frac{2\pi}{16}n - \frac{\pi}{3})$; (j) its spectrum

The DFT of the dc signal $x(n) = 3$, with N samples, is $X(k) = 3N\delta(k)$. That is $3 \Longleftrightarrow 3N\delta(k)$. As the dc signal has a constant value, its DFT evaluation essentially reduces to the summation of the sample values of the various complex exponentials. This sum is zero for all the complex exponentials with nonzero frequency index k.

For $k = 0$,

$$X(0) = 3 \sum_{n=0}^{N-1} e^{-j\frac{2\pi}{N}n0} = 3 \sum_{n=0}^{N-1} 1 = 3N$$

The complex exponential with $k = 0$ is the dc signal. The DFT of the dc signal $x(n) = 3$, with 16 samples, is shown in Figure 6.2(b).

The frequency range of the spectral components of a signal is called its bandwidth. The essential bandwidth of a signal is the frequency range of its spectral components containing most of its energy. The longer is the duration of a signal in the time-domain, the shorter is the essential bandwidth in its frequency-domain representation and vice versa. This is called reciprocal spreading and is well demonstrated in the case of the dc and impulse signals. The impulse signal is nonzero only at $n = 0$ in the time-domain and its spectrum is spread with significant values throughout the whole frequency range. The reverse is the case for the dc signal.

The complex exponential signal, although of no physical significance, is the standard unit in the frequency-domain representation and analysis of signals and systems, as it is easier to manipulate and the sum of its conjugate with itself is capable of representing a physical signal. Due to the orthogonality property, the complex exponential $x(n) = e^{j\frac{2\pi}{N}np}$ with frequency index p has the transform pair $e^{j\frac{2\pi}{N}np} \Longleftrightarrow N\delta(k - p)$. The dc case presented earlier is a specific case with $p = 0$. The complex exponential signal $x(n) = e^{j\frac{2\pi}{16}n}$ with $N = 16$ and its spectrum with $X(1) = 16$ are shown in Figures 6.2(c) and (d), respectively.

The complex exponential signal $x(n) = e^{j(2\frac{2\pi}{16}n - \frac{\pi}{6})}$ with $N = 16$ and its spectrum are shown in Figures 6.2(e) and (f), respectively. This signal can be expressed as $x(n) = e^{-j\frac{\pi}{6}} e^{j2\frac{2\pi}{16}n}$. Therefore, the DFT coefficient is that of $x(n) = e^{j2\frac{2\pi}{16}n}$, which is 16 at $k = 2$, multiplied by the complex constant $e^{-j\frac{\pi}{6}} = \frac{\sqrt{3}}{2} - j\frac{1}{2}$, as shown in Figure 6.2(f).

A real sinusoid, $x(n) = \cos(\frac{2\pi}{N}np + \theta)$, is the sum of two complex sinusoids,

$$x(n) = \cos\left(\frac{2\pi}{N}np + \theta\right) = \frac{1}{2}\left(e^{j\theta} e^{j\frac{2\pi}{N}np} + e^{-j\theta} e^{-j\frac{2\pi}{N}np}\right)$$

Using the DFT of complex exponentials, we get

$$\cos\left(\frac{2\pi}{N}np + \theta\right) \Longleftrightarrow \frac{N}{2}(e^{j\theta}\delta(k - p) + e^{-j\theta}\delta(k - (N - p)))$$

Note that, due to periodicity, $e^{-j\frac{2\pi}{N}np} = e^{j\frac{2\pi}{N}n(N-p)}$. We get the transform pairs for the cosine and sine waves, with $\theta = 0$ and $\theta = -\frac{\pi}{2}$, as

$$\cos\left(\frac{2\pi}{N}np\right) \Longleftrightarrow \frac{N}{2}(\delta(k - p) + \delta(k - (N - p)))$$

$$\sin\left(\frac{2\pi}{N}np\right) \Longleftrightarrow \frac{N}{2}(-j\delta(k-p)+j\delta(k-(N-p)))$$

The cosine and sine waves $x(n) = \cos(\frac{2\pi}{16}n)$ and $x(n) = \sin(2\frac{2\pi}{16}n)$ with $N = 16$, and their spectra are shown in Figures 6.2(g) and (h), respectively. The sinusoid $x(n) = \cos(2\frac{2\pi}{16}n - \frac{\pi}{3})$ with $N = 16$ and its spectrum are shown in Figures 6.2(i) and (j), respectively.

6.4 Properties of the Discrete Fourier Transform

In signal and system analysis, it is often required to carry out operations such as shifting, convolution etc., in both the domains. We know the effect, in the other domain, of carrying out an operation in one domain through properties. We repeatedly use the properties in applications of the DFT and in deriving DFT algorithms. In addition, new transform pairs can be derived from existing ones.

6.4.1 Linearity

If a sequence is a linear combination of a set of sequences, each of the same length N, then the DFT of that combined sequence is the same linear combination of the DFT of the individual sequences. That is,

$$x(n) \Longleftrightarrow X(k) \qquad y(n) \Longleftrightarrow Y(k) \qquad ax(n)+by(n) \Longleftrightarrow aX(k)+bY(k),$$

where a and b are arbitrary constants.

6.4.2 Periodicity

As the complex exponential W_N^{nk} is periodic in both the variables n and k with period N ($W_N^{nk} = W_N^{n(k+N)} = W_N^{(n+N)k}$), a sequence $x(n)$ of N samples and its DFT $X(k)$ are periodic with period N. By substituting $k + aN$ for k in the DFT equation and $n + aN$ for n in the IDFT equation, we get $X(k) = X(k + aN)$ and $x(n) = x(n + aN)$, where a is any integer.

6.4.3 Circular Shift of a Sequence

As any periodic sequence is completely specified by its elements over a period, the shifted version of a periodic sequence can be obtained by circularly shifting its elements over a period. As the time-domain sequence $x(n)$ and its DFT $X(k)$ are considered periodic, the shift of these sequences are called as circular shift. For example, the delayed sequence $(x - 1)$ is obtained by moving the last sample of $x(n)$ to the beginning of the sequence. Similarly, the advanced sequence $(x + 2)$ is obtained by moving the first two samples of $x(n)$ to the end of the sequence. Only $(N - 1)$ unique shifts are possible for a sequence with N samples.

The distance between two samples of a sinusoid completing k cycles in its period of N samples is $(2\pi/N)k$ radians. Therefore, a shift of the sinusoid by m sample intervals to the right amounts to changing its phase by $-(2\pi/N)mk$ radians, with its amplitude unchanged. The change in the phase is $(2\pi/N)mk$ radians for a left shift. Let $x(n) \Longleftrightarrow X(k)$. Then,

$$x(n \pm m) \Longleftrightarrow e^{\pm j\frac{2\pi}{N}mk} X(k) = W_N^{\mp mk} X(k)$$

The cosine waveform $x(n) = \cos(\frac{2\pi}{16}n)$ with $N = 16$ and its DFT are shown, respectively, in Figures 6.2(g) and (h). By shifting $x(n)$ to the right by two sample intervals, we get $x(n) = \cos(\frac{2\pi}{16}(n - 2))$. The spectral value $X(1)$ of the delayed waveform is obtained by multiplying $X(1) = 8$ in Figure 6.2(h) by $e^{-j\frac{2\pi}{16}(2)(1)} = e^{-j\frac{\pi}{4}} = \frac{1}{\sqrt{2}}(1 - j1)$. The result is $X(1) = \frac{8}{\sqrt{2}}(1 - j1)$. Similarly, $X(15) = \frac{8}{\sqrt{2}}(1 + j1)$.

6.4.4 Circular Shift of a Spectrum

The spectrum, $X(k)$, of a signal, $x(n)$, can be shifted by multiplying the signal by a complex exponential, $e^{\pm jk_0\frac{2\pi}{N}n}$, where k_0 is an integer and N is the length of $x(n)$. The new spectrum is $X(k \mp k_0)$, since a spectral component $X(k_a)e^{jk_a\frac{2\pi}{N}n}$ of the signal, multiplied by $e^{jk_0\frac{2\pi}{N}n}$, becomes $X(k_a)e^{j((k_a+k_0)\frac{2\pi}{N}n)}$ and the corresponding spectral value occurs at $k = (k_a + k_0)$, after a delay of k_0 samples. The spectrum is circularly shifted by k_0 sample intervals. For example, if $k_0 = 1$ or $k_0 = N + 1$, then the dc spectral value of the original signal appears at $k = 1$. With $k_0 = -1$ or $k_0 = N - 1$, it appears at $k = N - 1$. Let $x(n) \Longleftrightarrow X(k)$. Then,

$$e^{\mp j\frac{2\pi}{N}k_0 n} x(n) = W_N^{\pm k_0 n} x(n) \Longleftrightarrow X(k \pm k_0)$$

The complex exponential $x(n) = e^{(j\frac{2\pi}{16}n)}$ with $N = 16$ and its spectrum $X(1) = 16$ are shown, respectively, in Figures 6.2(c) and (d). By multiplying $x(n)$ with $e^{(j\frac{2\pi}{16}2n)}$, we get $x(n) = e^{(j\frac{2\pi}{16}3n)}$. Then, the spectrum becomes $X(3) = 16$.

6.4.5 Symmetry

Symmetry of a signal can be used to reduce its storage and computational requirements. The DFT symmetry properties for various types of signals are shown in Table 6.1. In this table, Re stands for 'real part of' and Im stands for 'imaginary part of'. The symbol * indicates the complex conjugation operation. Note that the even symmetry condition $x(n) = x(-n)$ is the same as $x(n) = x(N - n)$ for a periodic signal of period N.

Table 6.1. DFT symmetry properties

Signal $x(n), \quad n = 0, 1, \ldots, N-1$	DFT $X(k)$
Real, $\text{Im}(x(n)) = 0$	Hermitian $X(k) = X^*(N-k)$
Real and even	Real and even
$\text{Im}(x(n)) = 0$ and $x(n) = x(N-n)$	$\text{Im}(X(k)) = 0$ and $X(k) = X(N-k)$
Real and odd	Imaginary and odd
$\text{Im}(x(n)) = 0$ and $x(n) = -x(N-n)$	$\text{Re}(X(k)) = 0$ and $X(k) = -X(N-k)$
Real and even half-wave	Hermitian and even-indexed only
$\text{Im}(x(n)) = 0$ and $x(n) = x(n \pm \frac{N}{2})$	$X(k) = X^*(N-k)$ and $X(2k+1) = 0$
Real and odd half-wave	Hermitian and odd-indexed only
$\text{Im}(x(n)) = 0$ and $x(n) = -x(n \pm \frac{N}{2})$	$X(k) = X^*(N-k)$ and $X(2k) = 0$
Imaginary, $\text{Re}(x(n)) = 0$	Antihermitian $X(k) = -X^*(N-k)$
Imaginary and even	Imaginary and even
$\text{Re}(x(n)) = 0$ and $x(n) = x(N-n)$	$\text{Re}(X(k)) = 0$ and $X(k) = X(N-k)$
Imaginary and odd	Real and odd
$\text{Re}(x(n)) = 0$ and $x(n) = -x(N-n)$	$\text{Im}(X(k)) = 0$ and $X(k) = -X(N-k)$
Imaginary and even half-wave	Antihermitian and even-indexed only
$\text{Re}(x(n)) = 0$ and $x(n) = x(n \pm \frac{N}{2})$	$X(k) = -X^*(N-k)$ and $X(2k+1) = 0$
Imaginary and odd half-wave	Antihermitian and odd-indexed only
$\text{Re}(x(n)) = 0$ and $x(n) = -x(n \pm \frac{N}{2})$	$X(k) = -X^*(N-k)$ and $X(2k) = 0$
Complex and even, $x(n) = x(N-n)$	Even, $X(k) = X(N-k)$
Complex and odd, $x(n) = -x(N-n)$	Odd, $X(k) = -X(N-k)$
Complex and even half-wave	Even-indexed only
$x(n) = x(n \pm \frac{N}{2})$	$X(2k+1) = 0$
Complex and odd half-wave	Odd-indexed only
$x(n) = -x(n \pm \frac{N}{2})$	$X(2k) = 0$

6.4.6 Circular Convolution of Time-domain Sequences

Let $x(n)$ and $h(n)$ be two periodic time-domain sequences of the same period N. Then, the circular convolution of the sequences is defined as

$$y(n) = \sum_{m=0}^{N-1} x(m)h(n-m) = \sum_{m=0}^{N-1} h(m)x(n-m), \quad n = 0, 1, \ldots, N-1$$

The principal difference of this type of convolution from that of the linear convolution (Chapter 4) is that the range of the summation is restricted to a single period.

The convolution of $h(n)$ with a complex exponential $e^{jk_0\omega_0 n}$, $\omega_0 = 2\pi/N$ is given as

$$\sum_{m=0}^{N-1} h(m)e^{jk_0\omega_0(n-m)} = e^{jk_0\omega_0 n}\sum_{m=0}^{N-1} h(m)e^{-jk_0\omega_0 m} = H(k_0)e^{jk_0\omega_0 n}$$

As an arbitrary $x(n)$ is reconstructed by the IDFT as

$$x(n) = \frac{1}{N} \sum_{k=0}^{N-1} X(k) W_N^{-nk}$$

the convolution of $x(n)$ and $h(n)$ is given by

$$y(n) = \frac{1}{N} \sum_{k=0}^{N-1} X(k) H(k) W_N^{-nk},$$

where $X(k)$ and $H(k)$ are, respectively, the DFT of $x(n)$ and $h(n)$. The IDFT of $X(k)H(k)$ is the circular convolution of $x(n)$ and $h(n)$.

Example 6.1. Convolve $x(n) = \{2, 1, 3, 3\}$ and $h(n) = \{1, 0, 2, 4\}$.

Solution
$$X(k) = \{9, -1 + j2, 1, -1 - j2\} \text{ and } H(k) = \{7, -1 + j4, -1, -1 - j4\}$$
$$X(k)H(k) = \{63, -7 - j6, -1, -7 + j6\}$$

The product $X(k)H(k)$ is obtained by multiplying the corresponding terms in the two sequences. The IDFT of $X(k)H(k)$ is the convolution sum, $y(n) = \{12, 19, 19, 13\}$ ☐

6.4.7 Circular Convolution of Frequency-domain Sequences

Let $X(k)$ and $H(k)$ be two periodic frequency-domain sequences of the same period N. Then, the circular convolution of the sequences, divided by N, is given as

$$x(n)h(n) \Longleftrightarrow \frac{1}{N} \sum_{m=0}^{N-1} X(m) H(k-m) = \frac{1}{N} \sum_{m=0}^{N-1} H(m) X(k-m)$$

where $x(n)$ and $h(n)$, are the IDFT, respectively, of $X(k)$ and $H(k)$.

Example 6.2. Convolve $X(k) = \{9, -1 + j2, 1, -1 - j2\}$ and $H(k) = \{7, -1 + j4, -1, -1 - j4\}$.

Solution
$$x(n) = \{2, 1, 3, 3\} \text{ and } h(n) = \{1, 0, 2, 4\}$$
$$x(n)h(n) = \{2, 0, 6, 12\}$$

The product $x(n)h(n)$ is obtained by multiplying the corresponding terms in the two sequences. The DFT of $x(n)h(n)$ multiplied by four is the convolution sum of $X(k)$ and $H(k)$, $4\{20, -4 + j12, -4, -4 - j12\}$ ☐

6.4.8 Parseval's Theorem

This theorem expresses the power of a signal in terms of its DFT spectrum. Let $x(n) \Longleftrightarrow X(k)$ with sequence length N. The sum of the squared magnitude of the samples of a complex exponential with amplitude one, over the period N, is N. Remember that these samples occur on the unit-circle. The DFT decomposes a signal in terms of complex exponentials with coefficients $X(k)/N$. Therefore, the power of a complex exponential is $(|X(k)|^2/N^2)N = (|X(k)|^2)/N$. The power of the signal is the sum of the powers of its constituent complex exponentials and is given as

$$\sum_{n=0}^{N-1} |x(n)|^2 = \frac{1}{N} \sum_{k=0}^{N-1} |X(k)|^2$$

Example 6.3. Consider the DFT pair

$$\{2, 1, 3, 3\} \Longleftrightarrow \{9, -1 + j2, 1, -1 - j2\}$$

The sum of the squared magnitude of the data sequence is 23 and that of the DFT coefficients divided by 4 is also 23.   ☐

6.5 Applications of the Discrete Fourier Transform

The DFT is extensively used to approximate the forward and inverse transforms of the other versions of the Fourier analysis as described in other chapters. In addition, important operations such as convolution, interpolation, and decimation are carried out efficiently using the DFT as presented in this section.

6.5.1 Computation of the Linear Convolution Using the DFT

Circular convolution assumes two periodic sequences of the same period and results in a periodic sequence with that period. Using the DFT, circular convolution can be efficiently carried out, as the DFT assumes a finite length sequence is periodically extended. However, the linear convolution is of prime interest in LTI system analysis. The linear convolution of two finite sequences of length N and M is a sequence of length $(N + M - 1)$.

The basis of using the DFT to evaluate the linear convolution operation, as well as approximating other versions of Fourier analysis, is to make the period of the DFT so that all the essential information required is available in any one period with sufficient accuracy. Therefore, both the sequences to be convolved must be zero-padded to make them of length $(N + M - 1)$, at the least. This prevents the wrap-around effect of the circular convolution and makes one period output of the circular convolution the same as that of the linear convolution.

The linear convolution of $\{2, 1\}$ and $\{3, 4\}$ is $\{6, 11, 4\}$. The DFT of the sequences are, respectively, $\{3, 1\}$ and $\{7, -1\}$. The term by term product of these DFT is $\{21, -1\}$. The IDFT of this product yields the periodic convolution output $\{10, 11\}$. The last value 4 of the linear convolution is added to the first value 6 to make the first value of the circular convolution 10. The last value of the circular convolution is unaffected by aliasing in the time-domain. The DFT of the 4-point zero-padded sequences $\{2, 1, 0, 0\}$ and $\{3, 4, 0, 0\}$, respectively, are $\{3, 2 - j, 1, 2 + j\}$ and $\{7, 3 - j4, -1, 3 + j4\}$. The term-by-term product of these DFT is $\{21, 2 - j11, -1, 2 + j11\}$. The IDFT of this product yields the linear convolution output with one zero appended $\{6, 11, 4, 0\}$. We could have avoided the zero at the end by zero-padding the signals to make their length three. As fast DFT algorithms with high regularity are available only for data lengths those are an integral power of two, the input sequences are usually zero padded to make the length of the sequences an integral power of two. Of course, this length must be greater than or equal to the sum of the lengths of the two given sequences minus one.

6.5.2 Interpolation and Decimation

Changing the sampling rate of a signal is required for efficient signal processing. For example, reconstructing a signal is easier with a higher sampling rate while a lower sampling rate may be adequate for processing, requiring a shorter computation time. Changing the sampling rate of a signal by reconstructing the corresponding analog signal and resampling it at the new sampling rate introduces large errors. Therefore, sampling rate is usually changed in the discrete form itself. An analog signal sampled with an adequate sampling rate results in its proper discrete form. Sampling rate can be increased (interpolation) or decreased (decimation) to suit the processing requirements as long as the sampling theorem is not violated.

6.5.2.1 Interpolation

Increasing the sampling rate of a signal by a factor I is called interpolation or upsampling. First, the signal is zero padded with $(I - 1)$ samples with value zero between successive samples. In the frequency-domain, the operation of zero-padding corresponds to duplicating the spectrum of the given waveform $(I - 1)$ times, due to the periodicity of the complex exponential W_N^{nk}. This signal is passed through a lowpass filter with a cutoff frequency π/I radians and a passband gain I. The resulting spectrum corresponds to that of the interpolated version of the given waveform. Note that all the frequency components of the given signal lies in the range from zero to π/I

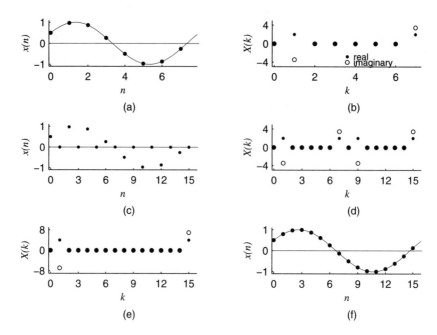

Figure 6.3 (a) A real signal; (b) its spectrum; (c) the signal shown in (a) with zero padding in between the samples; (d) its spectrum, which is the same as that shown in (b), but repeats; (e) the spectrum shown in (d) after lowpass filtering; (f) the corresponding time-domain signal, which is an interpolated version of that shown in (a)

radians of the duplicated spectrum. Frequency π corresponds to half the sampling frequency and the frequency with index $N/2$ in the DFT spectrum.

The signal, $x(n) = \cos(\frac{2\pi}{8}n - \frac{\pi}{3})$, is shown in Figure 6.3(a) and its spectrum is shown in Figure 6.3(b). With the interpolation factor $I = 2$, we want twice the number of samples in a cycle than that in Figure 6.3(a). This requires the insertion of one sample with zero value in between the samples, as shown in Figure 6.3(c). The DFT of the zero padded signal is shown in Figure 6.3(d). Except for the repetition, this spectrum is the same as that in Figure 6.3(b). This spectrum has two frequency components with frequency indices $k = 1$ and $k = 7$. We have to filter out the frequency component with $k = 7$. Therefore, lowpass filtering of this signal with the filter cutoff frequency $\pi/2$ radians and gain two yields the the spectrum shown in Figure 6.3(e) and the corresponding interpolated signal, $x(n) = \cos(\frac{2\pi}{16}n - \frac{\pi}{3})$, is shown in Figure 6.3(f). The spectrum in Figure 6.3(b) is the DFT of the sinusoid with 8 samples in a cycle, whereas that in Figure 6.3(e) is the DFT of the sinusoid with 16 samples in a cycle.

6.5.2.2 Decimation

Reducing the sampling rate of a signal by a factor D is called decimation or down-sampling. As we reduce the sampling rate, we have to filter the high-frequency

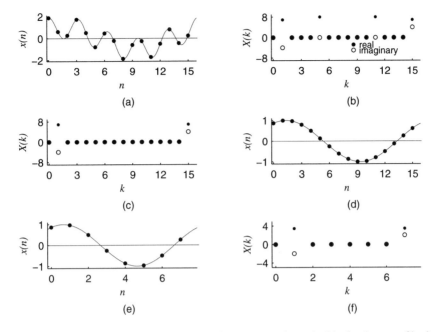

Figure 6.4 (a) A real signal; (b) its spectrum; (c) the spectrum shown in (b) after lowpass filtering; (d) the corresponding time-domain signal; (e) the signal shown in (d) with decimation of alternate samples; (f) its spectrum, which is the same as that shown in (c), but compressed

components of the signal first, by a filter with a cutoff frequency π/D and a passband gain 1, to eliminate aliasing. Then, we take every Dth sample. It is assumed that the filtered out high-frequency components are of no interest. The signal, $x(n) = \cos(\frac{2\pi}{16}n - \frac{\pi}{6}) + \cos(5\frac{2\pi}{16}n)$, is shown in Figure 6.4(a) and its spectrum is shown in Figure 6.4(b). With the decimation factor $D = 2$, we want half the number of samples in a cycle than that in Figure 6.4(a). The signal is passed through a lowpass filter with cutoff frequency $\pi/2$ and gain of 1. The spectrum of the filter output is shown in Figure 6.4(c) and the filtered signal, $x(n) = \cos(\frac{2\pi}{16}n - \frac{\pi}{6})$, is shown in Figure 6.4(d). Now, the decimated signal, $x(n) = \cos(\frac{2\pi}{8}n - \frac{\pi}{6})$, is obtained by taking every second sample. The decimated signal is shown in Figure 6.4(e) and its spectrum is shown in Figure 6.4(f).

6.5.2.3 Interpolation and Decimation

A sampling rate converter, which is a cascade of an interpolator and a decimator, can be used to convert the sampling rate by any rational factor, I/D. A single lowpass filter, with a cutoff frequency that is the smaller of π/I and π/D, and gain of I, is adequate. The signal, $x(n) = \cos(\frac{2\pi}{4}n - \frac{\pi}{3})$, is shown in Figure 6.5(a) and its spectrum is shown in Figure 6.5(b). With $I = 3$, $D = 2$, and $I/D = 3/2$ we want one and a half times the number of samples in a cycle than that in Figure 6.5(a). The insertion of

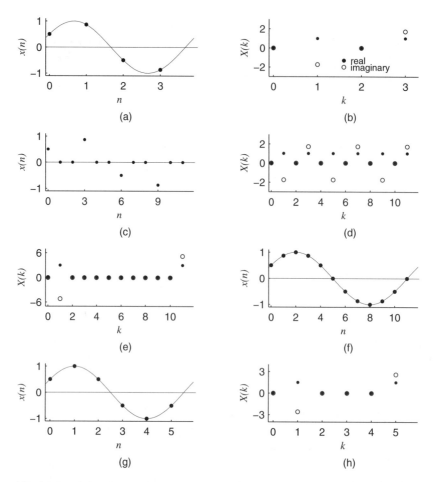

Figure 6.5 (a) A real signal; (b) its spectrum; (c) the signal shown in (a) with zero padding in between the samples; (d) its spectrum, which is the same as that shown in (b), but repeats twice; (e) the spectrum shown in (d) after lowpass filtering; (f) the corresponding time-domain signal, which is an interpolated version of that shown in (a); (g) the signal shown in (f) with decimation of alternate samples; (h) its spectrum, which is the same as that shown in (e), but compressed

two samples with zero value is required, as shown in Figure 6.5(c). The spectrum of this signal, which repeats twice, is shown in Figure 6.5(d). A lowpass filter, with cutoff frequency $\pi/3$ and gain of 3, eliminates the two high-frequency components. The resulting spectrum is shown in Figure 6.5(e) and the interpolated signal, $x(n) = \cos(\frac{2\pi}{12}n - \frac{\pi}{3})$, is shown in Figure 6.5(f). Now, by taking alternate samples, we get the decimated signal, $x(n) = \cos(\frac{2\pi}{6}n - \frac{\pi}{3})$, shown in Figure 6.5(g). Its spectrum is shown in Figure 6.5(h). Sampling rate conversion by a factor 3/2 resulted in six samples in a cycle, as shown in Figure 6.5(g), compared with four samples in a cycle in Figure 6.5(a).

6.6 Summary

- In this chapter, the DFT, its properties, and some of its applications have been presented.
- Frequency-domain analysis uses sinusoids or complex exponentials as basis signals to represent signals and systems, in contrast to impulse in the time-domain analysis.
- The basis functions used in Fourier analysis are constant amplitude sinusoids or exponentials with pure imaginary arguments. Fourier analysis has different versions, each version suitable for different type of signals. The sinusoidal basis functions differ, in each case, in characteristics such as discrete or continuous and finite or infinite in number.
- In all versions of Fourier analysis, the signal is represented with respect to the least-squares error criterion.
- The DFT version of Fourier analysis uses a finite number of harmonically related discrete sinusoids as basis functions. Therefore, both the input data and its spectrum are periodic and discrete. This fact makes it naturally suitable for numerical computation.
- The input to the DFT is a finite sequence of samples and it is assumed to be periodically extended. The DFT coefficients are the coefficients of the basis complex exponentials whose superposition sum yields the periodically extended discrete signal. The IDFT carries out this sum.
- The DFT is extensively used in the approximation of the other versions of Fourier analysis, in addition to efficient evaluation of important operations such as convolution, interpolation, and decimation.
- The periodicity property of the DFT is the key factor in deriving fast algorithms for its computation. These algorithms make the use of the DFT more efficient in most applications compared with alternate methods.

Further Reading

1. Sundararajan, D., *Discrete Fourier Transform, Theory, Algorithms, and Applications*, World Scientific, Singapore, 2001.
2. Sundararajan, D., *Digital Signal Processing, Theory and Practice*, World Scientific, Singapore, 2003.

Exercises

6.1 Given the DFT spectrum $X(k)$, express the corresponding time-domain signal $x(n)$ in terms of its constituent real sinusoids.

6.1.1 $\{X(0) = 3, X(1) = \frac{1}{\sqrt{2}} - j\frac{1}{\sqrt{2}}, X(2) = -2, X(3) = \frac{1}{\sqrt{2}} + j\frac{1}{\sqrt{2}}\}$.

6.1.2 $\{X(0) = -2, X(1) = \sqrt{3} + j1, X(2) = 3, X(3) = \sqrt{3} - j1\}$.

*6.1.3 $\{X(0) = 1, X(1) = 2 - j2\sqrt{3}, X(2) = -3, X(3) = 2 + j2\sqrt{3}\}$.

6.1.4 $\{X(0) = 3, X(1) = 4, X(2) = 1, X(3) = 4\}$.

6.1.5 $\{X(0) = -5, X(1) = j8, X(2) = 2, X(3) = -j8\}$.

6.2 Find the four samples of $x(n)$ over one period, and then use the DFT matrix equation to compute the spectrum $X(k)$.

6.2.1 $x(n) = 2 + 3\sin\left(\frac{2\pi}{4}n - \frac{\pi}{6}\right) - \cos(\pi n)$.

*6.2.2 $x(n) = -1 - 2\cos\left(\frac{2\pi}{4}n + \frac{\pi}{3}\right) + 2\cos(\pi n)$.

6.2.3 $x(n) = 3 + \cos\left(\frac{2\pi}{4}n - \frac{\pi}{4}\right) - 3\cos(\pi n)$.

6.2.4 $x(n) = 1 - 2\sin\left(\frac{2\pi}{4}n + \frac{\pi}{4}\right) + 4\cos(\pi n)$.

6.2.5 $x(n) = -2 + 3\cos\left(\frac{2\pi}{4}n + \frac{\pi}{6}\right) - 2\cos(\pi n)$.

6.3 Find the IDFT of the given spectrum $X(k)$ using the IDFT matrix equation.

6.3.1 $\{X(0) = -12, X(1) = 2 - j2\sqrt{3}, X(2) = 8, X(3) = 2 + j2\sqrt{3}\}$.

6.3.2 $\{X(0) = 4, X(1) = -4\sqrt{3} + j4, X(2) = -4, X(3) = -4\sqrt{3} - j4\}$.

6.3.3 $\{X(0) = 8, X(1) = 3 - j3\sqrt{3}, X(2) = 8, X(3) = 3 + j3\sqrt{3}\}$.

*6.3.4 $\{X(0) = -16, X(1) = -3\sqrt{2} - j3\sqrt{2}, X(2) = 8, X(3) = -3\sqrt{2} + j3\sqrt{2}\}$.

6.3.5 $\{X(0) = 12, X(1) = -2 - j2\sqrt{3}, X(2) = -12, X(3) = -2 + j2\sqrt{3}\}$.

6.4 Find the sample values of the waveform over one period first, and then use the matrix equation to find its DFT spectrum. Verify that the spectral values are the same as the corresponding coefficients of the exponentials multiplied by four.

6.4.1 $x(n) = (1 + j\sqrt{3})e^{j0\frac{2\pi}{4}n} + (2 - j2\sqrt{3})e^{j\frac{2\pi}{4}n} + (1 - j1)e^{j2\frac{2\pi}{4}n} - (1 + j1)e^{j3\frac{2\pi}{4}n}$.

6.4.2 $x(n) = (2 + j1)e^{j0\frac{2\pi}{4}n} + (3 - j2)e^{j\frac{2\pi}{4}n} + (4 - j1)e^{j2\frac{2\pi}{4}n} - (3 + j2)e^{j3\frac{2\pi}{4}n}$.

6.4.3 $x(n) = (1 - j2)e^{j0\frac{2\pi}{4}n} + (2 + j2)e^{j\frac{2\pi}{4}n} + (3 - j3)e^{j2\frac{2\pi}{4}n} + (1 - j4)e^{j3\frac{2\pi}{4}n}$.

6.4.4 $x(n) = (1 + j2)e^{j0\frac{2\pi}{4}n} + (2 + j3)e^{j\frac{2\pi}{4}n} + (4 + j4)e^{j2\frac{2\pi}{4}n} + (3 - j2)e^{j3\frac{2\pi}{4}n}$.

6.4.5 $x(n) = (2 - j2)e^{j0\frac{2\pi}{4}n} + (1 - j4)e^{j\frac{2\pi}{4}n} + (2 + j1)e^{j2\frac{2\pi}{4}n} + (1 - j2)e^{j3\frac{2\pi}{4}n}$.

6.5 Find the IDFT of the given spectrum $X(k)$ using the matrix IDFT equation.

*6.5.1 $\{X(0) = 1 - j1, X(1) = 3 - j2, X(2) = 4 + j1, X(3) = 1 + j2\}$.

6.5.2 $\{X(0) = 3 + j3, X(1) = 1 - j1, X(2) = 2 + j3, X(3) = 1 - j4\}$.

6.5.3 $\{X(0) = 2 - j3, X(1) = 1 + j5, X(2) = 2 + j3, X(3) = 2 + j4\}$.

6.5.4 $\{X(0) = 1 - j4, X(1) = 4 + j2, X(2) = 3 + j1, X(3) = 2 + j2\}$.

6.5.5 $\{X(0) = 3 - j4, X(1) = 2 + j5, X(2) = 1 - j3, X(3) = 2 - j4\}$.

6.6 Find the DFT $X(k)$ of the given $x(n)$. Using the periodicity property of the DFT and the IDFT, find the required $x(n)$ and $X(k)$.

6.6.1 $x(n) = \{2 + j3, 1 - j2, 2 + j1, 3 + j4\}$. Find $x(13)$, $x(-22)$, $X(10)$, and $X(-28)$.

*6.6.2 $x(n) = \{1 + j2, 2 - j3, 2 + j2, 1 - j4\}$. Find $x(-14)$, $x(43)$, $X(12)$, and $X(-7)$.

6.7 Find the DFT $X(k)$ of $x(n) = \{2 - j2, 1 + j3, 4 + j2, 1 - j2\}$. Using the time-domain shift property and $X(k)$, deduce the DFT of $x(n + 1)$, $x(n - 2)$, and $x(n + 3)$.

6.8 Find the IDFT $x(n)$ of $X(k) = \{12 + j4, 8 - j4, 4 + j4, 4 + j8\}$. Using the frequency-domain shift property, deduce the IDFT of $X(k + 1)$, $X(k - 2)$, and $X(k + 3)$.

6.9 Find the circular convolution of two frequency-domain sequences $X(k)$ and $H(k)$ using the DFT and the IDFT.

6.9.1 $X(k) = \{8 - j4, 4 + j4, 12 - j8, 8 - j12\}$ and $H(k) = \{12 - j4, 8 - j4, 4 + j8, 2 + j12\}$.

*6.9.2 $X(k) = \{8, 4, 8, 4\}$ and $H(k) = \{12, 8, 4, 12\}$.

6.9.3 $X(k) = \{0, j4, 0, -j4\}$ and $H(k) = \{0, 4, 0, -4\}$.

6.10 Find the DFT of $x(n)$ and verify the Parseval's theorem.

6.10.1 $x(n) = \{2, 4, 3, 1\}$.

6.10.2 $x(n) = \{-2, 4, 2, 5\}$.

6.10.3 $x(n) = \{4, -1, 3, 1\}$.

6.11 Find the linear convolution of the sequences $x(n)$ and $y(n)$ using the DFT and the IDFT.

6.11.1 $\{x(0) = 2, x(1) = 4, x(2) = 3\}$ and $\{y(0) = 1, y(1) = -2\}$.

6.11.2 $\{x(0) = 2, x(1) = -4, x(2) = 3\}$ and $\{y(0) = 1, y(1) = 2\}$.

*6.11.3 $\{x(0) = 1, x(1) = 4, x(2) = -3\}$ and $\{y(0) = -4, y(1) = 3\}$.

7

Fourier Series

Continuous periodic signals are analyzed using an infinite set of harmonically related sinusoids and a dc component in the FS frequency-domain representation. Increasing the number of samples in a period, by decreasing the sampling interval, results in a densely sampled time-domain waveform and a broader DFT periodic spectrum. As the sampling interval tends to zero, the time-domain waveform becomes a continuous function and the discrete spectrum becomes aperiodic. As the period of the waveform remains the same, the fundamental frequency and the harmonic spacing of the spectrum is fixed. Therefore, the discrete nature of the spectrum is unchanged. In Section 7.1, we derive the exponential form of the FS, starting from the defining equations of the DFT and the IDFT. Then, two equivalent trigonometric forms of the FS are deduced from the expressions of the exponential form. The properties of the FS are described in Section 7.2. The approximation of the FS coefficients by the DFT is presented in Section 7.3. Typical applications of the FS are presented in Section 7.4.

7.1 Fourier Series

A continuous periodic signal, $x(t)$, with period T is expressed as a sum of an infinite set of harmonically related sinusoids and a dc component in the FS. The frequency of the fundamental or first harmonic is the frequency of the waveform under analysis. That is, $\omega_0 = 2\pi/T$. The frequency of the second harmonic is $2\omega_0$, that of the third harmonic is $3\omega_0$, and so on.

7.1.1 FS as the Limiting Case of the DFT

While the FS can be derived using the orthogonality property of sinusoids, it is instructive to consider it as the limiting case of the DFT with the sampling interval tending to zero. Consider the continuous periodic signal $x(t)$, with period $T = 5$ s, and its five

A Practical Approach to Signals and Systems   D. Sundararajan
© 2008 John Wiley & Sons (Asia) Pte Ltd

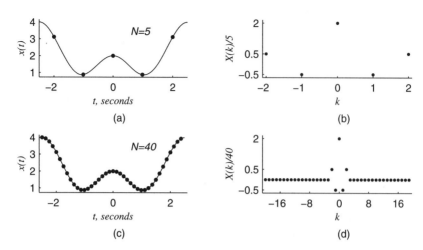

Figure 7.1   (a) Samples, at intervals of 1 s, of a periodic continuous signal $x(t)$ with period 5 s; (b) its scaled DFT spectrum; (c) samples of $x(t)$ at intervals of 0.125 s; (d) its scaled DFT spectrum

sample values, with a sampling interval of $T_s = 1$ s, shown in Figure 7.1(a). The scaled DFT spectrum of this sample set is shown in Figure 7.1(b). The frequency increment of the spectrum is $\omega_0 = 2\pi/5$ radians per second and one period of the DFT spectrum corresponds to $(2\pi/5) \times 5 = 2\pi$ radians. The samples of the same signal, with a sampling interval of $T_s = 0.125$ s, are shown in Figure 7.1(c). Reducing the sampling interval results in a densely sampled time-domain waveform. The scaled DFT spectrum of this sample set is shown in Figure 7.1(d). As the frequency increment is fixed at $2\pi/5$, the larger number of spectral values corresponds to a broader spectrum of width $(2\pi/5) \times 40 = 16\pi$ radians. Eventually, as the sampling interval tends to zero, the time-domain waveform becomes continuous and the discrete spectrum becomes aperiodic.

The mathematical presentation of the foregoing argument is as follows. The IDFT of $X(k)$, $-N \leq k \leq N$ is defined as

$$x(n) = \frac{1}{2N+1} \sum_{k=-N}^{N} X(k) e^{j\frac{2\pi}{(2N+1)}nk}, \quad n = 0, \pm 1, \pm 2, \ldots, \pm N$$

Substituting the DFT expression for $X(k)$, we get

$$x(n) = \frac{1}{2N+1} \sum_{k=-N}^{N} \left(\sum_{l=-N}^{N} x(l) e^{-j\frac{2\pi}{(2N+1)}lk} \right) e^{j\frac{2\pi}{(2N+1)}nk}$$

The frequency index k represents the discrete frequency $[2\pi/(2N+1)]k$. If the periodic signal, with period T, is sampled with a sampling interval of T_s seconds in order to get the samples, then the corresponding continuous frequency is given by

$[2\pi/(2N+1)T_s]k = (2\pi/T)k = k\omega_0$. The time index n corresponds to nT_s s. The number of samples in a period is $(2N+1) = T/T_s$. With these substitutions, we get

$$x(nT_s) = \sum_{k=-N}^{N} \left(\frac{1}{T} \sum_{l=-N}^{N} x(lT_s) e^{-j\frac{2\pi}{T} lT_s k} T_s \right) e^{j\frac{2\pi}{T} nT_s k}$$

As T_s is reduced, the number of samples $(2N+1)$ increases, but the product $(2N+1)T_s = T$ remains constant. Hence, the fundamental frequency ω_0 also remains constant. As $T_s \to 0$, nT_s and lT_s become continuous time variables t and τ, respectively, the inner summation becomes an integral over the period T, $N \to \infty$, and differential $d\tau$ formally replaces T_s. Therefore, we get

$$x(t) = \sum_{k=-\infty}^{\infty} \left(\frac{1}{T} \int_{-\frac{T}{2}}^{\frac{T}{2}} x(\tau) e^{-j\frac{2\pi}{T} \tau k} d\tau \right) e^{j\frac{2\pi}{T} tk}$$

The exponential form of the FS for a signal $x(t)$ is

$$x(t) = \sum_{k=-\infty}^{\infty} X_{cs}(k) e^{jk\omega_0 t} \tag{7.1}$$

where the FS coefficients $X_{cs}(k)$ are given as

$$X_{cs}(k) = \frac{1}{T} \int_{t_1}^{t_1+T} x(t) e^{-jk\omega_0 t} \, dt, \quad k = 0, \pm 1, \pm 2, \ldots \tag{7.2}$$

and t_1 is arbitrary. Because of periodicity of the FS with period T, the integral from t_1 to $(t_1 + T)$ will have the same value for any value of t_1. Since sinusoids are represented in terms of exponentials, Equation (7.1) is called the exponential form of the FS.

7.1.2 The Compact Trigonometric Form of the FS

The form of the FS, with sinusoids represented in polar form, is called the compact trigonometric form. Equation (7.1) can be rewritten as

$$x(t) = X_{cs}(0) + \sum_{k=1}^{\infty} (X_{cs}(k) e^{jk\omega_0 t} + X_{cs}(-k) e^{-jk\omega_0 t})$$

Since $X_{cs}(k)e^{jk\omega_0 t}$ and $X_{cs}(-k)e^{-jk\omega_0 t}$ form complex conjugate pair for real signals and their sum is twice the real part of either of the terms, we get

$$x(t) = X_p(0) + \sum_{k=1}^{\infty} X_p(k)\cos(k\omega_0 t + \theta(k)), \tag{7.3}$$

where

$$X_p(0) = X_{cs}(0), \quad X_p(k) = 2|X_{cs}(k)|, \quad \theta(k) = \angle(X_{cs}(k)), \ k = 1, 2, \ldots, \infty$$

7.1.3 The Trigonometric Form of the FS

The form of the FS, with sinusoids represented in rectangular form, is called the trigonometric form. Expressing the sinusoid in Equation (7.3) in rectangular form, we get

$$x(t) = X_c(0) + \sum_{k=1}^{\infty} (X_c(k)\cos(k\omega_0 t) + X_s(k)\sin(k\omega_0 t)), \tag{7.4}$$

where, $X_c(0) = X_p(0) = X_{cs}(0)$, $X_c(k) = X_p(k)\cos(\theta(k)) = 2\,\mathrm{Re}(X_{cs}(k))$, and $X_s(k) = -X_p(k)\sin(\theta(k)) = -2\,\mathrm{Im}(X_{cs}(k))$.

7.1.4 Periodicity of the FS

The FS is a periodic waveform of period that is the same as that of the fundamental, $T = 2\pi/\omega_0$. Replacing t by $t + T$ in Equation (7.3), we get

$$x(t) = X_p(0) + \sum_{k=1}^{\infty} X_p(k)\cos(k\omega_0(t + T) + \theta_k)$$

$$= X_p(0) + \sum_{k=1}^{\infty} X_p(k)\cos(k\omega_0 t + 2k\pi + \theta_k) = x(t)$$

If the waveform to be analyzed is defined only over the interval T, the FS represents the waveform only in that interval. On the other hand, if the waveform is periodic of period T, then the FS is valid for all t.

7.1.5 Existence of the FS

Any signal satisfying the Dirichlet conditions, which are a set of sufficient conditions, can be expressed in terms of a FS. The first of these conditions is that the signal $x(t)$

is absolutely integrable over one period, that is $\int_0^T |x(t)|dt < \infty$. From the definition of the FS, we get

$$|X_{cs}(k)| \leq \frac{1}{T} \int_{t_1}^{t_1+T} |x(t)e^{-jk\omega_0 t}|\, dt = \frac{1}{T} \int_{t_1}^{t_1+T} |x(t)||e^{-jk\omega_0 t}|\, dt$$

Since $|e^{-jk\omega_0 t}| = 1$,

$$|X_{cs}(k)| \leq \frac{1}{T} \int_{t_1}^{t_1+T} |x(t)|\, dt$$

The second condition is that the number of finite maxima and minima in one period of the signal must be finite. The third condition is that the number of finite discontinuities in one period of the signal must be finite. Most signals of practical interest satisfy these conditions.

Example 7.1. Find the three forms of the FS for the signal

$$x(t) = -1 - 2\cos\left(\frac{2\pi}{6}t - \frac{\pi}{3}\right)$$

Solution
As this signal can be expressed in terms of sinusoids easily, we do not need to evaluate any integral. The fundamental frequency of the waveform is $\omega_0 = 2\pi/6$, which is the same as that of the sinusoid. Note that the dc component is periodic with any period.

Compact trigonometric form

$$x(t) = -1 - 2\cos\left(\frac{2\pi}{6}t - \frac{\pi}{3}\right) = -1 + 2\cos\left(\frac{2\pi}{6}t + \frac{2\pi}{3}\right)$$

Comparing this expression with the definition, Equation (7.3), we get the compact trigonometric form of the FS coefficients as

$$X_p(0) = -1, \quad X_p(1) = 2, \quad \theta(1) = \frac{2\pi}{3}$$

A plot of the amplitude $X_p(k)$ of the constituent sinusoids of a signal $x(t)$ versus k or $k\omega_0$ is called the amplitude spectrum of $x(t)$. Similarly, the plot of the phase $\theta(k)$ is called the phase spectrum. The FS amplitude spectrum and the phase spectrum of the signal in compact trigonometric form are shown, respectively, in Figures 7.2(a) and (b).

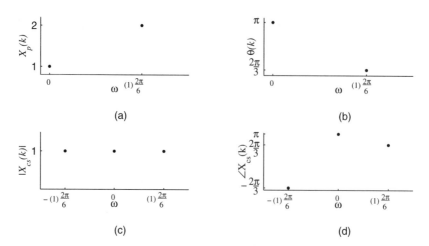

Figure 7.2 (a) The FS amplitude spectrum; (b) the phase spectrum of the signal in compact trigonometric form; (c) the FS amplitude spectrum; (d) the phase spectrum of the signal in exponential form

Trigonometric form

$$x(t) = -1 - 2\cos\left(\frac{2\pi}{6}t - \frac{\pi}{3}\right) = -1 - \cos\left(\frac{2\pi}{6}t\right) - \sqrt{3}\sin\left(\frac{2\pi}{6}t\right)$$

Comparing this expression with the definition, Equation (7.4), we get the trigonometric form of the FS coefficients as

$$X_c(0) = -1 \qquad X_c(1) = -1 \qquad X_s(1) = -\sqrt{3}$$

Exponential form

$$x(t) = -1 - 2\cos\left(\frac{2\pi}{6}t - \frac{\pi}{3}\right) = -1 + 2\cos\left(\frac{2\pi}{6}t + \frac{2\pi}{3}\right)$$

$$= -1 + e^{j\left(\frac{2\pi}{6}t + \frac{2\pi}{3}\right)} + e^{-j\left(\frac{2\pi}{6}t + \frac{2\pi}{3}\right)}$$

Comparing this expression with the definition, Equation (7.1), we get the exponential form of the FS coefficients as

$$X_{cs}(0) = -1, \quad X_{cs}(1) = 1\angle\frac{2\pi}{3} \qquad X_{cs}(-1) = 1\angle-\frac{2\pi}{3}$$

The FS amplitude spectrum and the phase spectrum of the signal in exponential form are shown, respectively, in Figures 7.2(c) and (d).   □

The frequencies of harmonically related continuous sinusoids must be rational numbers or rational multiples of the same transcendental or irrational number. Therefore, the ratio of frequencies of any two harmonically related sinusoids is a rational number. The fundamental frequency (of which the harmonic frequencies are integral multiples) of a combination of sinusoids is found as follows: (i) any common factors of the numerators and denominators of each of the frequencies are cancelled; and (ii) the greatest common divisor of the numerators is divided by the least common multiple of the denominators of the frequencies.

Example 7.2. Find the exponential form of the FS for the signal

$$x(t) = 2 + 4\sin\left(\frac{1}{2}t + \frac{\pi}{6}\right) + 3\cos\left(\frac{3}{5}t - \frac{\pi}{4}\right)$$

Solution
The frequency of the waveforms are $1/2$ and $3/5$. There are no common factors of the numerators and denominators. The least common multiple of the denominators (2,5) is 10. The greatest common divisor of the numerators (1,3) is one. Therefore, the fundamental frequency is $\omega_0 = 1/10$ radians per second. The fundamental period is $T = 2\pi/\omega_0 = 2\pi 10/1 = 20\pi$. The first sinusoid, the fifth harmonic shown in Figure 7.3 (dashed line), completes five cycles and the second sinusoid (dotted line), the sixth harmonic, completes six cycles in the period. The combined waveform (solid line) completes one cycle in the period.

$$x(t) = 2 + 2e^{j\left(\frac{1}{2}t - \frac{\pi}{3}\right)} + 2e^{-j\left(\frac{1}{2}t - \frac{\pi}{3}\right)} + \frac{3}{2}e^{j\left(\frac{3}{5}t - \frac{\pi}{4}\right)} + \frac{3}{2}e^{-j\left(\frac{3}{5}t - \frac{\pi}{4}\right)}$$

Comparing this expression with the definition, Equation (7.1), we get the exponential form of the FS coefficients as $X_{cs}(0) = 2$, $X_{cs}(\pm 5) = 2\angle\mp\frac{\pi}{3}$, $X_{cs}(\pm 6) = \frac{3}{2}\angle\mp\frac{\pi}{4}$.   □

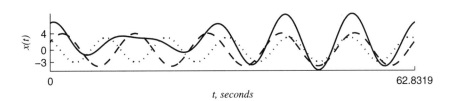

Figure 7.3   The harmonics and the combined waveform

Example 7.3. Find the FS for a square wave defined over one period as

$$x(t) = \begin{cases} 1 & \text{for } |t| < \frac{1}{4} \\ 0 & \text{for } \frac{1}{4} < |t| < \frac{1}{2} \end{cases}$$

Solution

The period of the waveform is one and the fundamental frequency is 2π. The waveform is even-symmetric and odd half-wave symmetric with a dc bias. Therefore, in addition to the dc component, the waveform is composed of odd-indexed cosine waves only.

$$X_c(0) = 2 \int_0^{\frac{1}{4}} dt = \frac{1}{2}$$

$$X_c(k) = 4 \int_0^{\frac{1}{4}} \cos(2\pi k\, t)dt = \begin{cases} \frac{2}{k\pi} \sin\left(\frac{\pi}{2}k\right) & \text{for } k \text{ odd} \\ 0 & \text{for } k \text{ even and } k \neq 0 \end{cases}$$

$$x(t) = \frac{1}{2} + \frac{2}{\pi}(\cos(2\pi t) - \frac{1}{3}\cos(3(2\pi t)) + \frac{1}{5}\cos(5(2\pi t)) - \cdots) \qquad (7.5)$$

The FS amplitude spectrum and the phase spectrum of the signal in exponential form are shown, respectively, in Figures 7.4(a) and (b). ☐

7.1.6 Gibbs Phenomenon

The FS converges uniformly for waveforms with no discontinuity. At any discontinuity of a waveform, the FS converges to the average of the left- and right-hand limits with overshoots and undershoots in the vicinity of the discontinuity. As the basis waveforms of the Fourier series are continuous sinusoids, they can never exactly add up to a discontinuity. This inability of the FS is referred as the Gibbs phenomenon.

Figures 7.5(a), (b), (c), and (d) show the FS for the square wave, using up to the first, third, seventh, and fifteenth harmonics, respectively. Consider the FS for the

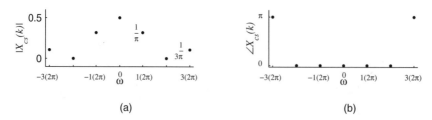

(a)   (b)

Figure 7.4 (a) The FS amplitude spectrum; (b) the phase spectrum of the square wave in exponential form

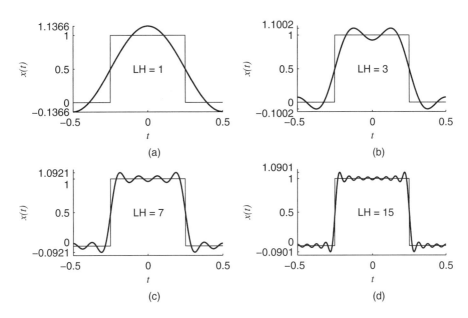

Figure 7.5 The FS for the square wave: (a) using up to the first harmonic (LH = 1); (b) using up to the third harmonic (LH = 3); (c) using up to the seventh harmonic (LH = 7); (d) using up to the fifteenth harmonic (LH = 15)

square wave using up to the first harmonic, $x(t) = \frac{1}{2} + \frac{2}{\pi}\cos(2\pi t)$. By differentiating this expression with respect to t and equating it to zero, we get $\sin(2\pi t) = 0$. The expression evaluates to zero for $t = 0$. Substituting $t = 0$ in the expression for $x(t)$, we get the value of the peak as 1.1366, as shown in Figure 7.5(a). We can find the maximum overshoots in other cases similarly.

As we use more harmonics, the frequency of oscillations increases and the oscillations are confined more closely to the discontinuity. But, the largest amplitude of the oscillations settles at 1.0869 for relatively small number of harmonics. Therefore, even if we use an infinite number of harmonics to represent a waveform with discontinuity, there will be deviations of 8.69% of the discontinuity for a moment. Of course, the area under the deviation tends to zero.

Example 7.4. Find the three forms of the FS for the periodic impulse train, shown in Figure 7.6(a), with period T s defined as

$$x(t) = \sum_{n=-\infty}^{\infty} \delta(t - nT)$$

Figure 7.6 (a) Impulse train with period T s; (b) its FS spectrum

Solution

$$X_{cs}(k) = \frac{1}{T}\int_{-\frac{T}{2}}^{\frac{T}{2}} \delta(t)e^{-jk\omega_0 t}\, dt = \frac{1}{T}\int_{-\frac{T}{2}}^{\frac{T}{2}} \delta(t)\, dt = \frac{1}{T} \qquad -\infty < k < \infty$$

The spectrum, shown in Figure 7.6(b), is also a periodic impulse train with period $\omega_0 = 2\pi/T$ and constant amplitude $1/T$. Note that the impulses in the time-domain are of continuous type (as $x(t)$ is a function of the continuous variable t), while those of the spectrum are of discrete type (as $X_{cs}(k)$ is a function of the discrete variable k). The FS for the impulse train, in exponential form, is given by

$$x(t) = \sum_{k=-\infty}^{\infty} X_{cs}(k)e^{jk\omega_0 t} = \frac{1}{T}\sum_{k=-\infty}^{\infty} e^{jk\omega_0 t} \qquad \omega_0 = \frac{2\pi}{T}$$

The FS coefficients, in compact trigonometric form, are

$$X_p(0) = X_{cs}(0) = \frac{1}{T} \qquad X_p(k) = 2|X_{cs}(k)| = \frac{2}{T} \qquad \theta(k) = 0 \qquad k = 1, 2, 3, \ldots$$

The FS is given by

$$x(t) = \frac{1}{T}(1 + 2(\cos(\omega_0 t) + \cos(2\omega_0 t) + \cos(3\omega_0 t) + \cdots)) \qquad \omega_0 = \frac{2\pi}{T}$$

As the phase $\theta(k) = 0$ is zero, the trigonometric form of the FS is the same as this form.

An alternate way of obtaining this FS is to consider the FS for a train of unit-area rectangular pulses of width a and height $1/a$, with the width a of one pulse including the point $t = 0$. In the limiting case of $a \to 0$, the train of pulses degenerates into an impulse train and the limiting form of its FS is the FS for the impulse train. ☐

7.2 Properties of the Fourier Series

For each operation in one domain, the properties establish the corresponding operation in the other domain, making evident the simpler relationship between variables for a particular operation. For example, the convolution operation in the time-domain corresponds to the much simpler multiplication operation in the frequency-domain. In addition, we can derive the FS coefficients for functions from that for related functions more easily than deriving them from the definition.

7.2.1 Linearity

The FS coefficients for a linear combination of a set of periodic signals, with the same period, is the same linear combination of their individual FS coefficients. That is,

$$x(t) \Longleftrightarrow X_{cs}(k) \qquad y(t) \Longleftrightarrow Y_{cs}(k) \qquad ax(t) + by(t) \Longleftrightarrow aX_{cs}(k) + bY_{cs}(k),$$

where a and b are arbitrary constants. For example, the FS coefficients for $\cos(t)$ and $\sin(t)$ are $X_{cs}(\pm 1) = 1/2$ and $X_{cs}(\pm 1) = \mp j/2$, respectively. The FS coefficients for $\cos(t) + j\sin(t) = e^{jt}$ are $X_{cs}(\pm 1) = 1/2 + (j)(\mp j/2)$. That is, the only nonzero FS coefficient is $X_{cs}(1) = 1$.

7.2.2 Symmetry

The symmetry properties simplify the evaluation of the FS coefficients. If the signal is real, then the real part of its spectrum is even and the imaginary part is odd, called the conjugate symmetry. The FS for a real signal $x(t)$, with period T, is given by

$$X_{cs}(k) = \frac{1}{T} \int_0^T x(t) e^{-jk\frac{2\pi}{T}t} dt = \frac{1}{T} \int_0^T x(t) \left(\cos\left(k\frac{2\pi}{T}t \right) - j\sin\left(k\frac{2\pi}{T}t \right) \right) dt$$

Conjugating both sides, we get

$$X_{cs}^*(k) = \frac{1}{T} \int_0^T x(t) \left(\cos\left(k\frac{2\pi}{T}t \right) + j\sin\left(k\frac{2\pi}{T}t \right) \right) dt$$

Replacing k by $-k$, we get $X_{cs}^*(-k) = X_{cs}(k)$. For example, the FS coefficients for $4\cos(t + \pi/3)$ are $X_{cs}(\pm 1) = 1 \pm j\sqrt{3}$.

7.2.2.1 Even Symmetry

If the signal $x(t)$ is real and even, then its product with sine basis waveforms is odd and the $X_s(k)$ coefficients are, therefore, zero. That is, the signal is composed of cosine waveforms alone and its spectrum is real and even. As the product of the cosine basis waveforms and the signal is even, the FS defining integral can be evaluated over half the period. That is,

$$X_c(0) = \frac{2}{T} \int_{t_1}^{t_1 + \frac{T}{2}} x(t) dt$$

$$X_c(k) = \frac{4}{T} \int_{t_1}^{t_1+\frac{T}{2}} x(t)\cos(k\omega_0 t)\,dt, \quad k = 1, 2, \ldots, \infty$$

For example, the FS coefficients for $\cos(t)$ are $X_{cs}(\pm 1) = \frac{1}{2}$.

7.2.2.2 Odd Symmetry

If the signal $x(t)$ is real and odd, then its product with cosine basis waveforms is odd and the $X_c(k)$ coefficients are, therefore, zero. That is, the signal is composed of sine waveforms alone and its spectrum is imaginary and odd. As the product of the sine basis waveforms and the signal is even, the FS defining integral can be evaluated over half the period. That is,

$$X_s(k) = \frac{4}{T} \int_{t_1}^{t_1+\frac{T}{2}} x(t)\sin(k\omega_0 t)\,dt \quad k = 1, 2, \ldots, \infty$$

For example, the FS coefficients for $\sin(t)$ are $X_{cs}(\pm 1) = \mp\frac{j}{2}$.

As the FS coefficients for a real and even signal are real and even and that for a real and odd signal are imaginary and odd, it follows that the real part of the FS coefficients, $\mathrm{Re}(X_{cs}(k))$, of an arbitrary real signal $x(t)$ are the FS coefficients for its even component $x_e(t)$ and $j\,\mathrm{Im}(X_{cs}(k))$ are that for its odd component $x_o(t)$.

7.2.2.3 Half-wave Symmetry

Even half-wave symmetry. If a periodic signal of period T satisfies the property $x(t \pm T/2) = x(t)$, then it is said to have even half-wave symmetry. That is, it completes two cycles of a pattern in the interval T. The FS coefficients can be expressed as

$$X_{cs}(k) = \frac{1}{T} \int_{t_1}^{t_1+\frac{T}{2}} \left(x(t) + (-1)^k x \left(t + \frac{T}{2} \right) \right) e^{-jk\omega_0 t}\,dt \tag{7.6}$$

The odd-indexed FS coefficients are zero. The even-indexed FS coefficients are given by

$$X_{cs}(k) = \frac{2}{T} \int_{t_1}^{t_1+\frac{T}{2}} x(t)e^{-jk\omega_0 t}\,dt \quad k = 0, 2, 4, \ldots$$

Odd half-wave symmetry. If a periodic signal of period T satisfies the property $-x(t \pm T/2) = x(t)$, then it is said to have odd half-wave symmetry. That is, the values of the signal over any half period are the negatives of the values over the succeeding or preceding half period. It is obvious, from Equation (7.6), that the even-indexed FS coefficients are zero. The odd-indexed FS coefficients are given by

$$X_{cs}(k) = \frac{2}{T} \int_{t_1}^{t_1 + \frac{T}{2}} x(t) e^{-jk\omega_0 t} \, dt \qquad k = 1, 3, 5, \ldots$$

Any periodic signal $x(t)$, with period T, can be decomposed into its even and odd half-wave symmetric components $x_{eh}(t)$ and $x_{oh}(t)$, respectively. That is $x(t) = x_{eh}(t) + x_{oh}(t)$, where

$$x_{eh}(t) = \frac{1}{2}\left(x(t) + x\left(t \pm \frac{T}{2}\right)\right) \quad \text{and} \quad x_{oh}(t) = \frac{1}{2}\left(x(t) - x\left(t \pm \frac{T}{2}\right)\right)$$

7.2.3 Time Shifting

When we shift a signal, the shape remains the same, but the signal is relocated. The shift of a typical spectral component, $X_{cs}(k_0)e^{jk_0\omega_0 t}$, by t_0 to the right results in the exponential, $X_{cs}(k_0)e^{jk_0\omega_0(t-t_0)} = e^{-jk_0\omega_0 t_0} X_{cs}(k_0)e^{jk_0\omega_0 t}$. That is, a delay of t_0 results in changing the phase of the exponential by $-k_0\omega_0 t_0$ radians without changing its amplitude.

Therefore, if the FS spectrum for $x(t)$, with the fundamental frequency $\omega_0 = 2\pi/T$, is $X_{cs}(k)$, then

$$x(t \pm t_0) \iff e^{\pm jk\omega_0 t_0} X_{cs}(k)$$

Consider the FS coefficients $X_{cs}(\pm 1) = \mp\frac{j}{2}$ for $\sin(2t)$. The FS coefficients for $\sin(2t + \frac{\pi}{2}) = \sin(2(t + \frac{\pi}{4})) = \cos(2t)$, with $k = \pm 1$, $\omega_0 = 2$, and $t_0 = \frac{\pi}{4}$, are $X_{cs}(\pm 1) = e^{\pm j(1)(2)\frac{\pi}{4}}(\mp\frac{j}{2}) = \frac{1}{2}$.

7.2.4 Frequency Shifting

The spectrum, $X_{cs}(k)$, of a signal, $x(t)$, can be shifted by multiplying the signal by a complex exponential, $e^{\pm jk_0\omega_0 t}$, where k_0 is an integer and ω_0 is the fundamental frequency. The new spectrum is $X_{cs}(k \mp k_0)$, since a spectral component $X_{cs}(k_a)e^{jk_a\omega_0 t}$ of the signal, multiplied by $e^{jk_0\omega_0 t}$, becomes $X_{cs}(k_a)e^{j((k_a+k_0)\omega_0 t)}$ and the corresponding spectral value occurs at $k = (k_a + k_0)$, after a delay of k_0 samples. Therefore, we get

$$x(t)e^{\pm jk_0\omega_0 t} \iff X_{cs}(k \mp k_0)$$

For example, consider the FS coefficients $X_{cs}(\pm 1) = \mp j/2$ for $\sin(t)$. The FS coefficients for $\cos(2t)\sin(t)$ can be computed using this property. As $\cos(2t) = \frac{1}{2}(e^{j2t} + e^{-j2t})$, the FS coefficients for the new function is the sum of the FS coefficients for $\sin(t)$ shifted to the right and left by two, in addition to the scale factor $1/2$. That is,

$$X_{cs}(\pm 1) = \pm \frac{j}{4} \quad \text{and} \quad X_{cs}(\pm 3) = \mp \frac{j}{4}$$

This spectrum corresponds to the time-domain function $\frac{1}{2}(\sin(3t) - \sin(t))$, which is, of course, equal to $\cos(2t)\sin(t)$.

7.2.5 Convolution in the Time-domain

Using the FS, we get periodic or cyclic convolution as FS analyzes periodic time-domain signals. The periodic convolution operation is defined for two periodic signals, $x(t)$ of period T_1 and $h(t)$ of period T_2, as

$$y(t) = \int_0^T x(\tau)h(t - \tau)d\tau$$

where T (common period of $x(t)$ and $h(t)$) is the least common multiple of T_1 and T_2. The FS coefficients for $y(t)$ are to be expressed in terms of those of $x(t)$ and $h(t)$.

The convolution of $h(t)$ of period T with a complex exponential $e^{jk_0\omega_0 t}$, $\omega_0 = 2\pi/T$ is given as

$$\int_0^T h(\tau)e^{jk_0\omega_0(t-\tau)}d\tau = e^{jk_0\omega_0 t}\int_0^T h(\tau)e^{-jk_0\omega_0\tau}d\tau = TH_{cs}(k_0)e^{jk_0\omega_0 t}$$

As an arbitrary $x(t)$ of period T is reconstructed by the inverse FS as $x(t) = \sum_{k=-\infty}^{\infty} X_{cs}(k)e^{jk_0\omega_0 t}$, the convolution of $x(t)$ and $h(t)$ is given by $y(t) = \sum_{k=-\infty}^{\infty} TX_{cs}(k)H_{cs}(k)e^{jk\omega_0 t}$, where $X_{cs}(k)$ and $H_{cs}(k)$ are, respectively, the FS coefficients for $x(t)$ and $h(t)$. The inverse of the FS spectrum $TX_{cs}(k)H_{cs}(k)$ is the periodic convolution of $x(t)$ and $h(t)$. That is,

$$\int_0^T x(\tau)h(t - \tau)d\tau = \sum_{k=-\infty}^{\infty} TX_{cs}(k)H_{cs}(k)e^{jk\omega_0 t} \iff TX_{cs}(k)H_{cs}(k)$$

Consider the convolution of $x(t) = \cos(t)$ and $h(t) = \sin(t)$ with the FS coefficients $X_{cs}(\pm 1) = 1/2$ and $H_{cs}(\pm 1) = \mp j/2$, respectively. Then, with $T = 2\pi$, we get

$$TX_{cs}(\pm 1)H_{cs}(\pm 1) = 2\pi \left(\mp \frac{j}{4}\right)$$

These FS coefficients correspond to the time-domain function $\pi \sin(t)$. By directly evaluating the time-domain convolution, we get

$$\int_0^{2\pi} \cos(\tau)\sin(t-\tau)d\tau = \int_0^{2\pi} \cos(\tau)(\sin(t)\cos(\tau) - \cos(t)\sin(\tau))d\tau = \pi \sin(t)$$

An important application of this property is in modeling the truncation of the FS spectrum. The signal corresponding to the truncated spectrum has to be expressed in terms of the original signal $x(t)$ with fundamental frequency ω_0. The truncation operation can be considered as multiplying the spectrum of $x(t)$ by the spectrum that is one for $-N \leq k \leq N$ and zero otherwise. The signal corresponding to this spectrum is

$$y(t) = \frac{\sin\left(\frac{(2N+1)\omega_0 t}{2}\right)}{\sin\left(\frac{\omega_0 t}{2}\right)}$$

Therefore, the signal corresponding to the truncated spectrum of $x(t)$ (using only $2N + 1$ FS coefficients) is given by the convolution of $x(t)$ and $y(t)$, multiplied by the factor $1/T$, as

$$x_N(t) = \frac{1}{T}\int_0^T x(\tau)\frac{\sin\left(\frac{(2N+1)\omega_0(t-\tau)}{2}\right)}{\sin\left(\frac{\omega_0(t-\tau)}{2}\right)}d\tau$$

This expression is often used in explaining the Gibbs phenomenon. The alternating nature of the second function in the integrand, even in the limit as $N \to \infty$, does not change and produces deviations at any discontinuity of $x(t)$.

7.2.6 Convolution in the Frequency-domain

Consider the FS representations of $x(t)$ and $y(t)$ with a common fundamental frequency $\omega_0 = 2\pi/T$.

$$x(t) = \sum_{m=-\infty}^{\infty} X_{cs}(m)e^{jm\omega_0 t} \quad \text{and} \quad y(t) = \sum_{l=-\infty}^{\infty} Y_{cs}(l)e^{jl\omega_0 t}$$

The FS coefficients for $x(t)y(t)$ are to be expressed in terms of those of $x(t)$ and $y(t)$. The product of the two functions is given by

$$z(t) = x(t)y(t) = \sum_{m=-\infty}^{\infty}\sum_{l=-\infty}^{\infty} X_{cs}(m)Y_{cs}(l)e^{j(m+l)\omega_0 t}$$

Letting $m + l = k$, we get

$$z(t) = x(t)y(t) = \sum_{k=-\infty}^{\infty} \left(\sum_{m=-\infty}^{\infty} X_{cs}(m)Y_{cs}(k - m) \right) e^{jk\omega_0 t}$$

This is a FS for $z(t) = x(t)y(t)$ with coefficients $Z_{cs}(k) = \sum_{m=-\infty}^{\infty} X_{cs}(m)Y_{cs}$ $(k - m)$. The convolution of two frequency-domain functions, with a common fundamental frequency, corresponds to the multiplication of their inverse FS in the time-domain. That is,

$$x(t)y(t) \Longleftrightarrow \frac{1}{T} \int_0^T x(t)y(t)e^{-jk\omega_0 t} \, dt = \sum_{m=-\infty}^{\infty} X_{cs}(m)Y_{cs}(k - m)$$

The convolution is aperiodic as the FS spectra are aperiodic.

Consider the convolution of the FS spectra given as $X_{cs}(\pm 1) = 1/2$ and $Y_{cs}(\pm 2) = \mp j/2$, with $\omega_0 = 1$. The linear convolution of these spectra is $Z_{cs}(\pm 3) = \mp j/4$ and $Z_{cs}(\pm 1) = \mp j/4$. The corresponding time-domain function is

$$\frac{1}{2}(\sin(t) + \sin(3t)) = \cos(t)\sin(2t)$$

Note that the given FS spectra corresponds to the time-domain functions $\cos(t)$ and $\sin(2t)$.

7.2.7 Duality

The analysis equation of the FS is an integral and the synthesis equation is a summation. Therefore, there is no duality between these operations. However, as the synthesis equation of the DTFT is an integral and the analysis equation is a summation, there is duality between these two transforms. This will be presented in the next chapter.

7.2.8 Time Scaling

Scaling is the operation of replacing the independent variable t by at, where $a \neq 0$ is a constant. As we have seen in Chapter 3, the signal is compressed or expanded in the time-domain by this operation. As a consequence, the spectrum of the signal is expanded or compressed in the frequency-domain. The amplitude of the spectrum remains the same with the fundamental frequency changed to $a\omega_0$. Let the spectrum of a signal $x(t)$, with the fundamental frequency $\omega_0 = 2\pi/T$, be $X_{cs}(k)$. Then, $x(at) \Longleftrightarrow X_{cs}(k)$ with the fundamental frequency $a\omega_0$ and $a > 0$. If $a < 0$, the spectrum, with the fundamental frequency $|a|\omega_0$, is also frequency-reversed. For example, with $a = 0.2$, the signal $\cos(t)$ becomes $\cos(0.2t)$. The spectrum remains the same, that is $X_{cs}(\pm 1) =$

1/2, with the fundamental frequency changed to 0.2 radians from 1 radian. With $a = -3$, the signal $\sin(t)$ becomes $\sin(-3t) = -\sin(3t)$. The spectrum gets frequency-reversed, $X_{cs}(\pm 1) = \pm j/2$, and the fundamental frequency of the FS spectrum is changed to 3 radians from 1 radian.

7.2.9 Time Differentiation

As the signal is decomposed in terms of exponentials of the form $e^{jk_0\omega_0 t}$, this property is essentially finding the derivative of all the constituent exponentials of a signal. The derivative of $e^{jk_0\omega_0 t}$ is $jk_0\omega_0 e^{jk_0\omega_0 t}$. Therefore, if the FS spectrum for a time-domain function $x(t)$ is $X_{cs}(k)$, then the FS spectrum for its derivative is $jk\omega_0 X_{cs}(k)$, where ω_0 is the fundamental frequency. Note that the FS coefficient with $k = 0$ is zero, as the dc component is lost in differentiating a signal. In general,

$$\frac{d^n x(t)}{dt^n} \Longleftrightarrow (jk\omega_0)^n X_{cs}(k)$$

This property can be stated as the invariance of the exponentials with respect to the differentiation operation. That is, the derivative of a exponential is the same exponential multiplied by a complex scale factor. The exponentials are invariant with respect to integration and summation operations also. These properties change an integro-differential equation in the time-domain to an algebraic equation in the frequency-domain. Therefore, the analysis of systems is easier in the frequency-domain.

Another use of this property, in common with other properties, is to find FS spectra for signals from those of the related signals. This property can be used to find the FS for the functions represented by polynomials in terms of the FS for their derivatives. When a function is reduced to a sum of impulses, by differentiating it successively, the FS of the impulses can be found easily and this FS is used to find the FS of $x(t)$ using the differentiation property. Consider a periodic rectangular pulse defined over one period as $x(t) = 1$, $|t| < a$ and $x(t) = 0$, $a < |t| < T/2$. The derivative of this signal in a period are the impulses $\delta(t + a)$ and $-\delta(t - a)$. The FS spectrum for this pair is $(1/T)(e^{jk\omega_0 a} - e^{-jk\omega_0 a}) = (j2/T)\sin(k\omega_0 a)$, where $\omega_0 = 2\pi/T$. This spectrum is related to the spectrum of the rectangular pulse by the factor $1/jk\omega_0$, $k \neq 0$. Therefore, the FS spectrum for the periodic rectangular pulse is

$$X_{cs}(k) = 2\frac{\sin(k\omega_0 a)}{k\omega_0 T} = \frac{\sin(k\omega_0 a)}{\pi k}$$

As there is no dc bias, this expression gives the correct value of $X_{cs}(0)$. In general, use this property to obtain $X_{cs}(k)$ for $k \neq 0$ only and determine $X_{cs}(0)$ directly from the given waveform.

7.2.10 Time Integration

For a signal $x(t) \Longleftrightarrow X_{cs}(k)$ with fundamental frequency ω_0,

$$\int_{-\infty}^{t} x(\tau)d\tau \Longleftrightarrow \frac{1}{jk\omega_0} X_{cs}(k)$$

provided the dc component of $x(t)$ is zero ($X_{cs}(0) = 0$). Consider the function $\cos(2t)$ with FS coefficients $X_{cs}(\pm 1) = 1/2$. Then, its integral, $y(t) = \sin(2t)/2$, has the FS coefficients

$$Y_{cs}(\pm 1) = \frac{1}{j(\pm 1)(2)} \frac{1}{2} = \mp \frac{j}{4}$$

7.2.10.1 Rate of Convergence of the Fourier Series

In practical problems, only the sum of a finite number of the infinite terms of the FS are used to approximate a given function $x(t)$, as no physical device can generate harmonics of infinite order. The rate of convergence of a FS indicates how rapidly the partial sums converge to $x(t)$. A smoother function has an higher rate of convergence. According to the time-integration property, each time the FS is integrated term by term, the coefficients are divided by the factor k, the index. That is, the rate of convergence of the FS is increased by the factor k, as the function becomes smoother by the integration operation. The FS for an impulse train converges slowly, as all its coefficients are the same (no dependence on the index k). As the integral of a function with impulses is a function with discontinuities, the FS of such a function converges more rapidly as the magnitude of their coefficients decrease at the rate $1/k$, for large values of k. As the integral of a function with discontinuities results in a function with no discontinuity, the FS for such functions converges still more rapidly as the magnitude of their coefficients decrease at the rate $1/k^2$. The magnitude of the coefficients of a function, whose nth derivative contains impulses, decrease at the rate $1/k^n$.

7.2.11 Parseval's Theorem

As the frequency-domain representation of a signal is an equivalent representation, the power of a signal can also be expressed in terms of its spectrum. That is, the average power of a signal is the sum of the average powers of its frequency components. The average power of a complex exponential, $Ae^{j\omega_0 t}$, is

$$P = \frac{1}{T} \int_0^T |Ae^{j\omega_0 t}|^2 dt = |A|^2$$

since $|e^{j\omega_0 t}| = 1$. That is, the average power of a complex exponential is the magnitude squared of its complex amplitude (irrespective of its frequency and phase). Therefore, the total average power of a signal is

$$P = \frac{1}{T} \int_0^T |x(t)|^2 dt = \sum_{k=-\infty}^{\infty} |X_{cs}(k)|^2$$

Example 7.5. Verify the Parseval's theorem for the square wave of Example 7.3. Find the sum of the powers of the dc, first harmonic, and the third harmonic components of the signal.

Solution
The power using the time-domain representation is

$$P = \frac{1}{T} \int_0^T |x(t)|^2 dt = 2 \int_0^{\frac{1}{4}} dt = \frac{1}{2}$$

The power using the FS is

$$P = \sum_{k=-\infty}^{\infty} |X_{cs}(k)|^2 = \left(\frac{1}{2}\right)^2 + 2 \sum_{k=1,3,}^{\infty} \left(\frac{1}{k\pi}\right)^2 = \frac{1}{4} + \frac{2}{\pi^2} \frac{\pi^2}{8} = \frac{1}{2}$$

The sum of the power of the components of the signal up to the third harmonic is

$$\frac{1}{4} + \frac{2}{\pi^2} + \frac{2}{9\pi^2} = \frac{1}{4} + \frac{20}{9\pi^2} = 0.25 + 0.2252 = 0.4752$$

This example shows that the approximation of the signal by a few harmonics includes most of its power. □

7.3 Approximation of the Fourier Series

We approximate the integral in Equation (7.2) by the rectangular rule of numerical integration. We take N samples of the signal,

$$x(0), x\left(\frac{T}{N}\right), x\left(2\frac{T}{N}\right), \ldots, x\left((N-1)\frac{T}{N}\right)$$

by dividing the period T of the signal into N intervals. The sampling interval is $T_s = T/N$. Now, Equation (7.2) is approximated as

$$X_{cs}(k) = \frac{1}{T} \sum_{n=0}^{N-1} x(nT_s)e^{-jk\omega_0 nT_s} \frac{T}{N} = \frac{1}{N} \sum_{n=0}^{N-1} x(nT_s)e^{-jk\frac{2\pi}{N}n}$$

Dropping T_s from $x(nT_s)$, we get

$$X_{cs}(k) = \frac{1}{N}\sum_{n=0}^{N-1} x(n)e^{-j\frac{2\pi}{N}nk} \qquad k = 0, 1, \ldots, N-1 \qquad (7.7)$$

This is the analysis equation. The synthesis equation, Equation (7.1), is approximated as

$$x(n) = \sum_{k=0}^{N-1} X_{cs}(k)e^{j\frac{2\pi}{N}nk} \qquad n = 0, 1, \ldots, N-1.$$

Except for constant factors, the approximations of the analysis and synthesis equations are the same as the DFT and IDFT equations, respectively. Note that $X_{cs}(k)$ is aperiodic and periodicity of N selected values is assumed in the IDFT computation. For example, if we truncate the FS spectrum to $X_{cs}(-1)$, $X_{cs}(0)$, and $X_{cs}(1)$, then the periodic extension can be written, starting with $X_{cs}(0)$ and $N = 3$, as $X_{cs}(0)$, $X_{cs}(1)$, and $X_{cs}(-1)$. With $N = 4$, the periodic values are $X_{cs}(0)$, $X_{cs}(1)$, 0, and $X_{cs}(-1)$.

For N even, comparing the coefficients of the DFT with that in Equation (7.2), we get, for real signals,

$$X_c(0) = \frac{X(0)}{N} \qquad X_c\left(\frac{N}{2}\right) = \frac{X\left(\frac{N}{2}\right)}{N}$$

$$X_c(k) = \frac{2}{N}\text{Re}(X(k)) \qquad X_s(k) = -\frac{2}{N}\text{Im}(X(k)) \qquad k = 1, 2, \ldots, \frac{N}{2} - 1$$

$$X_{cs}(k) = \frac{X(k)}{N} \qquad k = 0, 1, \ldots, \frac{N}{2} - 1 \quad \text{and} \quad \text{Re}\left(X_{cs}\left(\frac{N}{2}\right)\right) = \frac{X\left(\frac{N}{2}\right)}{2N}$$

7.3.1 Aliasing Effect

Let us find the FS spectrum of a sampled signal. The sampling operation can be considered as multiplying the signal ($x(t)$ with spectrum $X_{cs}(k)$) by the sampling signal ($s(t) = \sum_{n=-\infty}^{\infty} \delta(t - nT_s)$ with spectrum $\frac{1}{T_s}$). The sampled signal is $\sum_{n=-\infty}^{\infty} x(nT_s)\delta(t - nT_s)$. In the frequency domain, sampling operation corresponds to the convolution of the spectra of the two signals. As the convolution of a signal with an impulse is just translation of the origin of the signal to the location of the impulse, we get the spectrum of the sampled signal $x(t)s(t)$ as the superposition sum of the infinite frequency-shifted spectrum of the signal, multiplied by the factor $1/T_s$. That is, the

FS spectrum for the sampled signal is

$$\frac{1}{T_s} \sum_{m=-\infty}^{\infty} X_{cs}\left(k - m\frac{T}{T_s}\right)$$

where T is the period of the signal. Let us construct a sequence $x(n)$ such that its nth element has the value that is the same as the strength of the impulse $x(nT_s)\delta(t - nT_s)$ of the sampled signal. We get the DFT spectrum of $x(n)$ by multiplying the spectrum of the sampled signal by T (Remember that there is no normalization factor in the definition of the DFT.) and noting that $T/T_s = N$. Therefore, we get

$$X(k) = N \sum_{m=-\infty}^{\infty} X_{cs}(k - mN) \qquad k = 0, 1, \dots, N - 1 \tag{7.8}$$

This equation shows how the DFT spectrum is corrupted due to aliasing. By sampling the signal, in order to use the DFT, to obtain a finite number of N samples in a period, we simultaneously reduce the number of distinct sinusoids and, hence, the number of distinct spectral coefficients to N. Therefore, if the signal is band-limited we can get the exact FS coefficients by computing the DFT of the samples of the signal. If the signal is not band-limited or the number of samples is inadequate, we get a corrupted FS spectrum using the DFT due to the aliasing effect.

Consider the sampling of the square wave (Example 3) with $N = 4$ samples, shown in Figure 7.7(a). Note that, at any discontinuity, the average of the left- and right-hand limits should be taken as the sample value. The sample values, starting from $n = 0$, are $\{1, 0.5, 0, 0.5\}$. The DFT of this set of samples is $\{2, 1, 0, 1\}$, shown in Figure 7.7(b).

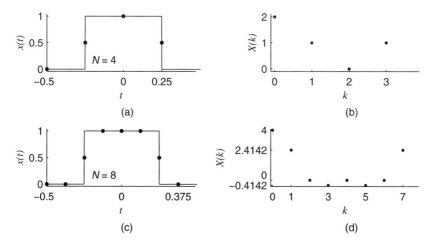

Figure 7.7 (a) Four samples of the square wave; (b) its DFT; (c) eight samples of the square wave; (d) its DFT

These values can be obtained from the FS coefficients using Equation (7.8). The dc
value is $2/4 = 0.5$, which is equal to the analytical value. The coefficient of the first
harmonic is $(1 + 1)/4 = 0.5$, which differs from the analytical value of 0.637. This is
due to the fact that, with only four samples, all the other odd harmonics alias as the
first harmonic.

$$X_c(1) = \frac{2}{\pi} \left(1 - \frac{1}{3} + \frac{1}{5} - \frac{1}{7} + \cdots \right)$$

The value of the summation can be obtained from Equation (7.5) by substituting
$t = 0$.

$$1 = \frac{1}{2} + \frac{2}{\pi} \left(1 - \frac{1}{3} + \frac{1}{5} - \frac{1}{7} + \cdots \right)$$

Therefore, we get $X_c(1) = 0.5$. As we double the number of samples, we get a better
approximation of the FS coefficients by the DFT. Figure 7.7(c) shows the square wave
with $N = 8$ samples and its DFT is shown in Figure 7.7(d). The value of the first
harmonic is $(2.4142 + 2.4142)/8 = 0.6036$, which is much closer to the actual value
of 0.637. The point is that DFT should be used to approximate the FS coefficients with
sufficient number of time-domain samples so that the accuracy of the approximation
is adequate.

7.4 Applications of the Fourier Series

The FS is used to analyze periodic waveforms, such as half- and full-wave rectified
waveforms. The steady-state response of stable LTI systems to periodic input signals
can also be found using the FS. The steady-state response is the response of a system
after the transient has decayed. The transient response of a stable system always decays
with time. The steady-state output of a LTI system to an input $e^{jk_0\omega_0 t}$ is the same
function multiplied by the complex scale factor, $H(jk_0\omega_0)$. Therefore, the output of
the system is $H(jk_0\omega_0)e^{jk_0\omega_0 t}$. The function $H(jk\omega_0)$ is the frequency response $H(j\omega)$
(Chapter 5) of the system, sampled at the discrete frequencies $\omega = k\omega_0$.

Consider the system governed by the differential equation

$$\frac{dy(t)}{dt} + y(t) = x(t)$$

The differential equation can be written, with the input $x(t) = e^{jk_0\omega_0 t}$, as

$$\frac{d(H(jk_0\omega_0)e^{jk_0\omega_0 t})}{dt} + H(jk_0\omega_0)e^{jk_0\omega_0 t} = e^{jk_0\omega_0 t}$$

Solving for $H(jk_0\omega_0)$, we get

$$H(jk_0\omega_0) = \frac{1}{1 + jk_0\omega_0}$$

For an arbitrary periodic input, as $x(t) = \sum_{k=-\infty}^{\infty} X_{cs}(k)e^{jk\omega_0 t}$, we get

$$y(t) = \sum_{k=-\infty}^{\infty} H(jk\omega_0)X_{cs}(k)e^{jk\omega_0 t} = \sum_{k=-\infty}^{\infty} \frac{X_{cs}(k\omega_0)}{1 + jk\omega_0}e^{jk\omega_0 t}$$

The more complex operation of solving a differential equation has been reduced to the evaluation of an algebraic operation.

7.5 Summary

- In this chapter, the FS has been studied. It is used to represent periodic time-domain signals by an aperiodic discrete spectrum in the frequency-domain.
- The three forms of the FS are the trigonometric, the compact trigonometric, and the exponential. In the trigonometric form, a time-domain function is expressed in terms of cosine and sine waveforms. In the compact trigonometric form, a time-domain function is expressed in terms of real sinusoids. In the exponential form, a time-domain function is expressed in terms of complex exponentials. Each of the three forms of the FS can be derived from the other forms.
- The Fourier representation fails to provide uniform convergence in the vicinity of a discontinuity of continuous waveforms in both the time- and frequency-domains (in the time-domain, in the case of the FS).
- In practice, the FS is often approximated by the DFT.

Further Reading

1. Sundararajan, D., *Discrete Fourier Transform, Theory, Algorithms, and Applications*, World Scientific, Singapore, 2001.
2. Lathi, B. P., *Linear Systems and Signals*, Oxford University Press, New York, 2004.
3. Guillemin, E. A., *The Mathematics of Circuit Analysis*, John Wiley, New York, 1952.

Exercises

7.1   The FS representation of a real periodic signal $x(t)$ of period T, satisfying the Dirichlet conditions, is given as

$$x(t) = X_c(0) + \sum_{k=1}^{\infty}(X_c(k)\cos(k\omega_0 t) + X_s(k)\sin(k\omega_0 t)),$$

where, $\omega_0 = 2\pi/T$ and $X_c(0)$, $X_c(k)$, and $X_s(k)$, the FS coefficients of the dc, cosine, and sine components of $x(t)$, respectively, are defined as

$$X_c(0) = \frac{1}{T} \int_{t_1}^{t_1+T} x(t)\,dt,$$

$$X_c(k) = \frac{2}{T} \int_{t_1}^{t_1+T} x(t)\cos(k\omega_0 t)\,dt \qquad k = 1, 2, \ldots, \infty$$

$$X_s(k) = \frac{2}{T} \int_{t_1}^{t_1+T} x(t)\sin(k\omega_0 t)\,dt \qquad k = 1, 2, \ldots, \infty$$

and t_1 is arbitrary. Derive the expressions for the coefficients using trigonometric identities.

7.2   Expand $x(t)$ and find the three forms of its FS coefficients without evaluating any integral. What is the fundamental frequency ω_0?

7.2.1 $x(t) = \cos(t)$.

7.2.2 $x(t) = \cos^2(t)$.

7.2.3 $x(t) = \cos^3(t)$.

*7.2.4 $x(t) = \cos^4(t)$.

7.2.5 $x(t) = \cos^5(t)$.

7.2.6 $x(t) = \sin(t)$.

7.2.7 $x(t) = \sin^2(t)$.

7.2.8 $x(t) = \sin^3(t)$.

7.2.9 $x(t) = \sin^4(t)$.

7.2.10 $x(t) = \sin^5(t)$.

7.3   Find the three forms of the FS coefficients of $x(t)$ without evaluating any integral. What is the fundamental frequency ω_0?

7.3.1 $x(t) = 3 + \cos\left(2\pi t - \frac{\pi}{3}\right) - 2\sin\left(4\pi t + \frac{\pi}{6}\right)$.

7.3.2 $x(t) = -1 - 2\sin\left(4\pi t - \frac{\pi}{6}\right) + 6\sin\left(8\pi t - \frac{\pi}{3}\right)$.

7.3.3 $x(t) = 2 - 3\sqrt{2}\cos\left(2\pi t - \frac{\pi}{4}\right) + 2\sin\left(6\pi t + \frac{\pi}{3}\right)$.

7.3.4 $x(t) = -3 + \sqrt{3}\cos(2\pi t) - \sin(2\pi t) + \sqrt{2}\cos(8\pi t) - \sqrt{2}\sin(8\pi t)$.

*7.3.5 $x(t) = 1 + \frac{1}{j2}e^{j(2\pi t + \frac{\pi}{3})} + e^{j(6\pi t - \frac{\pi}{6})} - \frac{1}{j2}e^{-j(2\pi t + \frac{\pi}{3})} + e^{-j(6\pi t - \frac{\pi}{6})}$.

7.4   Find the trigonometric form of the FS coefficients. What is the fundamental frequency ω_0?

7.4.1 $x(t) = 1 + 2\sin\left(\frac{4}{7}t\right) + 3\cos\left(\frac{2}{3}t\right)$.

*7.4.2 $x(t) = 2 - 5\cos\left(\frac{3}{7}t\right) - 2\sin\left(\frac{2}{9}t\right)$.

7.4.3 $x(t) = -3 + 2\cos\left(\frac{3}{7}t\right) - \sin\left(\frac{1}{3}t\right)$.

7.5   Find the FS of a periodic pulse train of period T, defined over one period as

$$x(t) = \begin{cases} \frac{1}{a} & \text{for } |t| < \frac{a}{2} \\ 0 & \text{for } \frac{a}{2} < |t| < \frac{T}{2} \end{cases}$$

Apply a limiting process, as $a \to 0$, to the pulse train and its FS to obtain the FS of the periodic impulse train of period T,

$$x(t) = \sum_{n=-\infty}^{\infty} \delta(t - nT)$$

7.6   Find the FS coefficients, using the time-domain convolution property, of $y(t) = x(t) * x(t)$, the convolution of $x(t)$ with itself, with $x(t)$ defined over a period as

$$x(t) = \begin{cases} A & 0 < t < \frac{T}{2} \\ 0 & \frac{T}{2} < t < T \end{cases}$$

7.7   Find the FS coefficients of $z(t) = x(t)y(t)$ with period equal to the common period of $x(t)$ and $y(t)$, where $x(t) = 2\sin(t) + 4\cos(3t)$ and $y(t) = 6\cos(2t)$, using the frequency-domain convolution property. Verify that the FS coefficients represent $z(t) = x(t)y(t)$.

7.8   Find the trigonometric FS representation of the periodic full-wave rectified sine wave defined over a period as

$$x(t) = \begin{cases} A\sin(\omega_0 t) & 0 \le t < \frac{T}{2} \\ A\sin(\omega_0(t - \frac{T}{2})) & \frac{T}{2} \le t < T \end{cases} \qquad \omega_0 = \frac{2\pi}{T}$$

using the time-differentiation property.

7.9   Using the time-differentiation property, find the FS coefficients of the periodic signal $x(t)$ defined over a period.

7.9.1

$$x(t) = \begin{cases} -A & 0 < t < \frac{T}{2} \\ A & -\frac{T}{2} < t < 0 \end{cases}$$

7.9.2

$$x(t) = \begin{cases} A + 2 & 0 < t < \frac{T}{2} \\ -A + 2 & -\frac{T}{2} < t < 0 \end{cases}$$

7.9.3

$$x(t) = \begin{cases} 0 & 0 < t < \frac{T}{2} \\ -(t + \frac{T}{2}) & -\frac{T}{2} < t < 0 \end{cases}$$

7.9.4

$$x(t) = \frac{t}{T}, \ 0 < t < T$$

7.9.5

$$x(t) = \begin{cases} \frac{2t}{T} & 0 \leq t \leq \frac{T}{2} \\ 2 - \frac{2t}{T} & \frac{T}{2} < t < T \end{cases}$$

*7.9.6

$$x(t) = \sin(2t), \ 0 < t < \frac{\pi}{2}$$

7.10 Find the trigonometric FS representation, using Equation (7.2), of the periodic signal defined over a period as $x(t) = \frac{3}{2}t$, $0 < t < 2$. Using the results, find the sum of the infinite series

$$1 - \frac{1}{3} + \frac{1}{5} - \frac{1}{7} + \cdots$$

Verify Parseval's theorem.
Find the power of the frequency components of the signal up to (i) the third harmonic ; (ii) the fifth harmonic.
Approximate the trigonometric FS coefficients using the DFT with $N = 4$. Verify that they are the same using Equation (7.8).
Find the location and the magnitude of a largest deviation due to Gibbs phenomenon if the signal is reconstructed using up to the third harmonic.
Deduce the trigonometric FS representation of the signals $x(t - 1)$ and $2x(t) - 3$.

*7.11 Find the trigonometric FS representation, using Equation (7.2), of the periodic signal defined over a period as

$$x(t) = \begin{cases} \frac{1}{2}t & 0 \leq t < 2 \\ 2(1 - \frac{t}{4}) & 2 \leq t < 4 \end{cases}$$

Using the results, find the sum of the infinite series

$$1 + \frac{1}{9} + \frac{1}{25} + \frac{1}{49} + \cdots$$

Verify Parseval's theorem.

Find the power of the frequency components of the signal up to : (i) the third harmonic; (ii) the fifth harmonic.

Approximate the trigonometric FS coefficients using the DFT with $N = 4$.

Verify that they are the same using Equation (7.8).

Deduce the trigonometric FS representation of the signals $x(t + 2)$ and $3x(t) - 2$.

7.12 Find the trigonometric FS representation of the periodic half-wave rectified sine wave defined over a period as

$$x(t) = \begin{cases} A \sin(\frac{2\pi}{T} t) & 0 \le t < \frac{T}{2} \\ 0 & \frac{T}{2} \le t < T \end{cases}$$

using the frequency-domain convolution property.

Verify Parseval's theorem.

Find the power of the frequency components of the signal up to the third harmonic.

Approximate the trigonometric FS coefficients using the DFT with $N = 4$.

Verify that they are the same using Equation (7.8).

Deduce the FS coefficients of $x(t) + x(t - \frac{T}{2})$ and $x(t) - x(t - \frac{T}{2})$.

7.13 Find the trigonometric FS representation of the periodic half inverted cosine wave defined over a period as

$$x(t) = \begin{cases} -A \cos(\frac{2\pi}{T} t) & 0 < t < \frac{T}{2} \\ 0 & \frac{T}{2} < t < T \end{cases}$$

using the frequency-domain convolution property.

Verify Parseval's theorem.

Find the power of the frequency components of the signal up to the fifth harmonic. Find the FS coefficients using the DFT with $N = 4$. Verify that they are the same using Equation (7.8).

Find the location and the magnitude of a largest deviation due to Gibbs phenomenon if the signal is reconstructed using up to the third harmonic.

Deduce the FS coefficients of $x(t) - x(t - \frac{T}{2})$.

7.14 Using the result of Exercise 7.13, deduce the FS representation of periodic two inverted half-cosine waves defined over a period as

$$x(t) = \begin{cases} -A\cos(\frac{2\pi}{T}t) & 0 < t < \frac{T}{2} \\ A\cos(\frac{2\pi}{T}t) & \frac{T}{2} < t < T \end{cases}$$

*7.15 Find the response of the system governed by the differential equation

$$\frac{dy(t)}{dt} + y(t) = e^{j2t} + e^{j3t}$$

7.16 Find the response of the system governed by the differential equation

$$\frac{dy(t)}{dt} + 2y(t) = 2 - 3\sin(t) + \cos\left(2t + \frac{\pi}{3}\right)$$

8

The Discrete-time Fourier Transform

A continuum of discrete sinusoids over a finite frequency range is used as the basis signals in the DTFT to analyze aperiodic discrete signals. Compared with the DFT, as the discrete aperiodic time-domain waveform contains an infinite number of samples, the frequency increment of the periodic spectrum of the DFT tends to zero and the spectrum becomes continuous. The period is not affected since it is determined by the sampling interval in the time-domain. An alternate view of the DTFT is that it is the same as the FS with the roles of time- and frequency-domain functions interchanged.

In Section 8.1, the DTFT and its inverse, and the dual relationship between the DTFT and the FS are derived. The properties of the DTFT are presented in Section 8.2. The approximation of the DTFT by the DFT is described in Section 8.3. Some typical applications of the DTFT are presented in Section 8.4.

8.1 The Discrete-time Fourier Transform

8.1.1 The DTFT as the Limiting Case of the DFT

In the last chapter, we found that the FS is the limiting case of the DFT as the sampling interval in the time-domain tends to zero with the period of the waveform fixed. In this chapter, we find that the DTFT is the limiting case of the DFT as the period in the time-domain tends to infinity with the sampling interval fixed. With a predetermined sampling interval, the effective frequency range of the spectrum is fixed.

Consider the DFT magnitude spectrum $|X(k)|$ of $x(n)$ with $N = 5$ samples, shown, respectively, in Figures 8.1(b) and (a). The frequency increment of the spectrum is $2\pi/5$. Even if a signal $x(n)$ is aperiodic, in the DFT computation, periodicity is assumed. Therefore, only a set of samples of the continuous spectrum of an aperiodic

A Practical Approach to Signals and Systems   D. Sundararajan
© 2008 John Wiley & Sons (Asia) Pte Ltd

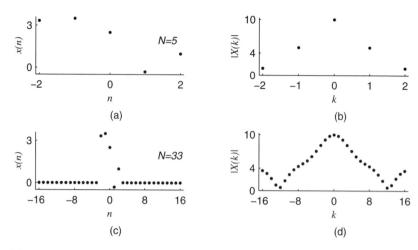

Figure 8.1 (a) The time-domain function $x(n)$ with $N = 5$; (b) the corresponding DFT magnitude spectrum, $|X(k)|$; (c) same as (a) with $N = 33$; (d) the corresponding DFT magnitude spectrum, $|X(k)|$

$x(n)$ is computed by the DFT. By zero-padding on either side of $x(n)$, we have made the signal longer with $N = 33$ samples, as shown in Figure 8.1(c). Its spectrum is shown in Figure 8.1(d), which is denser (frequency increment $2\pi/33$) compared with that in Figure 8.1(b). Eventually, as N tends to infinity, we get the aperiodic discrete signal and its periodic continuous spectrum. The spectrum is always periodic with the same period, 2π, as the sampling interval is fixed at $T_s = 1$.

The foregoing argument can be put mathematically, as follows. The IDFT of $X(k)$, $-N \le k \le N$ is defined as

$$x(n) = \frac{1}{2N + 1} \sum_{k=-N}^{N} X(k) e^{j \frac{2\pi}{(2N+1)} nk}, \quad n = 0, \pm 1, \pm 2, \ldots, \pm N$$

Substituting the DFT expression for $X(k)$, we get

$$x(n) = \frac{1}{2N + 1} \sum_{k=-N}^{N} \left(\sum_{l=-N}^{N} x(l) e^{-j \frac{2\pi}{2N+1} lk} \right) e^{j \frac{2\pi}{2N+1} nk}$$

As N tends to ∞, due to zero-padding of $x(n)$, $([2\pi/(2N + 1)]k)$ becomes a continuous variable ω, differential $d\omega$ formally replaces $2\pi/(2N + 1)$, and $2N + 1 = 2\pi/d\omega$. The outer summation becomes an integral with limits $-\pi$ and π (actually any continuous interval of 2π). The limits of the inner summation can be written as $-\infty$ and ∞. Therefore, the DTFT $X(e^{j\omega})$ of the signal $x(n)$ is defined as

$$X(e^{j\omega}) = \sum_{n=-\infty}^{\infty} x(n) e^{-j\omega n} \tag{8.1}$$

The DTFT is commonly written as $X(e^{j\omega})$ instead of $X(j\omega)$ to emphasize the fact that it is a periodic function of ω. The inverse DTFT $x(n)$ of $X(e^{j\omega})$ is defined as

$$x(n) = \frac{1}{2\pi} \int_{-\pi}^{\pi} X(e^{j\omega})e^{j\omega n} d\omega, \quad n = 0, \pm1, \pm2, \ldots \tag{8.2}$$

When deriving closed-form expressions for $x(n)$ or $X(e^{j\omega})$,

$$X(e^{j0}) = \sum_{n=-\infty}^{\infty} x(n), \quad X(e^{j\pi}) = \sum_{n=-\infty}^{\infty} (-1)^n x(n), \quad x(0) = \frac{1}{2\pi} \int_{-\pi}^{\pi} X(e^{j\omega}) d\omega$$

which can be easily evaluated, are useful to check their correctness.

The analysis equation of the DTFT is a summation and the synthesis equation is an integral. In these equations, it is assumed that the sampling interval of the time-domain signal T_s, is 1 s. For other values of T_s, only scaling of the frequency axis is required. However, the DTFT equations can also be expressed including T_s as

$$X(e^{j\omega T_s}) = \sum_{n=-\infty}^{\infty} x(nT_s)e^{-jn\omega T_s} \tag{8.3}$$

$$x(nT_s) = \frac{1}{\omega_s} \int_{-\omega_s/2}^{\omega_s/2} X(e^{j\omega T_s})e^{jn\omega T_s} d\omega \qquad n = 0, \pm1, \pm2, \ldots \tag{8.4}$$

where $\omega_s = 2\pi/T_s$. The DTFT represents a discrete aperiodic signal, $x(nT_s)$, with T_s s between consecutive samples, as integrals of a continuum of complex sinusoids $e^{jn\omega T_s}$ (amplitude $(1/\omega_s)X(e^{j\omega T_s})d\omega$) over the finite frequency range $-\omega_s/2$ to $\omega_s/2$ (over one period of $X(e^{j\omega T_s})$). $X(e^{j\omega T_s})$ is periodic of period $\omega_s = 2\pi/T_s$, since $e^{-jn\omega T_s} = e^{-jn(\omega+2\pi/T_s)T_s}$. Therefore, the integration in Equation (8.4) can be evaluated over any interval of length ω_s. As the amplitude, $(1/\omega_s)X(e^{j\omega T_s})d\omega$, of the constituent sinusoids of a signal is infinitesimal, the spectral density $X(e^{j\omega T_s})$, which is proportional to the spectral amplitude, represents the frequency content of a signal. Although the DTFT is the spectral density of a signal, it is still called the spectrum. Therefore, the DTFT spectrum is a relative amplitude spectrum.

The summation in Equation (8.3) converges uniformly to $X(e^{j\omega T_s})$, if $x(nT_s)$ is absolutely summable, that is $\sum_{n=-\infty}^{\infty} |x(nT_s)| < \infty$. The summation converges in the least-squares error sense, if $x(nT_s)$ is square summable, that is $\sum_{n=-\infty}^{\infty} |x(nT_s)|^2 < \infty$ (for example, $x(n)$ in Example 8.2). The Gibbs phenomenon is also common to all forms of Fourier analysis whenever reconstructing a continuous waveform, with one or more discontinuities, in either domain. In the case of the DTFT, the Gibbs phenomenon occurs in the frequency-domain as the spectrum is a continuous function.

Example 8.1. Find the DTFT of the unit impulse signal $x(n) = \delta(n)$.

$$X(e^{j\omega}) = \sum_{n=-\infty}^{\infty} \delta(n)e^{-j\omega n} = 1 \qquad \text{and} \qquad \delta(n) \Longleftrightarrow 1$$

That is, the unit impulse signal is composed of complex sinusoids of all frequencies from $\omega = -\pi$ to $\omega = \pi$ in equal proportion. $\qquad\square$

Example 8.2. One period of a DTFT spectrum, shown in Figure 8.2(a), is given as $X(e^{j\omega}) = u(\omega + \frac{\pi}{4}) - u(\omega - \frac{\pi}{4})$, $-\pi < \omega \le \pi$. Find the corresponding $x(n)$.

Solution
As the spectrum is even-symmetric,

$$x(n) = \frac{1}{\pi} \int_0^{\frac{\pi}{4}} \cos(\omega n)\, d\omega = \frac{\sin\left(\frac{\pi n}{4}\right)}{n\pi}, \qquad -\infty < n < \infty$$

The time-domain signal $x(n)$ is shown in Figure 8.2(b). $\qquad\square$

The function of the form $x(n) = \sin(\frac{\pi n}{4})/(n\pi)$, shown in Figure 8.2(b), is called the sinc function that occurs often in signal and system analysis. It is an even function of n. At $n = 0$, the peak value is $\frac{1}{4}$, as $\lim_{\theta \to 0} \sin(\theta) = \theta$. The zeros of the sinc function occur whenever the argument of the sine function in the numerator is equal to $\pm\pi, \pm 2\pi, \pm 3\pi, \ldots$. For the specific case, the zeros occur whenever n is an integral multiple of four. As $a \to 0$, $\sin(an)/(an)$ degenerates into a dc function with amplitude one, as the zeros move to infinity. The sinc function is an energy signal, as it is square summable. However, it is not absolutely summable.

Example 8.3. Find the DTFT of the signal $x(n) = a^n u(n)$, $|a| < 1$.

$$X(e^{j\omega}) = \sum_{n=-\infty}^{\infty} a^n u(n)e^{-j\omega n} = \sum_{n=0}^{\infty} (ae^{-j\omega})^n = \frac{1}{1 - ae^{-j\omega}}, \quad |a| < 1 \qquad\square$$

(a)   (b)

Figure 8.2   (a) One period of a DTFT spectrum; (b) the corresponding aperiodic discrete signal

The DTFT of some frequently used signals, which are neither absolutely nor square summable, such as the unit-step, is obtained by applying a limiting process to appropriate signals so that they degenerate into these signals in the limit. The limit of the corresponding transform is the transform of the signal under consideration, as presented in the next example.

Example 8.4.   Find the DTFT of the unit-step signal $x(n) = u(n)$.
As this signal is not absolutely or square summable, its DTFT is derived as that of the limiting form of the signal $a^n u(n)$, as $a \to 1$.

$$X(e^{j\omega}) = \lim_{a \to 1} \frac{1}{1 - ae^{-j\omega}} = \lim_{a \to 1} \left(\frac{1 - a\cos(\omega)}{1 - 2a\cos(\omega) + a^2} - j \frac{a\sin(\omega)}{1 - 2a\cos(\omega) + a^2} \right)$$

The real and imaginary parts of the DTFT spectrum of the signal $0.8^n u(n)$, shown in Figure 8.3(a), are shown, respectively in Figures 8.3(c) and (e). Figures 8.3(d) and (f) show the same for the signal $0.99^n u(n)$, shown in Figure 8.3(b). The real part of the spectrum is even and the imaginary part is odd. The area enclosed by the real part of the spectrum is a constant (2π) independent of the value a, the base of the

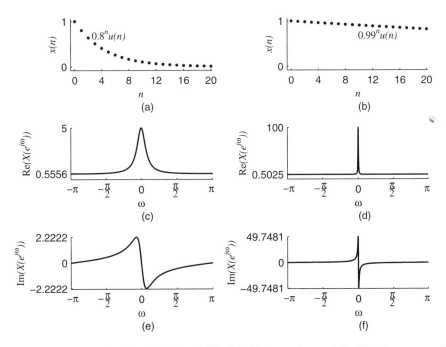

Figure 8.3   (a) $x(n) = 0.8^n u(n)$; (b) $x(n) = 0.99^n u(n)$; (c) the real part of the DTFT spectrum of the signal in (a); (e) its imaginary part; (d) the real part of the DTFT spectrum of the signal in (b); (f) its imaginary part

exponential signal. This is so because, from the inverse DTFT with $n = 0$,

$$x(0) = 1 = \frac{1}{2\pi} \int_{-\pi}^{\pi} \frac{1 - a\cos(\omega)}{1 - 2a\cos(\omega) + a^2} e^{j\omega 0} d\omega = \frac{1}{2\pi} \int_{-\pi}^{\pi} \frac{1 - a\cos(\omega)}{1 - 2a\cos(\omega) + a^2} d\omega$$

As can be seen from the figures, the real part of the spectrum becomes more peaked as $a \to 1$. Eventually, the spectrum consists of a strictly continuous component (except at $\omega = 0$) and an impulsive component. The constant area 2π is split up, as the function evaluates to 0.5 for $\omega \neq 0$ with $a \to 1$, between these components and the spectrum becomes

$$X(e^{j\omega}) = \pi\delta(\omega) + \frac{1}{1 - e^{-j\omega}} \quad \text{and} \quad u(n) \Longleftrightarrow \pi\delta(\omega) + \frac{1}{1 - e^{-j\omega}} \qquad \square$$

Example 8.5. Find the DTFT of the dc signal, $x(n) = 1$.
The dc signal can be written as $x(n) = u(n) + u(-n) - \delta(n)$. Due to the time-reversal property, if $x(n) \Longleftrightarrow X(e^{j\omega})$ then $x(-n) \Longleftrightarrow X(e^{-j\omega})$. The DTFT of $u(-n)$ is obtained from that of $u(n)$ by replacing ω by $-\omega$. Therefore, the DTFT of the dc signal is

$$\pi\delta(\omega) + \frac{1}{1 - e^{-j\omega}} + \pi\delta(-\omega) + \frac{1}{1 - e^{j\omega}} - 1 = 2\pi\delta(\omega)$$

Explicitly showing the periodicity of the DTFT spectrum, we get

$$1 \Longleftrightarrow 2\pi \sum_{k=-\infty}^{\infty} \delta(\omega + 2k\pi)$$

That is, the dc signal, which is the complex exponential $x(n) = e^{j\omega n}$ with $\omega = 0$, has, nonzero spectral component only at the single frequency $\omega = 0$. Note that 2π in the spectral value is a constant factor. $\qquad \square$

8.1.2 The Dual Relationship Between the DTFT and the FS

The DTFT is the same as the FS with the roles of time- and frequency-domain functions interchanged. The analysis equation, with period of the time-domain waveform T and the fundamental frequency $\omega_0 = 2\pi/T$, of the FS is

$$X_{cs}(k\omega_0) = \frac{1}{T} \int_{-\frac{T}{2}}^{\frac{T}{2}} x(t) e^{-jk\omega_0 t} \, dt \qquad k = 0, \pm 1, \pm 2, \ldots$$

Replacing ω_0 by T_s, T by $\omega_s = 2\pi/T_s$, ω by t, t by ω, and k by $-k$ in this equation we get

$$X_{cs}(-kT_s) = \frac{1}{\omega_s} \int_{-\frac{\omega_s}{2}}^{\frac{\omega_s}{2}} x(\omega) e^{jk\omega T_s} \, d\omega \qquad k = 0, \pm1, \pm2, \ldots$$

This equation is the same as the inverse DTFT with $x(kT_s) = X_{cs}(-kT_s)$ and $X(e^{j\omega T_s}) = x(\omega)$. Due to this similarity,

$$x(kT_s) \Longleftrightarrow X(e^{j\omega T_s}) \quad \text{implies} \quad X(e^{jtT_s}) = x(t) \Longleftrightarrow x(-kT_s) = X_{cs}(k\omega_0)$$

For the same periodic waveform, we get two sets of FS coefficients related by the time-reversal operation because the periodic waveform occurs in the frequency-domain in the case of the DTFT and in the time-domain in the case of the FS. By convention, we use a complex exponential with negative exponent in the forward transform definitions of the FS and the DTFT.

Consider the signal $x(t) = \sin(3t)$ shown in Figure 8.4(a) and the corresponding FS coefficients $X_{cs}(k\omega_0) = X_{cs}(\pm3) = \mp j0.5$ shown in Figure 8.4(b). From the FS synthesis equation, $-0.5\,je^{j3t} + 0.5\,je^{-j3t} = \sin(3t)$. Consider the spectrum $X(e^{j\omega T_s}) = \sin(3\omega)$ shown in Figure 8.4(c) and the corresponding $x(kT_s)$, $(x(\pm3) = \pm j0.5)$, shown in Figure 8.4(d). From the DTFT analysis equation, $0.5\,je^{-j3\omega} - 0.5\,je^{j3\omega} = \sin(3\omega)$.

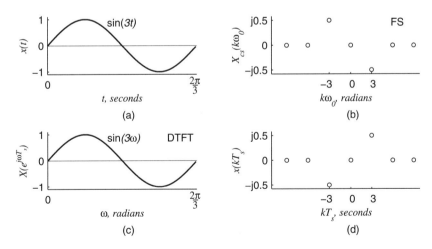

Figure 8.4 (a) One period of the periodic time-domain function $x(t) = \sin(3t)$; (b) the corresponding FS spectrum, $X_{cs}(k\omega_0)$; (c) one period of the periodic frequency-domain function $X(e^{j\omega T_s}) = \sin(3\omega)$; (d) the corresponding inverse DTFT, $x(kT_s)$, which is the time-reversal of $X_{cs}(k\omega_0)$ in (b)

8.1.3 The DTFT of a Discrete Periodic Signal

A periodic signal $x(n)$ is reconstructed using its DFT coefficients $X(k)$ as

$$x(n) = \frac{1}{N} \sum_{k=0}^{N-1} X(k)e^{jk\omega_0 n} \qquad \omega_0 = \frac{2\pi}{N}$$

Since the DTFT of $e^{jk\omega_0 n}$ is $2\pi\delta(\omega - k\omega_0)$, we get, from the linearity property of the DTFT, one period of the DTFT $X(e^{j\omega})$ of $x(n)$ as

$$X(e^{j\omega}) = \frac{2\pi}{N} \sum_{k=0}^{N-1} X(k)\delta(\omega - k\omega_0)$$

Therefore, the DTFT of a periodic signal is a periodic train of impulses with strength $\frac{2\pi}{N}X(k)$ at $\frac{2\pi}{N}k$ with period 2π.

For example, the DFT of $\cos((2\pi/4)n)$ is $\{X(0) = 0, X(1) = 2, X(2) = 0, X(3) = 2\}$ with $N = 4$. One period of the DTFT $X(e^{j\omega})$ is given as $\{X(e^{j0}) = 0, X(e^{j\frac{2\pi}{4}}) = \pi\delta(\omega - \frac{2\pi}{4}), X(e^{j2\frac{2\pi}{4}}) = 0, X(e^{j3\frac{2\pi}{4}}) = \pi\delta(\omega - 3\frac{2\pi}{4})\}$.

8.1.4 Determination of the DFT from the DTFT

The DTFT of a finite sequence $x(n)$, starting from $n = n_0$, of length N is given as

$$X(e^{j\omega}) = \sum_{n=n_0}^{n_0+N-1} x(n)e^{-jn\omega}$$

The DFT of $x(n)$ is given as

$$X(k) = \sum_{n=n_0}^{n_0+N-1} x(n)e^{-jk\omega_0 n} \qquad \omega_0 = \frac{2\pi}{N}$$

Comparing the DFT and DTFT definitions of the signal, we get

$$X(k) = X(e^{j\omega})|_{\omega=k\omega_0} = X(e^{jk\omega_0})$$

The DTFT spectrum is evaluated at all frequencies along the unit-circle in the complex plane, whereas the DFT spectrum is the set of N samples of the DTFT spectrum at intervals of $\frac{2\pi}{N}$.

Let the nonzero samples of a signal is defined as $\{x(-2) = 0, x(-1) = -1, x(0) = 0, x(1) = 1\}$. The DTFT of $x(n)$ is $X(e^{j\omega}) = -e^{j\omega} + e^{-j\omega} = -j2\sin(\omega)$. The set of samples of $X(e^{j\omega})$, $\{X(0) = 0, X(1) = -j2, X(2) = 0, X(3) = j2\}$, at $\omega = 0$, $\omega = \frac{2\pi}{4}$, $\omega = 2\frac{2\pi}{4}$, and $\omega = 3\frac{2\pi}{4}$ is the DFT of $x(n)$.

8.2 Properties of the Discrete-time Fourier Transform

Properties present the frequency-domain effect of time-domain characteristics and operations on signals and vice versa. In addition, they are used to find new transform pairs more easily.

8.2.1 Linearity

The DTFT of a linear combination of a set of signals is the same linear combination of their individual DTFT. That is,

$$x(n) \Longleftrightarrow X(e^{j\omega}), \quad y(n) \Longleftrightarrow Y(e^{j\omega}), \quad ax(n) + by(n) \Longleftrightarrow aX(e^{j\omega}) + bY(e^{j\omega}),$$

where a and b are arbitrary constants. This property follows from the linearity property of the summation operation defining the DTFT.

Consider the signal $x(n) = a^{|n|}$, $|a| < 1$. This signal can be decomposed as $x(n) = a^n u(n) + a^{-n} u(-n) - \delta(n)$. The DTFT of $a^n u(n)$, from Example 8.3, is $\frac{1}{1-ae^{-j\omega}}$. Due to time-reversal property, if $x(n) \Longleftrightarrow X(e^{j\omega})$ then $x(-n) \Longleftrightarrow X(e^{-j\omega})$. Therefore, the DTFT of $a^{-n} u(-n)$ is obtained from that of $a^n u(n)$ as $\frac{1}{1-ae^{j\omega}}$. The DTFT of the signal $x(n) = a^{|n|}$, $|a| < 1$, due to linearity property, is

$$X(e^{j\omega}) = \frac{1}{1 - ae^{-j\omega}} + \frac{1}{1 - ae^{j\omega}} - 1 = \frac{1 - a^2}{1 - 2a\cos(\omega) + a^2}$$

8.2.2 Time Shifting

When we shift a signal, the shape remains the same, but the signal is relocated. The shift of a typical spectral component, $X(e^{j\omega_a})e^{j\omega_a n}$, by an integral number of sample intervals, n_0, to the right results in the exponential, $X(e^{j\omega_a})e^{j\omega_a(n-n_0)} = e^{-j\omega_a n_0} X(e^{j\omega_a})e^{j\omega_a n}$. That is, a delay of n_0 results in changing the phase of the exponential by $-\omega_a n_0$ radians without changing its amplitude. Therefore, if the transform of a time-domain function $x(n)$ is $X(e^{j\omega})$, then the transform of $x(n \pm n_0)$ is given by $e^{\pm j\omega n_0} X(e^{j\omega})$. That is,

$$x(n \pm n_0) \Longleftrightarrow e^{\pm j\omega n_0} X(e^{j\omega})$$

Consider the transform pair

$$(0.8)^n u(n) \Longleftrightarrow \frac{1}{1 - 0.8e^{-j\omega}}$$

Then, due to this property, we get the transform pair

$$(0.8)^{(n-2)}u(n-2) \Longleftrightarrow \frac{e^{-j2\omega}}{1-0.8e^{-j\omega}}$$

8.2.3 Frequency Shifting

The spectrum, $X(e^{j\omega})$, of a signal, $x(n)$, can be shifted by multiplying the signal by a complex exponential, $e^{\pm j\omega_0 n}$. The new spectrum is $X(e^{j(\omega \mp \omega_0)})$, since a spectral component $X(e^{j\omega_a})e^{j\omega_a n}$ of the signal multiplied by $e^{j\omega_0 n}$ becomes $X(e^{j\omega_a})e^{j(\omega_a+\omega_0)n}$ and the spectral value $X(e^{j\omega_a})$ occurs at $(\omega_a + \omega_0)$, after a delay of ω_0 radians. Therefore, we get

$$x(n)e^{\pm j\omega_0 n} \Longleftrightarrow X(e^{j(\omega \mp \omega_0)})$$

The complex exponential $e^{j\omega_0 n}$ can be considered as the product of the dc signal $x(n) = 1$ and $e^{j\omega_0 n}$. From the frequency shift property, we get the transform pair $e^{j\omega_0 n} \Longleftrightarrow 2\pi\delta(\omega - \omega_0)$. The complex exponential is characterized by the single frequency ω_0 alone. Therefore, its spectrum is an impulse at ω_0 in the fundamental frequency range from $-\pi$ to π. As $\cos(\omega_0 n) = 0.5(e^{j\omega_0 n} + e^{-j\omega_0 n})$ and $\sin(\omega_0 n) = 0.5j(e^{-j\omega_0 n} - e^{j\omega_0 n})$,

$$\cos(\omega_0 n) \Longleftrightarrow \pi(\delta(\omega - \omega_0) + \delta(\omega + \omega_0))$$

$$\sin(\omega_0 n) \Longleftrightarrow j\pi(\delta(\omega + \omega_0) - \delta(\omega - \omega_0))$$

In Example 8.2, the frequency response of an ideal lowpass filter and its impulse response were presented. By shifting the frequency response, shown in Figure 8.2(a), by π radians, we get the frequency response of an ideal highpass filter with cutoff frequency $\pi - \pi/4 = 3\pi/4$, as shown in Figure 8.5(a). As the frequency response is shifted by π radians, we get the impulse response of the highpass filter by multiplying that of the lowpass filter by $e^{j\pi n} = (-1)^n$. That is, the impulse response of the highpass filter is $(-1)^n \sin(\frac{\pi n}{4})/n\pi$, shown in Figure 8.5(b).

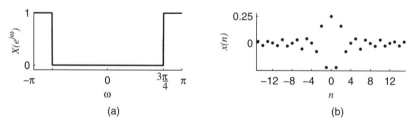

Figure 8.5 (a) One period of a DTFT spectrum of a highpass filter; (b) the corresponding impulse response

8.2.4 Convolution in the Time-domain

The convolution of signals $x(n)$ and $h(n)$ is defined, in Chapter 4, as

$$y(n) = \sum_{m=-\infty}^{\infty} x(m)h(n-m)$$

The convolution of $h(n)$ with a complex exponential $e^{j\omega_0 n}$ is given as

$$\sum_{m=-\infty}^{\infty} h(m)e^{j\omega_0(n-m)} = e^{j\omega_0 n} \sum_{m=-\infty}^{\infty} h(m)e^{-j\omega_0 m} = H(e^{j\omega_0})e^{j\omega_0 n}$$

As an arbitrary $x(n)$ is reconstructed by the inverse DTFT as

$$x(n) = \frac{1}{2\pi} \int_{-\pi}^{\pi} X(e^{j\omega})e^{j\omega n} d\omega$$

the convolution of $x(n)$ and $h(n)$ is given by

$$y(n) = \frac{1}{2\pi} \int_{-\pi}^{\pi} X(e^{j\omega})H(e^{j\omega})e^{j\omega n} d\omega$$

where $X(e^{j\omega})$ and $H(e^{j\omega})$ are, respectively, the DTFT of $x(n)$ and $h(n)$. The inverse DTFT of $X(e^{j\omega})H(e^{j\omega})$ is the convolution of $x(n)$ and $h(n)$. Therefore, we get the transform pair

$$\sum_{m=-\infty}^{\infty} x(m)h(n-m) = \frac{1}{2\pi} \int_{-\pi}^{\pi} X(e^{j\omega})H(e^{j\omega})e^{j\omega n} d\omega \iff X(e^{j\omega})H(e^{j\omega})$$

Consider the rectangular signal

$$x(n) = \begin{cases} 1 & \text{for } |n| \leq 2 \\ 0 & \text{for } |n| > 2 \end{cases}$$

shown in Figure 8.6(a) and its spectrum shown in Figure 8.6(b). The DTFT of the signal is $\sin(\frac{5\omega}{2})/\sin(\frac{\omega}{2})$. The DTFT of the convolution of this signal with itself is, due to the property, $(\sin(\frac{5\omega}{2})/\sin(\frac{\omega}{2}))^2$. As the convolution of a rectangular signal with itself is a triangular signal, this DTFT is that of a triangular signal. The triangular signal and its spectrum, which is positive for all ω, are shown, respectively, in Figures 8.6(c) and (d).

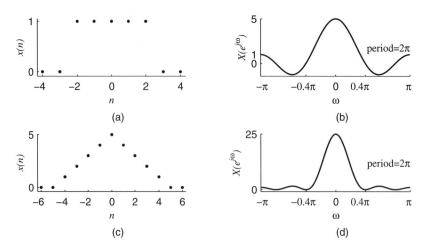

Figure 8.6 (a) The rectangular signal; (b) its spectrum; (c) the triangular signal, which is the convolution of the signal in (a) with itself; (d) its spectrum

8.2.5 Convolution in the Frequency-domain

The convolution of two functions in the frequency-domain corresponds to the multiplication of the inverse DTFT of the functions in the time-domain with a scale factor. That is,

$$x(n)y(n) \Longleftrightarrow \sum_{n=-\infty}^{\infty} x(n)y(n)e^{-j\omega n} = \frac{1}{2\pi} \int_0^{2\pi} X(e^{jv})Y(e^{j(\omega-v)})dv$$

Note that this convolution is periodic, since the DTFT spectrum is periodic.

Consider finding the DTFT of the product of the signal $\sin(n)/\pi n$ with itself. One period of the DTFT of the signal is the rectangular function

$$\begin{cases} 1 & \text{for } |\omega| < 1 \\ 0 & \text{for } 1 < |\omega| < \pi \end{cases}$$

The convolution of this function with itself divided by 2π is the periodic triangular function, one period of which is defined as

$$\begin{cases} \frac{\omega+2}{2\pi} & \text{for } -2 \le \omega \le 0 \\ \frac{2-\omega}{2\pi} & \text{for } 0 < \omega \le 2 \\ 0 & \text{for } -\pi \le \omega < -2 \text{ and } 2 < \omega < \pi \end{cases}$$

8.2.6 Symmetry

If a signal is real, then the real part of its spectrum $X(e^{j\omega})$ is even and the imaginary part is odd, called the conjugate symmetry. The DTFT of a real signal is given by

$$X(e^{j\omega}) = \sum_{n=-\infty}^{\infty} x(n)e^{-j\omega n} = \sum_{n=-\infty}^{\infty} x(n)(\cos(\omega n) - j\sin(\omega n))$$

Conjugating both sides, we get

$$X^*(e^{j\omega}) = \sum_{n=-\infty}^{\infty} x(n)(\cos(\omega n) + j\sin(\omega n))$$

Replacing ω by $-\omega$, we get $X^*(e^{-j\omega}) = X(e^{j\omega})$. For example, the DTFT of $\cos(\omega_a(n - \pi/4)) = \cos(\omega_a\frac{\pi}{4})\cos(\omega_a n) + \sin(\omega_a\frac{\pi}{4})\sin(\omega_a n)$ is

$$X(e^{j\omega}) = \pi \cos\left(\omega_a\frac{\pi}{4}\right)(\delta(\omega - \omega_a) + \delta(\omega + \omega_a))$$

$$+ j\pi \sin\left(\omega_a\frac{\pi}{4}\right)(\delta(\omega + \omega_a) - \delta(\omega - \omega_a))$$

If a signal is real and even, then its spectrum also is real and even. Since $x(n)\cos(\omega n)$ is even and $x(n)\sin(\omega n)$ is odd,

$$X(e^{j\omega}) = x(0) + 2\sum_{n=1}^{\infty} x(n)\cos(\omega n) \quad \text{and} \quad x(n) = \frac{1}{\pi}\int_0^{\pi} X(e^{j\omega})\cos(\omega n)d\omega$$

The DTFT of the cosine function is an example of this symmetry.

If a signal is real and odd, then its spectrum is imaginary and odd. Since $x(n)\cos(\omega n)$ is odd and $x(n)\sin(\omega n)$ is even,

$$X(e^{j\omega}) = -j2\sum_{n=1}^{\infty} x(n)\sin(\omega n) \quad \text{and} \quad x(n) = \frac{j}{\pi}\int_0^{\pi} X(e^{j\omega})\sin(\omega n)d\omega$$

The DTFT of the sine function is an example of this symmetry.

As the DTFT of a real and even signal is real and even and that of a real and odd is imaginary and odd, it follows that the real part of the DTFT, $\text{Re}(X(e^{j\omega}))$, of an arbitrary real signal $x(n)$ is the transform of its even component $x_e(n)$ and $j\,\text{Im}(X(e^{j\omega}))$ is that of its odd component $x_o(n)$.

8.2.7 Time Reversal

Let the spectrum of a signal $x(n)$ be $X(e^{j\omega})$. Then, $x(-n) \Longleftrightarrow X(e^{-j\omega})$. That is the time-reversal of a signal results in its spectrum also reflected about the vertical axis at the origin. This result is obtained if we replace n by $-n$ and ω by $-\omega$ in the DTFT definition.

8.2.8 Time Expansion

As we have seen in Chapter 2, a signal is compressed or expanded by scaling operation. The DTFT of the compressed signal, in the discrete case, is unrelated to the DTFT of the original signal, since part of the data is lost in contrast to merely speeding it up in the case of a continuous signal. Therefore, consider the case of signal expansion alone. Let the spectrum of a signal $x(n)$ be $X(e^{j\omega})$. If we pad $x(n)$ with zeros to get $y(n)$ defined as

$$y(an) = x(n) \text{ for } -\infty < n < \infty \quad \text{and} \quad y(n) = 0 \quad \text{otherwise}$$

where $a \neq 0$ is any positive integer, then,

$$Y(e^{j\omega}) = X(e^{ja\omega})$$

The DTFT of the sequence $y(n)$ is given by

$$Y(e^{j\omega}) = \sum_{n=-\infty}^{\infty} y(n)e^{-j\omega n}$$

Since we have nonzero input values only if $n = ak, \ k = 0, \pm1, \pm2, \ldots$, we get

$$Y(e^{j\omega}) = \sum_{k=-\infty}^{\infty} y(ak)e^{-j\omega ak} = \sum_{k=-\infty}^{\infty} x(k)e^{-j\omega ak} = X(e^{ja\omega})$$

Therefore,

$$y(n) \Longleftrightarrow X(e^{ja\omega})$$

The spectrum is compressed. That is, the spectral value at ω in the spectrum of the signal occurs at ω/a in the spectrum of its expanded version. If a is negative, the spectrum is also frequency-reversed.

For example, the DTFT of the signal $x(n)$ shown in Figure 8.7(a) with dots, with its only nonzero values given as $x(-1) = 1$ and $x(1) = 1$, is $X(e^{j\omega}) = e^{j\omega} + e^{-j\omega} = 2\cos(\omega)$. Using the theorem, we get the DTFT of $y(n)$ with $a = 2$, shown in Figure 8.7(a) with unfilled circles, as

$$Y(e^{j\omega}) = X(e^{j2\omega}) = 2\cos(2\omega)$$

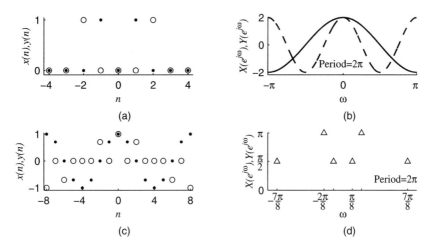

Figure 8.7 (a) Signal $x(n)$ (dots) and its expanded version $y(n)$ (unfilled circles) with $a = 2$; (b) the DTFT of $x(n)$ (solid line) and that of $y(n)$ (dashed line); (c) signal $x(n)$ (dots) and its expanded version $y(n)$ (unfilled circles) with $a = 2$; (d) the DTFT of $x(n)$ (two impulses of strength π) and that of $y(n)$ (four impulses of strength $\pi/2$)

This result can be verified from the DTFT definition. The DTFT of the signal (solid line) and that of its expanded version (dashed line) are shown in Figure 8.7(b). Since the signal is expanded by a factor of two, the spectrum is compressed by a factor of two.

As another example, consider the cosine signal $x(n) = \cos((2\pi/8)n)$, shown in Figure 8.7(c) by dots, and its DTFT

$$X(e^{j\omega}) = \pi \sum_{k=-\infty}^{\infty} \left(\delta\left(\omega - \frac{2\pi}{8} + 2\pi k\right) + \delta\left(\omega + \frac{2\pi}{8} + 2\pi k\right) \right)$$

shown in Figure 8.7(d) with two impulses of strength π. The DTFT of $y(n)$ with $a = 2$, shown in Figure 8.7(c) by unfilled circles, is

$$Y(e^{j\omega}) = \pi \sum_{k=-\infty}^{\infty} \left(\delta\left(2\omega - \frac{2\pi}{8} + 2\pi k\right) + \delta\left(2\omega + \frac{2\pi}{8} + 2\pi k\right) \right)$$

$$= \frac{\pi}{2} \sum_{k=-\infty}^{\infty} \left(\delta\left(\omega - \frac{\pi}{8} + \pi k\right) + \delta\left(\omega + \frac{\pi}{8} + \pi k\right) \right)$$

$$= \frac{\pi}{2} \left(\left(\delta\left(\omega - \frac{\pi}{8}\right) + \delta(\omega + \frac{\pi}{8}) \right) + \left(\delta\left(\omega - \frac{7\pi}{8}\right) + \delta\left(\omega + \frac{7\pi}{8}\right) \right) \right),$$

$-\pi < \omega \le \pi$, shown in Figure 8.7(d) with four impulses of strength $\pi/2$ in the fundamental frequency range from $-\pi$ to π. The expanded time-domain signal is reconstructed from its spectrum as follows.

$$y(n) = 0.5 \cos \left(\frac{\pi}{8}n\right) + 0.5 \cos \left(\left(\pi - \frac{\pi}{8}\right)n\right) = 0.5 \cos \left(\frac{\pi}{8}n\right)(1 + (-1)^n)$$

$$= \cos \left(\frac{\pi}{8}n\right)$$

for n even and $y(n)$ is zero otherwise.

8.2.9 Frequency-differentiation

Differentiating both sides of the DTFT defining equation, with respect to ω, we get the transform pair

$$(-jn)x(n) \Longleftrightarrow \frac{dX(e^{j\omega})}{d\omega} \quad \text{or} \quad (n)x(n) \Longleftrightarrow (j)\frac{dX(e^{j\omega})}{d\omega}$$

In general,

$$(-jn)^m x(n) \Longleftrightarrow \frac{d^m X(e^{j\omega})}{d\omega^m} \quad \text{or} \quad (n)^m x(n) \Longleftrightarrow (j)^m \frac{d^m X(e^{j\omega})}{d\omega^m}$$

This property is applicable only if the resulting signal satisfies the existence conditions of the DTFT. Consider the transform pair

$$\delta(n - 2) \Longleftrightarrow e^{-j2\omega}$$

Using the property, we get the transform pair

$$n\delta(n - 2) \Longleftrightarrow (j)(-j2)e^{-j2\omega} = 2e^{-j2\omega}$$

8.2.10 Difference

The derivative of a function is approximated by differences in the discrete case.

$$y(n) = x(n) - x(n - 1) \Longleftrightarrow Y(e^{j\omega}) = (1 - e^{-j\omega})X(e^{j\omega})$$

using the time-shifting property.

8.2.11 Summation

The summation of a time-domain function, $x(n)$, can be expressed, in terms of its DTFT $X(e^{j\omega})$, as

$$y(n) = \sum_{l=-\infty}^{n} x(l) \Longleftrightarrow Y(e^{j\omega}) = \frac{X(e^{j\omega})}{(1 - e^{-j\omega})} + \pi X(e^{j0})\delta(\omega), \quad -\pi < \omega \leq \pi$$

The transform $[X(e^{j\omega})/(1 - e^{-j\omega})] + \pi X(e^{j0})\delta(\omega)$ is the product of the transforms of $x(n)$ and $u(n)$ and corresponds to the convolution of $x(n)$ and $u(n)$ in the time-domain, which, of course, is equivalent to the sum of the values of $x(n)$ from $-\infty$ to n. The time-summation operation can be considered as the inverse of the time-differencing operation, if $X(e^{j0}) = 0$. This justifies the strictly continuous component of the spectrum. The impulsive component is required to take into account of the dc component of $x(n)$. This property is applicable only if the resulting signal satisfies the existence conditions of the DTFT.

Since the DTFT of unit-impulse is one and the unit-step function is a summation of the impulse, we get, using this property, the DTFT of $u(n)$, over one period, as

$$u(n) = \sum_{l=0}^{n} \delta(l) \Longleftrightarrow \frac{1}{(1 - e^{-j\omega})} + \pi\delta(\omega) \quad -\pi < \omega \leq \pi$$

As another example, consider the signal, shown in Figure 8.8(a), and the resulting signal, shown in Figure 8.8(b), obtained by summing it. The DTFT of the given signal is, from the DTFT definition, $1 + e^{-j\omega}$. Using the property, we get the DTFT of its summation as

$$\frac{1 + e^{-j\omega}}{1 - e^{-j\omega}} + 2\pi\delta(\omega), \quad -\pi < \omega \leq \pi$$

The summation of $x(n)$ is $y(n)$, shown in Figure 8.8(b) along with its two components corresponding to the two terms of the transform.

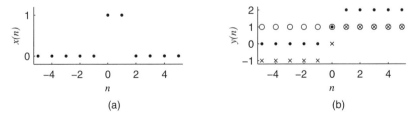

(a)   (b)

Figure 8.8 (a) Signal $x(n) = u(n) - u(n - 2)$; (b) $y(n) = \sum_{l=-\infty}^{n} x(l)$ (dotted line) and its two components

8.2.12 Parseval's Theorem and the Energy Transfer Function

As the frequency-domain representation of a signal is an equivalent representation, the energy of a signal can also be expressed in terms of its spectrum.

$$E = \sum_{n=-\infty}^{\infty} |x(n)|^2 = \frac{1}{2\pi} \int_0^{2\pi} |X(e^{j\omega})|^2 d\omega$$

Since $x(n)$ can be considered as the FS coefficients of $X(e^{j\omega})$, this expression is the same as that corresponding to the FS with the roles of the domains interchanged. The quantity $|X(e^{j\omega})|^2$ is called the energy spectral density of the signal, since $(1/2\pi)|X(e^{j\omega})|^2 d\omega$ is the signal energy over the infinitesimal frequency band ω to $\omega + d\omega$.

Consider the signal, shown in Figure 8.8(a), and its DTFT $1 + e^{-j\omega}$. The energy of the signal, from its time-domain representation, is $1^2 + 1^2 = 2$. The energy of the signal, from its frequency-domain representation, is

$$E = \frac{1}{2\pi} \int_0^{2\pi} |1 + e^{-j\omega}|^2 d\omega = \frac{1}{2\pi} \int_0^{2\pi} (2 + 2\cos(\omega)) d\omega = 2$$

The input and output of a LTI system, in the frequency-domain, is related by the transfer function $H(e^{j\omega})$ as

$$Y(e^{j\omega}) = H(e^{j\omega})X(e^{j\omega})$$

where $X(e^{j\omega})$, $Y(e^{j\omega})$, and $H(e^{j\omega})$ are the DTFT of the input, output, and impulse response of the system. The output energy spectrum is given by

$$|Y(e^{j\omega})|^2 = Y(e^{j\omega})Y^*(e^{j\omega})$$
$$= H(e^{j\omega})X(e^{j\omega})H^*(e^{j\omega})X^*(e^{j\omega}) = |H(e^{j\omega})|^2|X(e^{j\omega})|^2$$

The quantity $|H(e^{j\omega})|^2$ is called the energy transfer function, as it relates the input and output energy spectral densities of the input and output of a system.

8.3 Approximation of the Discrete-time Fourier Transform

In the computation of the DFT, we usually use the time-domain range from $n = 0$ to $n = N - 1$. Due to periodicity of the DFT, we can always get the samples in this interval even though the data is defined in other intervals. Replacing ω by $(2\pi/N)k$ in the DTFT definition, we get

$$X(e^{j\frac{2\pi}{N}k}) = \sum_{n=0}^{N-1} x(n)e^{-j\frac{2\pi}{N}nk} = \sum_{n=0}^{N-1} x(n)W^{nk} \qquad k = 0, 1, \dots, N-1$$

Let us approximate the samples of the DTFT spectrum shown in Figure 8.2(a) using the DFT. The time-domain signal, shown in Figure 8.2(b), is of infinite duration and, therefore, we have to truncate it. For example, let us take the fifteen samples $x(-7), x(-6), \dots, x(6), x(7)$. The record length of the truncated signal should be such that most of the energy of the signal is retained in the truncated signal. As the most efficient and regular DFT algorithms are of length that is an integral power of two, the truncated data is usually zero-padded. With one zero added and $N = 16$, the data for the DFT computation becomes $x(0), x(1), \dots, x(7), 0, x(-7), \dots, x(-2), x(-1)$, as shown in Figure 8.9(a). The corresponding DFT spectrum is shown in Figure 8.9(b). As the spectrum is even-symmetric, only the positive frequency half of the spectrum is shown. As the number of samples is increased, the spectral samples become more accurate, as shown in Figures 8.9(c) and (d) with $N = 64$. Note the Gibbs phenomenon in the vicinity of the discontinuity of the spectrum.

The spectral samples obtained using the DFT are not exact because of the truncation of the input data. In effect, the actual data are multiplied by a rectangular window. Therefore, the desired spectrum is convolved with that of the rectangular window (a sinc function). This results in the distortion of the spectrum. As the level of truncation is reduced, the distortion also gets reduced. In the end, with no truncation (a rectangular window of infinite length), we get an undistorted spectrum. As an infinite data length

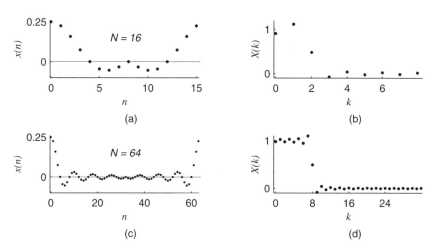

Figure 8.9 (a), (c) One period of the periodic extension of the truncated and zero padded aperiodic discrete signal, shown in Figure 8.2(b), with $N = 16$ and $N = 64$ samples, respectively; (b), (d) The DFT of the signals in (a) and (c), respectively

is unacceptable for DFT computation, we start with some finite data length and keep increasing it until the difference between two successive spectra becomes negligible.

8.3.1 Approximation of the Inverse DTFT by the IDFT

Replacing ω by $(2\pi/N)k$ and dω by $2\pi/N$ in the inverse DTFT definition, we get

$$x(n) = \frac{1}{N}\sum_{k=0}^{N-1} X\left(e^{j\frac{2\pi}{N}k}\right) e^{j\frac{2\pi}{N}nk} = \frac{1}{N}\sum_{k=0}^{N-1} X\left(e^{j\frac{2\pi}{N}k}\right) W^{-nk} \qquad n = 0, 1, \dots, N-1$$

Let us approximate the inverse DTFT of the spectrum shown in Figure 8.2(a) by the IDFT. As always, at points of discontinuity, the average of the left- and right-hand limits should be taken as the sample value in Fourier analysis. The sample values of the spectrum with $N = 8$ are shown in Figure 8.10(a). The IDFT of these samples is shown in Figure 8.10(b) along with the exact values. Only half of the signal is shown, as it is even-symmetric. As the number of samples is increased, as shown in Figure 8.10(c), the time-domain values become more accurate, as shown in Figure 8.10(d). As the time-domain data length is infinite, the necessary sampling interval of the spectrum is zero radians. However, as that interval is not practical with numerical analysis, we use some finite sample interval. That results in time-domain aliasing. As mentioned earlier, practical signals, with an adequate sampling interval and a sufficient record length, can be considered as both time-limited and band-limited with a desired accuracy. This fact enables the use of the DFT and IDFT, which can be computed using fast algorithms, to approximate the other versions of Fourier analysis.

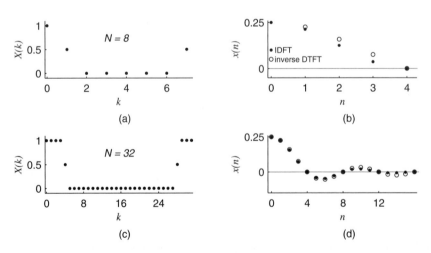

Figure 8.10 (a), (c) Samples of one period of the periodic DTFT spectrum, shown in Figure 8.2(a), with $N = 8$ and $N = 32$ samples, respectively; (b), (d) The IDFT of the spectra in (a) and (c), respectively

8.4 Applications of the Discrete-time Fourier Transform

8.4.1 Transfer Function and the System Response

The input–output relationship of a LTI system is given by the convolution operation in the time-domain. Since convolution corresponds to multiplication in the frequency-domain, we get

$$y(n) = \sum_{m=-\infty}^{\infty} x(m)h(n-m) \iff Y(e^{j\omega}) = X(e^{j\omega})H(e^{j\omega})$$

where $x(n)$, $h(n)$, and $y(n)$ are, respectively the system input, impulse response, and output, and $X(e^{j\omega})$, $H(e^{j\omega})$, and $Y(e^{j\omega})$ are their respective transforms. As input is transferred to output by multiplication with $H(e^{j\omega})$, $H(e^{j\omega})$ is called the transfer function of the system. The transfer function, which is the transform of the impulse response, characterizes a system in the frequency-domain just as the impulse response does in the time-domain.

Since the impulse function, whose DTFT is one (a uniform spectrum), is composed of complex exponentials, $e^{j\omega n}$, of all frequencies from $\omega = -\pi$ to $\omega = \pi$ with equal magnitude and zero phase, the transform of the impulse response, the transfer function, is also called the frequency response of the system. Therefore, an exponential $Ae^{j(\omega_a n + \theta)}$ is changed to $(|H(e^{j\omega_a})|A)e^{j(\omega_a n + (\theta + \angle(H(e^{j\omega_a}))))}$ at the output. A real sinusoidal input signal $A\cos(\omega_a n + \theta)$ is also changed at the output by the same amount of amplitude and phase of the complex scale factor $H(e^{j\omega_a})$. That is, $A\cos(\omega_a n + \theta)$ is changed to $(|H(e^{j\omega_a})|A)\cos(\omega_a n + (\theta + \angle(H(e^{j\omega_a}))))$. The steady-state response of a stable system to the input $Ae^{j(\omega_a n + \theta)}u(n)$ is also the same.

As $H(e^{j\omega}) = Y(e^{j\omega})/X(e^{j\omega})$, the transfer function can also be described as the ratio of the transform $Y(e^{j\omega})$ of the response $y(n)$ to an arbitrary signal $x(n)$ to that of its transform $X(e^{j\omega})$, provided $|X(e^{j\omega})| \neq 0$ for all frequencies and the system is initially relaxed.

Since the transform of a delayed signal is its transform multiplied by a factor, we can just as well find the transfer function by taking the transform of the difference equation characterizing a system. Consider the difference equation of a causal LTI discrete system.

$$y(n) + a_{K-1}y(n-1) + a_{K-2}y(n-2) + \cdots + a_0 y(n-K)$$
$$= b_M x(n) + b_{M-1}x(n-1) + \cdots + b_0 x(n-M)$$

Taking the transform of both sides, we get, assuming initial conditions are all zero,

$$Y(e^{j\omega})(1 + a_{K-1}e^{-j\omega} + a_{K-2}e^{-j2\omega} + \cdots + a_0 e^{-jK\omega})$$
$$= X(e^{j\omega})(b_M + b_{M-1}e^{-j\omega} + \cdots + b_0 e^{-jM\omega})$$

The transfer function $H(e^{j\omega})$ is obtained as

$$H(e^{j\omega}) = \frac{Y(e^{j\omega})}{X(e^{j\omega})} = \frac{b_M + b_{M-1}e^{-j\omega} + \cdots + b_0 e^{-jM\omega}}{1 + a_{K-1}e^{-j\omega} + a_{K-2}e^{-j2\omega} + \cdots + a_0 e^{-jK\omega}}$$

Example 8.6. Find the response, using the DTFT, of the system governed by the difference equation

$$y(n) = x(n) + 0.6y(n-1)$$

to the input $x(n) = \cos(\frac{2\pi}{6}n + \frac{\pi}{6})$.

Solution

$$H(e^{j\omega}) = \frac{e^{j\omega}}{e^{j\omega} - 0.6}$$

Substituting $\omega = \frac{2\pi}{6}$, we get

$$H\left(e^{j\frac{2\pi}{6}}\right) = \frac{e^{j\frac{2\pi}{6}}}{e^{j\frac{2\pi}{6}} - 0.6} = 1.1471\angle(-0.6386)$$

The response of the system to the input $x(n) = \cos(\frac{2\pi}{6}n + \frac{\pi}{6})$ is $y(n) = 1.1471\cos(\frac{2\pi}{6}n + \frac{\pi}{6} - 0.6386)$. ☐

Example 8.7. Find the impulse response $h(n)$, using the DTFT, of the system governed by the difference equation

$$y(n) = x(n) - x(n-1) + 2x(n-2) + \frac{7}{12}y(n-1) - \frac{1}{12}y(n-2)$$

Solution

$$H(e^{j\omega}) = \frac{1 - e^{-j\omega} + 2e^{-j2\omega}}{\left(1 - \frac{7}{12}e^{-j\omega} + \frac{1}{12}e^{-j2\omega}\right)} = \frac{1 - e^{-j\omega} + 2e^{-j2\omega}}{\left(1 - \frac{1}{3}e^{-j\omega}\right)\left(1 - \frac{1}{4}e^{-j\omega}\right)}$$

Expanding into partial fractions, we get

$$H(e^{j\omega}) = 24 + \frac{64}{\left(1 - \frac{1}{3}e^{-j\omega}\right)} - \frac{87}{\left(1 - \frac{1}{4}e^{-j\omega}\right)}$$

Taking the inverse DTFT, we get the impulse response as

$$h(n) = 24\delta(n) + \left(64\left(\frac{1}{3}\right)^n - 87\left(\frac{1}{4}\right)^n\right)u(n)$$

The first four values of the impulse response $h(n)$ are

$$h(0) = 1, \quad h(1) = -0.4167, \quad h(2) = 1.6736, \quad h(3) = 1.011 \qquad \square$$

Example 8.8. Find the zero-state response, using the DTFT, of the system governed by the difference equation

$$y(n) = 2x(n) - x(n-1) + 3x(n-2) + \frac{9}{20}y(n-1) - \frac{1}{20}y(n-2)$$

with the input $x(n) = u(n)$, the unit-step function.

Solution

$$H(e^{j\omega}) = \frac{2 - e^{-j\omega} + 3e^{-j2\omega}}{\left(1 - \frac{9}{20}e^{-j\omega} + \frac{1}{20}e^{-j2\omega}\right)} = \frac{2 - e^{-j\omega} + 3e^{-j2\omega}}{\left(1 - \frac{1}{5}e^{-j\omega}\right)\left(1 - \frac{1}{4}e^{-j\omega}\right)}$$

With $X(e^{j\omega}) = \frac{1}{(1-e^{-j\omega})} + \pi\delta(\omega)$,

$$Y(e^{j\omega}) = H(e^{j\omega})X(e^{j\omega}) = \frac{2 - e^{-j\omega} + 3e^{-j2\omega}}{(1 - e^{-j\omega})\left(1 - \frac{1}{5}e^{-j\omega}\right)\left(1 - \frac{1}{4}e^{-j\omega}\right)} + \frac{20}{3}\pi\delta(\omega)$$

Expanding into partial fractions, we get

$$Y(e^{j\omega}) = \frac{\frac{20}{3}}{(1 - e^{-j\omega})} + \frac{72}{\left(1 - \frac{1}{5}e^{-j\omega}\right)} - \frac{\frac{230}{3}}{\left(1 - \frac{1}{4}e^{-j\omega}\right)} + \frac{20}{3}\pi\delta(\omega)$$

Taking the inverse DTFT, we get the zero-state response.

$$y(n) = \left(\frac{20}{3} + 72\left(\frac{1}{5}\right)^n - \frac{230}{3}\left(\frac{1}{4}\right)^n\right)u(n)$$

The steady-state response is $(20/3)u(n)$, the response after the transient response has died out completely. The first four values of the sequence $y(n)$ are

$$y(0) = 2, \quad y(1) = 1.9, \quad y(2) = 4.755, \quad y(3) = 6.0448 \qquad \square$$

The transfer function concept can still be used, even if the initial conditions of a system are nonzero. In that case, we have to assume that additional inputs are applied to the system at the instant the system is turned on, which will produce the same response as do the initial conditions. However, the z-transform is relatively easier for system analysis. In addition, it can handle a larger class of signals and systems than can be analyzed by the DTFT. Wherever the DTFT is more suitable, it is better for numerical analysis as it can be approximated by the DFT using fast algorithms.

8.4.2 Digital Filter Design Using DTFT

Digital filters are widely used in signal processing applications. Usually, the specification of a filter is given in terms of its frequency response. As the filter, which is a system, is characterized by its impulse response, the design of a filter is to determine its impulse response. Therefore, one way of finding the impulse response is to find inverse DTFT of its frequency response. For example, the frequency response and the corresponding impulse response of an ideal lowpass filter are shown, respectively, in Figures 8.2(a) and (b) and those of a highpass filter are shown, respectively, in Figures 8.5(a) and (b). A system with this type of impulse response is not practically implementable because: (i) As the impulse response is not absolutely summable, it is an unstable system; (ii) the impulse response is noncausal. The first problem is overcome by truncating part of the impulse response. The second problem is solved by shifting the impulse response to the right so that it becomes causal. With these modifications of the impulse response, of course, the filter response will not be ideal.

We prefer the response of the actual filter to converge uniformly to that of the ideal filter. But, in Fourier analysis, the convergence criteria is with respect to the square error. That is, there is a 9% deviation of the frequency response at the band edges (discontinuities) of the filter. This problem can be reduced by using window functions to smooth the truncated impulse response. This time the price that is paid is of longer transition bands.

8.4.3 Digital Differentiator

In this subsection, we derive the impulse response of the digital differentiator from its frequency response. This differentiator takes the samples of a continuous signal $x(t)$ and produces the samples of its derivative. The periodic frequency response, shown in Figure 8.11(a) over one period, of the ideal digital differentiator is defined as

$$H(e^{j\omega}) = j\omega, \quad -\pi < \omega < \pi$$

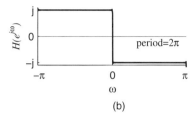

(a)   (b)

Figure 8.11   (a) The frequency response of the ideal digital differentiator; (b) the frequency response of the ideal Hilbert transformer

For example, the input and the output of the differentiator are

$$\sin(\omega_0 n) \iff j\pi(\delta(\omega + \omega_0) - \delta(\omega - \omega_0))$$

$$j\pi(j)(-\omega_0\delta(\omega + \omega_0) - \omega_0\delta(\omega - \omega_0)) \iff \omega_0 \cos(\omega_0 n)$$

The impulse response of the ideal differentiator is obtained by finding the inverse DTFT of its frequency response.

$$h(n) = \frac{1}{2\pi} \int_{-\pi}^{\pi} j\omega e^{j\omega n} d\omega = \frac{\cos(\pi n)}{n} = \begin{cases} \frac{(-1)^n}{n} & \text{for } n \neq 0 \\ 0 & \text{for } n = 0 \end{cases} \quad -\infty < n < \infty$$

As the frequency response of the differentiator is imaginary and odd-symmetric, the impulse response is real and odd-symmetric.

8.4.4 Hilbert Transform

Although most practical signals are real-valued, we need, in applications such as the sampling of bandpass signals and single-sideband amplitude modulation, a complex signal whose real part is the given real signal $x(n)$ and the imaginary part is the Hilbert transform of $x(n)$. In the Hilbert transform, every real frequency component of a real signal $x(n)$ is shifted to the right by $-\pi/2$ radians. That is, a phase of $-\pi/2$ radians is added. For example, the Hilbert transform of $\sin(\omega n)$ is $\sin(\omega n - \pi/2) = -\cos(\omega n)$. Most of the transforms have two domains, whereas there is only one domain in the Hilbert transform. Consider the complex signal formed with the real part being a real signal and the imaginary part being its Hilbert transform. The spectral values of this complex signal are zero for negative frequencies (a one-sided spectrum). The complex signal formed by the sine signal and its Hilbert transform is

$$\sin(\omega n) - j\cos(\omega n) = -je^{j\omega n}$$

The DFT of $\sin(\omega n)$, with N samples in a cycle, is $-jN/2$ at ω and $jN/2$ at $-\omega$ whereas that of $-je^{j\omega n}$ is $-jN$ at ω only. Similarly, a transform with its imaginary part being the Hilbert transform of its real part, for example the transfer function of a causal system, corresponds to a one-sided time-domain signal. In this subsection, the impulse response of the Hilbert transformer is derived from its frequency response.

Figures 8.12(a) and (b) show the signal

$$x(n) = 0.3 + \sin\left(\frac{2\pi}{16}n\right) + \cos\left(3\frac{2\pi}{16}n\right) + \sin\left(5\frac{2\pi}{16}n - \frac{\pi}{3}\right) + \cos(\pi n)$$

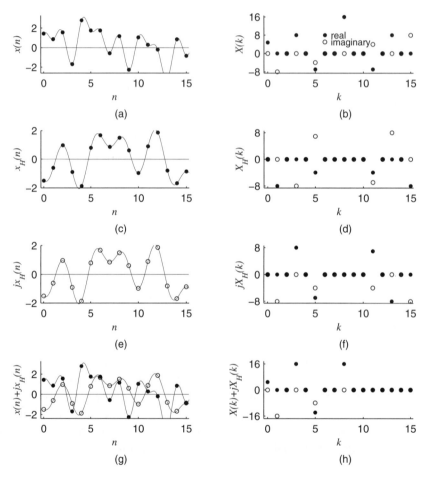

Figure 8.12   (a) An arbitrary signal; (b) its DFT; (c) the Hilbert transform of signal in (a) and (d) its DFT; (e) the signal in (c) multiplied by j and (f) its DFT; (g) the sum of signals in (a) and (e), and (h) its one-sided DFT spectrum

and its DFT spectrum, respectively. Figures 8.12(c) and (d) show the Hilbert transform
of the signal in (a)

$$x_H(n) = \sin\left(\frac{2\pi}{16}n - \frac{\pi}{2}\right) + \cos\left(3\frac{2\pi}{16}n - \frac{\pi}{2}\right) + \sin\left(5\frac{2\pi}{16}n - \frac{\pi}{3} - \frac{\pi}{2}\right)$$

and its spectrum, respectively. The dc component 0.3 and the component with frequency π, $\cos(\pi n)$, become sine terms with frequencies 0 and π radians. At these
frequencies, the samples of the sine wave are all zero. The differences between the
spectra in (b) and (d) are that the values at index $k = 0$ and at $k = \frac{N}{2} = 8$ are zero in
(d), the values of the other positive frequency components in (b) are multiplied by $-j$,
and those of the negative frequency components in (b) are multiplied by j. Therefore,
the spectrum of a real signal modified in this way is the DFT of its Hilbert transform
and its IDFT gives the Hilbert transform of the signal.

The signal $jx_H(n)$ and its spectrum are shown in Figures 8.12(e) and (f), respectively.
Compared with the spectrum in Figure 8.12(b), the coefficients at index $k = 0$ and
at $k = N/2 = 8$ are zero, the coefficients of the positive frequency components are
modified by $j(-j) = 1$, and those of the negative frequency components are modified
by $j(j) = -1$. Therefore, the spectrum is the same as in (b) with the values of the
negative frequency components negated and the values with indices 0 and 8 zero. The
complex signal $x(n) + jx_H(n)$ and its spectrum are shown in Figures 8.12(g) and (h),
respectively. The spectral values in (h) with indices from 1 to 7 are twice of those in
the first half of (b). Values with indices 0 and 8 are the same and the rest of the values
are zero.

The periodic frequency response, shown in Figure 8.11(b) over one period, of the
ideal Hilbert transformer is defined as

$$H(e^{j\omega}) = \begin{cases} -j & \text{for } 0 < \omega < \pi \\ j & \text{for } -\pi < \omega < 0 \end{cases}$$

The impulse response of the ideal Hilbert transformer is obtained by finding the inverse
DTFT of its frequency response.

$$h(n) = \frac{1}{2\pi}\int_0^\pi -je^{j\omega n}\,d\omega + \frac{1}{2\pi}\int_{-\pi}^0 je^{j\omega n}\,d\omega$$

$$= \begin{cases} \frac{2\sin^2\left(\frac{\pi n}{2}\right)}{\pi n} & \text{for } n \neq 0 \\ 0 & \text{for } n = 0 \end{cases} \quad -\infty < n < \infty$$

8.5 Summary

- In this chapter, the DTFT, its properties, its applications, and its approximation by the DFT have been presented.
- The DTFT analyzes aperiodic discrete signals in terms of a continuum of discrete sinusoids over a finite frequency range. Due to the discrete nature of the signal with an infinite range, the DTFT spectrum is periodic and continuous.
- There is a dual relationship between the FS and the DTFT.
- The spectral analysis of discrete signals, design of filters, and LTI discrete system analysis are typical applications of the DTFT.
- As is the case with the other versions of the Fourier analysis, the DTFT is also approximated by the DFT.

Further Reading

1. Sundararajan, D., *Digital Signal Processing, Theory and Practice*, World Scientific, Singapore, 2003.
2. Lathi, B. P., *Linear Systems and Signals*, Oxford University Press, New York, 2004.
3. Guillemin, E. A., *Theory of Linear Physical Systems*, John Wiley, New York, 1963.

Exercises

8.1 Find the DTFT of

$$x(n) = \begin{cases} 1 & \text{for } 0 \le n < N \\ 0 & \text{otherwise} \end{cases}$$

With $N = 5$, compute the values of $X(e^{j\omega})$ of $x(n)$ at $\omega = 0, \pi$.

*8.2 Find the DTFT of

$$x(n) = \begin{cases} 1 & \text{for } -N \le n \le N \\ 0 & \text{otherwise} \end{cases}$$

With $N = 5$, compute the values of $X(e^{j\omega})$ of $x(n)$ at $\omega = 0, \pi$.

8.3 Find the DTFT of $x(n) = (a)^n \cos(\omega_0 n)u(n)$, $a < 1$. With $\omega_0 = \pi/2$ and $a = 0.9$, compute the values of $X(e^{j\omega})$ of $x(n)$ at $\omega = 0, \pi$.

*8.4 Find the DTFT of $x(n) = (a)^n \sin(\omega_0 n)u(n)$, $a < 1$. With $\omega_0 = \pi/2$ and $a = 0.7$, compute the values of $X(e^{j\omega})$ of $x(n)$ at $\omega = 0, \pi$.

8.5 Apply a limiting process, as $N \to \infty$, so that

$$x(n) = \begin{cases} \cos(\omega_0 n) & \text{for } |n| \le N \\ 0 & \text{for } |n| > N \end{cases}$$

degenerates into the cosine function and, hence, derive the DTFT of the signal $\cos(\omega_0 n)$, $-\infty < n < \infty$.

8.6   Apply a limiting process, as $a \to 1$, so that $a^{|n|} \cos(\omega_0 n)$, $a < 1$ degenerates into $\cos(\omega_0 n)$ and, hence, derive the DTFT of the signal $\cos(\omega_0 n)$, $-\infty < n < \infty$.

8.7   Apply a limiting process so that $x(n)$ degenerates into the dc function and, hence, derive the DTFT of the dc function, $x(n) = 1$.

8.7.1 $x(n) = \begin{cases} 1 & \text{for } |n| \leq M \\ 0 & \text{for } |n| > M \end{cases}$ as $M \to \infty$.

8.7.2 $x(n) = a^{|n|}$, $0 < a < 1$ as $a \to 1$.

8.7.3 $x(n) = \frac{\sin(an)}{an}$ as $a \to 0$.

8.8   Given the description of the periodic signal $x(t)$ over one period, find its FS. Then, using the duality property, find the corresponding DTFT pair. Verify the DTFT pair using the inverse DTFT equation.

8.8.1 $x(t) = \begin{cases} 2 & \text{for } 0 < t < 2 \\ -2 & \text{for } 2 < t < 4 \end{cases}$

8.8.2 $x(t) = 1.5t$, $0 \leq t < 2$.

8.8.3 $x(t) = \begin{cases} \frac{4}{3}t & \text{for } 0 \leq t < 1.5 \\ \frac{4}{3}(3 - t) & \text{for } 1.5 \leq t < 3 \end{cases}$

8.9   Find the DTFT of $x(n)$.

8.9.1 $x(n) = 2\cos\left(\frac{2\pi}{8}n + \frac{\pi}{3}\right)$.

8.9.2 $x(n) = j4\sin\left(\frac{2\pi}{6}n - \frac{\pi}{6}\right)$.

8.9.3 $x(n) = 2e^{j\left(\frac{2\pi}{9}n + \frac{\pi}{4}\right)}$.

8.9.4 $x(n) = u(n - 2)$.

*8.9.5 $x(n) = (0.6)^n u(n - 2)$.

8.10 Given the sample values over a period of a periodic sequence, find its DTFT using the DFT.

8.10.1 $\{x(0) = 2, x(1) = 3, x(2) = 1, x(3) = 4\}$.

*8.10.2 $\{x(0) = 4, x(1) = 1, x(2) = 2, x(3) = 3\}$.

8.10.3 $\{x(0) = 3, x(1) = 4, x(2) = -2, x(3) = 1\}$.

8.11 Find the DTFT, $X(e^{j\omega})$, of $x(n)$. Find also the DFT, $X(k)$, of $x(n)$ with $N = 4$. Verify that the DFT values correspond to the samples of $X(e^{j\omega})$ at $\omega = 0, \pi/2, \pi, 3\pi/2$.

8.11.1 $\{x(n), n = 0, 1, 2, 3\} = \{2, 3, -1, 4\}$ and $x(n) = 0$ otherwise.

8.11.2 $\{x(n), n = 0, 1, 2, 3\} = \{4, 0, 0, 0\}$ and $x(n) = 0$ otherwise.

8.11.3 $\{x(n), n = 0, 1, 2, 3\} = \{0, -2, 0, 0\}$ and $x(n) = 0$ otherwise.

8.11.4 $\{x(n), n = 0, 1, 2, 3\} = \{3, 3, 3, 3\}$ and $x(n) = 0$ otherwise.

8.11.5 $\{x(n), n = 0, 1, 2, 3\} = \{2, -2, 2, -2\}$ and $x(n) = 0$ otherwise.

8.12 Find the DTFT of the signal

$$x(n) = \begin{cases} 1 & \text{for } n \geq 0 \\ -1 & \text{for } n < 0 \end{cases}$$

using the linearity property.

8.13   Find the DTFT of the signal $x(n) = (n + 1)(0.7)^n u(n)$ using the linearity prop-
       erty. Find the spectral values at $\omega = 0, \pi$.

8.14   Find the DTFT of the signal $x(n) = 0, \ n < 0, x(0) = 2, x(1) = 2$, and $x(n) =$
       $5, \ n > 1$ using the transform of $u(n)$, and linearity and time-shifting properties.

8.15   Find the DTFT of the signal with its nonzero values defined as $x(n) =$
       $(0.6)^n, \ 0 \le n \le 7$ using the transform of $(0.6)^n u(n)$, and the linearity and time-
       shifting properties.

8.16   Find the inverse DTFT of $X(e^{j\omega})$ using the linearity property.
       8.16.1 $X(e^{j\omega}) = \frac{1}{(1-0.5e^{-j\omega})(1-0.4e^{-j\omega})}$.
       *8.16.2 $X(e^{j\omega}) = \frac{1}{(1-0.5e^{-j\omega})(1-0.25e^{-j\omega})}$.

8.17   Find the impulse response $h_l(n)$ of an ideal lowpass filter with cutoff frequency
       $\pi/3$ radians. Using the frequency-shifting property and the $h_l(n)$ obtained, find
       the impulse response $h_h(n)$ of an ideal highpass filter with cutoff frequency
       $2\pi/3$ radians.

8.18   Find the DTFT of the signal

$$x(n) = \begin{cases} 1 & \text{for } -N \le n \le N \\ 0 & \text{otherwise} \end{cases}$$

       using the DTFT of shifted unit-step signals.

8.19   Using the frequency-shifting property, find the inverse DTFT of

$$X(e^{j\omega}) = \frac{1}{(1 - 0.6e^{-j(\omega - \frac{\pi}{3})})}$$

8.20   Find the DTFT of the signal $x(n) = e^{j\omega_0 n} u(n)$ using the frequency-shifting
       property.

8.21   Find the convolution of the finite sequences $x(n)$ and $h(n)$ using the DTFT.
       *8.21.1 $\{x(n), \ n = 0, 1, 2, 3\} = \{1, 0, 2, 3\}$ and $\{h(n), \ n = 1, 2, 3\} = \{-2,$
       $1, -4\}$.
       8.21.2 $\{x(n), \ n = -4, -3, -2, -1\} = \{3, 1, 0, -4\}$ and $\{h(n), \ n = -4,$
       $-3, -2, -1\} = \{1, 0, -1, 3\}$.

       8.21.3 $\{x(n), \ n = -1, 0, 1\} = \{2, 0, 3\}$ and $\{h(n), \ n = -1, 0, 1\} = \{-3,$
       $2, 2\}$.

8.22   Using the time-domain convolution property, find the DTFT of the convolution
       of $x(n)$ and $h(n)$.
       8.22.1 $x(n) = (0.5)^n u(n)$ and $h(n) = x(n)$.
       8.22.2 $x(n) = (0.6)^n u(n)$ and $h(n) = u(n)$.
       8.22.3 $x(n) = (0.7)^n u(n)$ and $h(n) = (0.3)^n u(n)$.

8.23 Using the frequency-domain convolution property, find the DTFT of the product of $x(n)$ and $h(n)$.
  8.23.1 $x(n) = 2\sin(n)$ and $h(n) = \cos(n)$.
  8.23.2 $x(n) = e^{(j\omega_0 n)}$ and $h(n) = u(n)$.

8.24 Using the time-expansion property, find the DTFT of the signal $y(n)$ defined as

$$y(an) = x(n) \text{ for } -\infty < n < \infty \quad \text{and} \quad y(n) = 0 \quad \text{otherwise}$$

  8.24.1 $x(n) = 3$, $|n| \le 2$ and $x(n) = 0$ otherwise, and $a = 2$.
  *8.24.2 $x(n) = (0.6)^n u(n)$ and $a = -4$.
  8.24.3 $x(n) = \frac{\sin(\frac{\pi n}{3})}{\pi n}$ and $a = 2$.
  8.24.4 $x(n) = u(n)$ and $a = 3$.
  8.24.5 $x(n) = u(n-2)$ and $a = 2$.
  8.24.6 $x(n) = \cos\left(\frac{2\pi}{8}(n-1)\right)$ and $a = 2$.

8.25 Find the DTFT of the signal $x(n) = n(0.8)^n u(n)$ using the multiplication by n property.

8.26 Using the time-summation property, find the DTFT of the summation

$$y(n) = \sum_{l=-\infty}^{n} x(l)$$

  8.26.1 $x(-1) = 2$, $x(1) = -2$ and $x(n) = 0$ otherwise.
  8.26.2 $x(n) = \delta(n+2)$.
  8.26.3 $x(n) = u(n+2)$.
  *8.26.4 $x(n) = (0.6)^n u(n)$.

8.27 Verify Parseval's theorem.
  8.27.1 $x(-1) = 1$, $x(1) = -1$, and $x(n) = 0$ otherwise.
  8.27.2 $x(n) = \frac{\sin(n)}{\pi n}$.

8.28 Find the DTFT of $x(n) = (0.4)^n u(n)$. Compute the samples of $X(e^{j\omega})$ of $x(n)$ using the DFT with $N = 4$. Compare the DFT values with the exact sample values of $X(e^{j\omega})$.

*8.29 Find the DTFT of $x(n) = (0.3)^n u(n)$. Approximate the values of $x(n)$, using the IDFT with $N = 4$, from the samples of the DTFT of $x(n)$. Compare the IDFT values with the exact values of $x(n)$.

8.30 Using the DTFT, find the impulse response $h(n)$ of the system governed by the difference equation

$$y(n) = 2x(n) - 3x(n-1) + 2x(n-2) + \frac{5}{6}y(n-1) - \frac{1}{6}y(n-2)$$

with input $x(n)$ and output $y(n)$. List the first four values of $h(n)$.

*8.31 Using the DTFT, find the impulse response $h(n)$ of the system governed by the difference equation

$$y(n) = x(n) - 4x(n-1) - \frac{11}{12}y(n-1) - \frac{1}{6}y(n-2)$$

with input $x(n)$ and output $y(n)$. List the first four values of $h(n)$.

8.32 Using the DTFT, find the impulse response $h(n)$ of the system governed by the difference equation

$$y(n) = x(n) + \frac{11}{15}y(n-1) - \frac{2}{15}y(n-2)$$

with input $x(n)$ and output $y(n)$. List the first four values of $h(n)$.

8.33 Using the DTFT, find the frequency response of the system governed by the difference equation

$$y(n) = x(n) + 0.8y(n-1)$$

Deduce the steady-state response of the system to the input $x(n) = \cos(\frac{2\pi}{8}n - \frac{\pi}{6})u(n)$.

8.34 Using the DTFT, find the zero-state response of the system governed by the difference equation

$$y(n) = x(n) - 2x(n-1) + 3x(n-2) + \frac{7}{12}y(n-1) - \frac{1}{12}y(n-2)$$

with the input $x(n) = u(n)$, the unit-step function.

*8.35 Using the DTFT, find the zero-state response of the system governed by the difference equation

$$y(n) = 3x(n) + 2x(n-1) + x(n-2) + \frac{8}{15}y(n-1) - \frac{1}{15}y(n-2)$$

with the input $x(n) = (\frac{1}{2})^n u(n)$.

8.36 Find the Hilbert transform $x_H(n)$ of the signal.
    8.36.1 $x(n) = 2 - \cos^2(n)$.
    *8.36.2 $x(n) = (-1)^n + \sin^2(0.5n)$.

8.37 Find the Hilbert transform $x_H(n)$ of the signal $\{x(n), n = 0, 1, 2, 3\} = \{4, 5, 4, 3\}$ using the DFT and the IDFT.

9

The Fourier Transform

The FT is the frequency-domain representation of continuous aperiodic signals in terms of a continuum of sinusoids over an infinite frequency range. Compared with the FS, as the period of the periodic waveform tends to infinity, the waveform becomes aperiodic and the interval between the spectral points tends to zero, resulting in a continuous aperiodic spectrum. Compared with the DTFT, as the sampling interval of the time-domain waveform tends to zero, the waveform becomes continuous and the period of the spectrum tends to infinity, resulting in the continuous periodic spectrum of the DTFT becoming a continuous aperiodic spectrum.

In Section 9.1, we derive the FT, starting from the definition of the DTFT. The properties of the FT are presented in Section 9.2. The FT of mixed class of signals is derived in Section 9.3. In Section 9.4, the approximation of the samples of the FT by those of the DFT is described. Some typical applications of the FT are presented in Section 9.5.

9.1 The Fourier Transform

9.1.1 The FT as a Limiting Case of the DTFT

The FT is the same as the DTFT with the sampling interval of the time-domain waveform tending to zero. Consider the samples of the continuous sinc function, $\sin(\frac{\pi}{3}t)/\pi t$, with sampling interval $T_s = 1$ s and its DTFT spectrum, multiplied by T_s, shown, respectively, in Figures 9.1(a) and (b). The DTFT spectrum is periodic with period $2\pi/T_s = 2\pi/1$ radians.

Reducing the sampling interval by a factor of two results in the doubling of the period of the spectrum, as shown in Figures 9.1(c) and (d). As the number of samples is increased, the amplitude of the spectrum will also increase. But the product of the amplitude and the sampling interval approaches a finite limiting function. As the

A Practical Approach to Signals and Systems   D. Sundararajan
© 2008 John Wiley & Sons (Asia) Pte Ltd

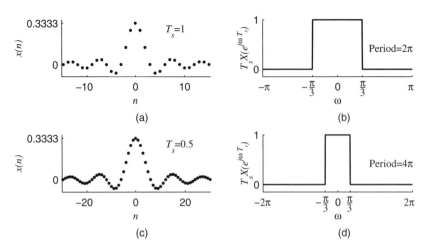

Figure 9.1 (a) Samples of the sinc function $\sin\left(\frac{\pi}{3}t\right)/\pi t$, with $T_s = 1$ s; (b) its DTFT spectrum, multiplied by T_s, with period 2π radians; (c) sinc function with $T_s = 0.5$ s; (d) its DTFT spectrum, multiplied by T_s, with period 4π radians

sampling interval tends to zero, the time-domain waveform becomes continuous with a corresponding aperiodic spectrum.

The foregoing argument can be, mathematically, put as follows. Substituting for $X(e^{j\omega T_s})$ and $1/\omega_s$ replaced by $T_s/2\pi$ in Equation (8.4), we get

$$x(nT_s) = \frac{T_s}{2\pi} \int_{-\frac{\omega_s}{2}}^{\frac{\omega_s}{2}} e^{j\omega n T_s} \left(\sum_{l=-\infty}^{\infty} x(lT_s)e^{-j\omega l T_s} \right) d\omega$$

As T_s tends to 0, ω_s tends to ∞, nT_s and lT_s become, respectively, continuous time variables t and τ, differential $d\tau$ formally replaces T_s, and the summation becomes an integral. Therefore, we get

$$x(t) = \frac{1}{2\pi} \int_{-\infty}^{\infty} \left(\int_{-\infty}^{\infty} x(\tau)e^{-j\omega\tau} d\tau \right) e^{j\omega t} d\omega = \frac{1}{2\pi} \int_{-\infty}^{\infty} X(j\omega)e^{j\omega t} d\omega$$

The FT $X(j\omega)$ of $x(t)$ is defined as

$$X(j\omega) = \int_{-\infty}^{\infty} x(t)e^{-j\omega t} dt \tag{9.1}$$

The inverse FT $x(t)$ of $X(j\omega)$ is defined as

$$x(t) = \frac{1}{2\pi} \int_{-\infty}^{\infty} X(j\omega)e^{j\omega t} d\omega \tag{9.2}$$

The FT represents a continuous aperiodic signal $x(t)$ as integrals of a continuum of complex sinusoids (amplitude $(1/2\pi)X(j\omega)d\omega$) over an infinite frequency range. Although the amplitudes are infinitesimal, the spectrum $X(j\omega)$ (actually the spectral density) gives the relative variations of the amplitudes of the constituent complex sinusoids of a signal. When deriving closed-form expressions for $X(j\omega)$ or $x(t)$,

$$X(j0) = \int_{-\infty}^{\infty} x(t)dt \quad \text{and} \quad x(0) = \frac{1}{2\pi} \int_{-\infty}^{\infty} X(j\omega)d\omega$$

which can be easily evaluated, are useful to check their correctness. By replacing ω by $2\pi f$ and since $d\omega = 2\pi d f$, Equations (9.1) and (9.2) can be expressed in terms of the cyclic frequency f as

$$X(j2\pi f) = \int_{-\infty}^{\infty} x(t)e^{-j2\pi ft}dt \quad \text{and} \quad x(t) = \int_{-\infty}^{\infty} X(j2\pi f)e^{j2\pi ft}d f$$

The Gibbs phenomenon is common to all forms of Fourier analysis, whenever a continuous function, with one or more discontinuities, is reconstructed in either domain.

9.1.2 Existence of the FT

Any signal satisfying the Dirichlet conditions, which are a set of sufficient conditions, can be expressed in terms of a FT. The first of these conditions is that the signal $x(t)$ is absolutely integrable, that is $\int_{-\infty}^{\infty} |x(t)|dt < \infty$. From the definition of the FT, we get

$$|X(j\omega)| \le \int_{-\infty}^{\infty} |x(t)e^{-j\omega t}| \, dt = \int_{-\infty}^{\infty} |x(t)||e^{-j\omega t}| \, dt$$

Since $|e^{-j\omega t}| = 1$,

$$|X(j\omega)| \le \int_{-\infty}^{\infty} |x(t)| \, dt$$

Hence, the condition $\int_{-\infty}^{\infty} |x(t)|dt < \infty$ implies that $X(j\omega)$ will exist. The second condition is that the number of finite maxima and minima of $x(t)$ in any finite interval must be finite. The third condition is that the number of finite discontinuities of $x(t)$ in any finite interval must be finite. Most signals of practical interest satisfy these conditions.

As Fourier analysis approximates a signal in the least-squares error sense,

$$\int_{-\infty}^{\infty} |x(t)|^2 dt - \frac{1}{2\pi} \int_{-\infty}^{\infty} |X(j\omega)|^2 d\omega = 0$$

the FT $X(j\omega)$ of a square integrable signal, $\int_{-\infty}^{\infty} |x(t)|^2 dt < \infty$, also exists.

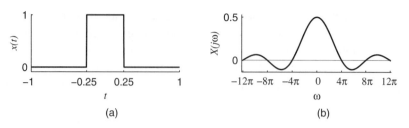

Figure 9.2   (a) The pulse $x(t) = u(t + 0.25) - u(t - 0.25)$; (b) its FT spectrum

Example 9.1.   Find the FT of the rectangular pulse $x(t) = u(t + a) - u(t - a)$.

Solution

As $x(t)$ is even,

$$X(j\omega) = \int_{-a}^{a} e^{-j\omega t}\,\mathrm{d}t = 2\int_{0}^{a} \cos(\omega t)\mathrm{d}t = \frac{2\sin(\omega a)}{\omega}$$

$$u(t + a) - u(t - a) \Longleftrightarrow \frac{2\sin(\omega a)}{\omega}$$

The pulse and its FT are shown, respectively, in Figures 9.2(a) and (b) with $a = 0.25$.   ☐

The function of the form $\sin(\omega a)/\omega$, a specific case shown in Figure 9.2(b), is called the sinc function that occurs often in signal and system analysis. It is an even function of ω. At $\omega = 0$, the peak value is a, as $\lim_{\theta\to 0}\sin(\theta) = \theta$. The zeros of the sinc function occur whenever the numerator argument (ωa) of the sine function is equal to $\pm\pi, \pm 2\pi, \dots$. That is, at $\omega = \pm\pi/a, \pm 2\pi/a, \dots$. For the specific case, the zeros occur whenever ω equals a multiple of 4π. The area enclosed by the sinc function is π irrespective of the value of a, as, by finding the inverse FT of $X(j\omega)$ in Example 9.1 with $t = 0$,

$$x(0) = \frac{1}{2\pi}\int_{-\infty}^{\infty} \frac{2\sin(\omega a)}{\omega}\,\mathrm{d}\omega = 1$$

It is also known that the area enclosed by the function is equal to the area of the triangle inscribed within its main hump. The sinc function is not absolutely integrable. But, it is square integrable and, hence, is an energy signal.

As $a \to 0$, the function $\sin(\omega a)/a\omega$ is expanded and, eventually, degenerates into a dc function. The first pair of zeros at $\omega = \pm\pi/a$ move to infinity and the function becomes a horizontal line with amplitude one. As a becomes larger, the numerator sine function $\sin(\omega a)$ of $\sin(\omega a)/\pi\omega$ alone is compressed (frequency of oscillations is increased). As a consequence, the amplitudes of all the ripples along with that of the main hump increase with fixed ratios to one another. While the ripples and the main hump become taller and narrower, the area enclosed by each and the total area enclosed by the function remains fixed. In the limit, as $a \rightarrow \infty$, the main hump and all the ripples of significant amplitude are concentrated at $\omega = 0$ and $\sin(\omega a)/\pi\omega$ degenerates into a unit-impulse.

Example 9.2. Find the FT $X(j\omega)$ of the real, causal, and decaying exponential signal $x(t) = e^{-at}u(t)$, $a > 0$. Find the value of $x(0)$ from $X(j\omega)$.

Solution

$$X(j\omega) = \int_0^\infty e^{-at}e^{-j\omega t}dt = \int_0^\infty e^{-(a+j\omega)t}dt = -\frac{e^{-(a+j\omega)t}}{a+j\omega}\bigg|_0^\infty = \frac{1}{a+j\omega}$$

$$e^{-at}u(t), \ a > 0 \Longleftrightarrow \frac{1}{a+j\omega}$$

$$x(0) = \frac{1}{2\pi}\int_{-\infty}^\infty \frac{1}{a+j\omega}d\omega = \frac{1}{2\pi}\int_{-\infty}^\infty \frac{a}{\omega^2+a^2}d\omega - \frac{j}{2\pi}\int_{-\infty}^\infty \frac{\omega}{\omega^2+a^2}d\omega$$

As the imaginary part of $X(j\omega)$ is odd, its integral evaluates to zero. Therefore,

$$x(0) = \frac{1}{2\pi}\int_{-\infty}^\infty \frac{a}{\omega^2+a^2}d\omega = \frac{1}{2\pi}\int_{-\infty}^\infty \frac{d\left(\frac{\omega}{a}\right)}{\left(\frac{\omega}{a}\right)^2+1} = \frac{1}{2\pi}\tan^{-1}\left(\frac{\omega}{a}\right)\bigg|_{-\infty}^\infty = \frac{1}{2}$$

The value of $x(t)$ at $t = 0$ is always $1/2$ for any value of a. Note that the Fourier reconstructed waveform converges to the average of the right- and left-hand limits at any discontinuity.   □

For some signals, such as a step signal or a sinusoid, which are neither absolutely nor square integrable, the FT is obtained by applying a limiting process to appropriate signals so that they degenerate into these signals in the limit. The limit of the corresponding transform is the transform of the signal under consideration, as presented in the next example.

Example 9.3. Find the FT of $x(t) = u(t)$, the unit-step function.

Solution
As $u(t)$ is not absolutely or square integrable, we consider it as the limiting form of the decaying exponential, $e^{-at}u(t)$, $a > 0$, as $a \rightarrow 0$. Therefore, as the FT of the

exponential is $1/(j\omega + a)$, the FT of $u(t)$ is given by

$$X(j\omega) = \lim_{a \to 0} \frac{1}{j\omega + a} = \lim_{a \to 0} \frac{a}{\omega^2 + a^2} - \lim_{a \to 0} \frac{j\omega}{\omega^2 + a^2} = \lim_{a \to 0} \frac{a}{\omega^2 + a^2} + \frac{1}{j\omega}$$

The area under the real part of $X(j\omega)$ is π regardless of the value of a, as found in Example 9.2. As $a \to 0$, the value of this function tends to zero at all values of ω except when $\omega = 0$, where its area is π. Therefore

$$\lim_{a \to 0} \frac{a}{\omega^2 + a^2} = \pi\delta(\omega)$$

and

$$u(t) \Longleftrightarrow \pi\delta(\omega) + \frac{1}{j\omega}$$

That is, the spectrum of the unit-step function is composed of an impulsive component $\pi\delta(\omega)$ (an impulse of strength π at $\omega = 0$) and a strictly continuous (except at $\omega = 0$) component $1/j\omega$. The real part of the transform $\pi\delta(\omega)$ corresponds to the even component $u_e(t) = 0.5$ of $u(t)$ and the imaginary part $1/j\omega$ corresponds to the odd component $u_o(t) = -0.5$, $t < 0$ and $u_o(t) = 0.5$, $t > 0$.

Figure 9.3 depicts the limiting process by which a real exponential function degenerates into a unit-step function. Figures 9.3(a) and (c) show, respectively, the signal $e^{-at}u(t)$ with $a = 1$ and $a = 0.1$. Figures 9.3(b) and (d) show, respectively, their corresponding spectra. The real part of the spectrum (continuous line) is an even function

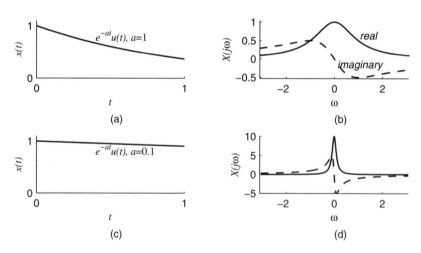

Figure 9.3 (a) $x(t) = e^{-at}u(t)$, $a = 1$; (b) its FT spectrum (the real part of the FT is shown by the continuous line and the imaginary part is shown by the dashed line); (c) $x(t) = e^{-at}u(t)$, $a = 0.1$; (d) its FT spectrum

with a peak value of $1/a$ at $\omega = 0$ and the imaginary part (dashed line) is an odd function with peaks of value $\pm 1/2a$ at $\omega = \mp a$. As $a \to 0$, the real part becomes more peaked and, eventually, degenerates into an impulse of strength π, that is $\pi\delta(\omega)$. The imaginary part becomes a rectangular hyperbola in the limit. $\square$

Example 9.4. Find the FT of the unit impulse signal $x(t) = \delta(t)$.

Solution

Using the sampling property of the impulse, we get

$$X(j\omega) = \int_{-\infty}^{\infty} \delta(t)e^{-j\omega t}\,\mathrm{d}t = e^{-j\omega 0} \int_{-\infty}^{\infty} \delta(t)\mathrm{d}t = 1 \qquad \text{and} \qquad \delta(t) \Longleftrightarrow 1$$

The unit impulse signal is composed of complex sinusoids, with zero phase shift, of all frequencies from $\omega = -\infty$ to $\omega = \infty$ in equal proportion. That is,

$$\delta(t) = \frac{1}{2\pi} \int_{-\infty}^{\infty} e^{j\omega t}\,\mathrm{d}\omega = \frac{1}{2\pi} \int_{-\infty}^{\infty} \cos(\omega t)\mathrm{d}\omega = \frac{1}{\pi} \int_{0}^{\infty} \cos(\omega t)\mathrm{d}\omega \qquad \square$$

Example 9.5. Find the inverse FT of $X(j\omega) = \delta(\omega)$.

Solution

$$x(t) = \frac{1}{2\pi} \int_{-\infty}^{\infty} \delta(\omega)e^{j\omega t}\,\mathrm{d}\omega = \frac{1}{2\pi} \qquad \text{and} \qquad 1 \Longleftrightarrow 2\pi\delta(\omega)$$

An impulse at $\omega = 0$ properly represents the dc signal, since it is characterized by the single frequency $\omega = 0$ alone. That is $x(t) = e^{j\omega_0 t}$ with $\omega_0 = 0$. Similar to the DFT, the scale factor $1/2\pi$ is included in the inverse transform. Therefore, the spectrum of dc is an impulse at $\omega = 0$ with strength 2π rather than one. The placement of the constant in the forward or inverse definition of a transform is a matter of convention. $\square$

Example 9.6. Find the inverse FT of $X(j\omega) = \delta(\omega - \omega_0)$.

Solution

$$x(t) = \frac{1}{2\pi} \int_{-\infty}^{\infty} \delta(\omega - \omega_0)e^{j\omega t}\,\mathrm{d}\omega = \frac{1}{2\pi}e^{j\omega_0 t} \qquad \text{and} \qquad e^{j\omega_0 t} \Longleftrightarrow 2\pi\delta(\omega - \omega_0)$$

That is, the spectrum of the complex sinusoid $e^{j\omega_0 t}$ is an impulse at $\omega = \omega_0$ with strength 2π. $\square$

9.2 Properties of the Fourier Transform

Properties present the frequency-domain effect of time-domain characteristics and operations on signals and vice versa. In addition, they are used to find new transform pairs more easily.

9.2.1 Linearity

The FT of a linear combination of a set of signals is the same linear combination of their individual FT. That is,

$$x(t) \Longleftrightarrow X(j\omega) \qquad y(t) \Longleftrightarrow Y(j\omega) \qquad ax(t) + by(t) \Longleftrightarrow aX(j\omega) + bY(j\omega),$$

where a and b are arbitrary constants. This property follows from the linearity property of the integral defining the FT. Consider the signal defined as

$$x(t) = \begin{cases} 1 & \text{for } t > 0 \\ -1 & \text{for } t < 0 \end{cases}$$

This signal can be expressed as $(2u(t) - 1)$. Substituting the respective FT, we get the FT of $x(t)$ as $2(\pi\delta(\omega) + (1/j\omega)) - 2\pi\delta(\omega) = 2/j\omega$.

9.2.2 Duality

The forward and inverse FT definitions differ only by the reversed algebraic sign in the exponent of the complex exponential, the interchange of the variables t and ω, and the constant $1/2\pi$ in the inverse FT. Due to this similarity, there exists a dual relationship between time- and frequency-domain functions. Consider the inverse FT defined as

$$x(t) = \frac{1}{2\pi} \int_{-\infty}^{\infty} X(j\omega)e^{j\omega t} d\omega$$

By replacing t by $-t$, we get

$$x(-t) = \frac{1}{2\pi} \int_{-\infty}^{\infty} X(j\omega)e^{-j\omega t} d\omega \quad \text{and} \quad 2\pi x(-t) = \int_{-\infty}^{\infty} X(j\omega)e^{-j\omega t} d\omega$$

This is a forward transform with $2\pi x(-t)$ being the FT of $X(j\omega)$. To put it another way, we get $2\pi x(-t)$ by taking the FT of $x(t)$ twice in succession, $2\pi x(-t) = \text{FT}(\text{FT}(x(t)))$. Let $x(t) \Longleftrightarrow X(j\omega)$. If we replace the variable ω in the frequency-domain function by $\pm t$, then the corresponding frequency-domain function is obtained by replacing the variable t by $\mp\omega$ in the original time-domain function multiplied by 2π. For an even $x(t)$, as $X(j\omega)$ is also even, the sign change of either t or ω is not required. For example, consider the FT pairs

$$\cos(2t) \Longleftrightarrow \pi(\delta(\omega + 2) + \delta(\omega - 2)) \quad \text{and} \quad \sin(3t) \Longleftrightarrow j\pi(\delta(\omega + 3) - \delta(\omega - 3))$$

Using the property, we get the transform pairs

$$2\cos(2(-\omega)) \quad = \quad 2\cos(2\omega) \Longleftrightarrow (\delta(t + 2) + \delta(t - 2))$$

$$2\sin(3\omega) \Longleftrightarrow j(\delta(-t + 3) - \delta(-t - 3)) = j(\delta(t - 3) - \delta(t + 3))$$

9.2.3 Symmetry

If a signal $x(t)$ is real, then the real part of its spectrum $X(j\omega)$ is even and the imaginary part is odd, called the conjuagte symmetry. The FT of $x(t)$ is given by

$$X(j\omega) = \int_{-\infty}^{\infty} x(t)e^{-j\omega t}dt = \int_{-\infty}^{\infty} x(t)(\cos(\omega t) - j\sin(\omega t))dt$$

Conjugating both sides, we get

$$X^*(j\omega) = \int_{-\infty}^{\infty} x(t)(\cos(\omega t) + j\sin(\omega t))dt$$

Replacing ω by $-\omega$, we get $X^*(-j\omega) = X(j\omega)$. An example is

$$x(t) = e^{-t}u(t) \Longleftrightarrow X(j\omega) = \frac{1}{j\omega + 1} = \frac{1}{\omega^2 + 1} - \frac{j\omega}{\omega^2 + 1}$$

If a signal $x(t)$ is real and even, then its spectrum also is real and even. Since $x(t)\cos(\omega t)$ is even and $x(t)\sin(\omega t)$ is odd,

$$X(j\omega) = 2\int_{0}^{\infty} x(t)\cos(\omega t)dt \quad \text{and} \quad x(t) = \frac{1}{\pi}\int_{0}^{\infty} X(j\omega)\cos(\omega t)d\omega$$

The FT $\pi(\delta(\omega + 1) + \delta(\omega - 1))$ of $\cos(t)$ is an example of the FT of an even function. Similarly, if a signal $x(t)$ is real and odd, then its spectrum is imaginary and odd.

$$X(j\omega) = -j2\int_{0}^{\infty} x(t)\sin(\omega t)dt \quad \text{and} \quad x(t) = \frac{j}{\pi}\int_{0}^{\infty} X(j\omega)\sin(\omega t)d\omega$$

The FT $j\pi(\delta(\omega + 1) - \delta(\omega - 1))$ of $\sin(t)$ is an example of the FT of an odd function.
As the FT of a real and even signal is real and even and that of a real and odd is imaginary and odd, it follows that the real part of the FT, $\text{Re}(X(j\omega))$, of an arbitrary

real signal $x(t)$ is the transform of its even component $x_e(t)$ and $j\,\text{Im}(X(j\omega))$ is that of its odd component $x_o(t)$.

9.2.4 Time Shifting

When we shift a signal, the shape remains the same but the signal is relocated. The shift of a typical spectral component, $X(j\omega_a)e^{j\omega_a t}$, by t_0 to the right results in the exponential, $X(j\omega_a)e^{j\omega_a(t-t_0)} = e^{-j\omega_a t_0}X(j\omega_a)e^{j\omega_a t}$. That is, a delay of t_0 results in changing the phase of the exponential by $-\omega_a t_0$ radians without changing its amplitude. Therefore, if the FT of $x(t)$ is $X(j\omega)$, then

$$x(t \pm t_0) \Longleftrightarrow e^{\pm j\omega t_0}X(j\omega)$$

Consider the FT of $\cos(2t)$, $\pi(\delta(\omega + 2) + \delta(\omega - 2))$. Now, the FT of $\cos(2(t - \frac{\pi}{4})) = \cos(2t - \frac{\pi}{2}) = \sin(2t)$ is

$$\pi(e^{-j(-2)\frac{\pi}{4}}\delta(\omega + 2) + e^{-j2\frac{\pi}{4}}\delta(\omega - 2)) = j\pi(\delta(\omega + 2) - \delta(\omega - 2))$$

9.2.5 Frequency Shifting

The spectrum, $X(j\omega)$, of a signal, $x(t)$, can be shifted by multiplying the signal by a complex exponential, $e^{\pm j\omega_0 t}$. The new spectrum is $X(j(\omega \mp \omega_0))$, since a spectral component $X(j\omega_a)e^{j\omega_a t}$ of the signal multiplied by $e^{j\omega_0 t}$ becomes $X(j\omega_a)e^{j(\omega_a + \omega_0)t}$ and the spectral value $X(j\omega_a)$ occurs at $(\omega_a + \omega_0)$, after a delay of ω_0 radians. That is,

$$x(t)e^{\pm j\omega_0 t} \Longleftrightarrow X(j(\omega \mp \omega_0))$$

Duality applies for both transform pairs and properties. This property is the dual of the time-shifting property.

Consider the FT pair $e^{-2t}u(t) \Longleftrightarrow \frac{1}{2+j\omega}$. The FT of

$$e^{-2t}\cos(3t)u(t) = e^{-2t}\frac{(e^{j3t} + e^{-j3t})}{2}u(t)$$

is

$$\frac{1}{2}\left(\frac{1}{2 + j(\omega - 3)} + \frac{1}{2 + j(\omega + 3)}\right) = \frac{2 + j\omega}{(2 + j\omega)^2 + 9}$$

9.2.6 Convolution in the Time-domain

The convolution $x(t) * h(t)$ of signals $x(t)$ and $h(t)$ is defined, in Chapter 5, as

$$y(t) = x(t) * h(t) = \int_{-\infty}^{\infty} x(\tau)h(t - \tau)d\tau$$

The convolution of $h(t)$ with a complex exponential $e^{j\omega_0 t}$ is given as

$$\int_{-\infty}^{\infty} h(\tau)e^{j\omega_0(t-\tau)}d\tau = e^{j\omega_0 t} \int_{-\infty}^{\infty} h(\tau)e^{-j\omega_0 \tau}d\tau = H(j\omega_0)e^{j\omega_0 t}$$

As an arbitrary signal $x(t)$ is reconstructed by the inverse FT as $x(t) = \frac{1}{2\pi} \int_{-\infty}^{\infty} X(j\omega)e^{j\omega t}d\omega$, the convolution of $x(t)$ and $h(t)$ is given by $y(t) = \frac{1}{2\pi} \int_{-\infty}^{\infty} X(j\omega)H(j\omega)e^{j\omega t}d\omega$. The inverse FT of $X(j\omega)H(j\omega)$ is the convolution of $x(t)$ and $h(t)$. That is,

$$\int_{-\infty}^{\infty} x(\tau)h(t - \tau)d\tau = \frac{1}{2\pi} \int_{-\infty}^{\infty} X(j\omega)H(j\omega)e^{j\omega t}d\omega \iff X(j\omega)H(j\omega)$$

Therefore, convolution in the time-domain corresponds to multiplication in the frequency-domain. This property is one of the major reasons for the dominant role of the frequency-domain analysis in the study of signals and systems.

The convolution of a rectangular pulse, centered at the origin, of width a and height $1/a$ with itself yields a triangular waveform, centered at the origin, with width $2a$ and height $1/a$. Figures 9.4(a) and (b) show, respectively, these waveforms with $a = 2$. Since convolution in the time-domain corresponds to multiplication in the frequency-domain and the FT of the rectangular pulse is $2\sin(\frac{\omega a}{2})/a\omega$, we get the FT of the triangular waveform as

$$X(j\omega) = \frac{2\sin\left(\frac{\omega a}{2}\right)}{a\omega} \frac{2\sin\left(\frac{\omega a}{2}\right)}{a\omega} = \left(\frac{2\sin\left(\frac{\omega a}{2}\right)}{a\omega}\right)^2$$

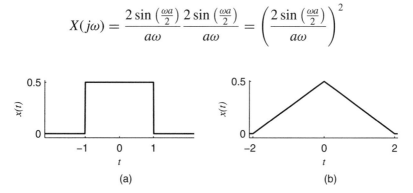

Figure 9.4 (a) The rectangular pulse with width two and height 0.5; (b) the triangular waveform with width four and height 0.5, which is the convolution of the pulse in (a) with itself

9.2.7 Convolution in the Frequency-domain

Similar to the time-domain convolution, we find that the convolution of two frequency-domain functions corresponds to the multiplication of the inverse FT of the functions in the time-domain with a scale factor. That is,

$$x(t)y(t) \Longleftrightarrow \int_{-\infty}^{\infty} x(t)y(t)e^{-j\omega t}\,dt = \frac{1}{2\pi}\int_{-\infty}^{\infty} X(jv)Y(j(\omega-v))dv$$

The FT of $\sin(t)\cos(t)$ is the convolution of the FT of $\sin(t)$ and $\cos(t)$ divided by 2π. That is,

$$\frac{1}{2\pi}(j\pi(\delta(\omega+1)-\delta(\omega-1)) * \pi(\delta(\omega+1)+\delta(\omega-1)))$$

$$= \frac{j\pi^2}{2\pi}((\delta(\omega+1)*\delta(\omega+1))-(\delta(\omega-1)*\delta(\omega-1)))$$

$$= \frac{j\pi}{2}(\delta(\omega+2)-\delta(\omega-2)) \Longleftrightarrow \frac{1}{2}\sin(2t) = \sin(t)\cos(t)$$

9.2.8 Conjugation

Let $x(t) \Longleftrightarrow X(j\omega)$. Then, $x^*(\pm t) \Longleftrightarrow X^*(\mp j\omega)$. This result is obtained if we replace t by $-t$ or ω by $-\omega$, in addition to conjugating both sides of the FT definition. For example,

$$e^{-(1+j2)t}u(t) \Longleftrightarrow \frac{1}{(1+j2)+j\omega} \quad \text{and} \quad e^{(1-j2)t}u(-t) \Longleftrightarrow \frac{1}{(1-j2)-j\omega}$$

9.2.9 Time Reversal

Let $x(t) \Longleftrightarrow X(j\omega)$. Then, $x(-t) \Longleftrightarrow X(-j\omega)$. That is the time-reversal of a signal results in its spectrum also reflected about the vertical axis at the origin. This result is obtained if we replace t by $-t$ and ω by $-\omega$ in the FT definition. For example,

$$e^{-3t}u(t) \Longleftrightarrow \frac{1}{3+j\omega} \quad \text{and} \quad e^{3t}u(-t) \Longleftrightarrow \frac{1}{3-j\omega}$$

9.2.10 Time Scaling

Scaling is the operation of replacing the independent variable t by at, where $a \neq 0$ is a real constant. As we have seen in Chapter 3, the signal is compressed ($|a| > 1$)

or expanded ($|a| < 1$) in the time-domain by this operation. As a consequence, the spectrum of the signal is expanded or compressed in the frequency-domain. With a negative, the signal is also time-reversed.

Let the spectrum of a signal $x(t)$ be $X(j\omega)$. By replacing at by τ, t by τ/a and dt by $d\tau/a$, with $a > 0$, in the FT definition of $x(at)$, we get

$$\int_{-\infty}^{\infty} x(at)e^{-j\omega t}\,dt = \frac{1}{a}\int_{-\infty}^{\infty} x(\tau)e^{-j\omega\frac{\tau}{a}}\,d\tau = \frac{1}{a}X\left(j\left(\frac{\omega}{a}\right)\right)$$

The FT of $x(-at)$, due to the time-reversal property, becomes

$$\frac{1}{a}X\left(j\left(\frac{-\omega}{a}\right)\right) = \frac{1}{a}X\left(j\left(\frac{\omega}{-a}\right)\right)$$

By combining both the results, we get

$$x(at) \Longleftrightarrow \frac{1}{|a|}X\left(j\left(\frac{\omega}{a}\right)\right), \qquad a \neq 0$$

The factor $1/|a|$ ensures that the scaled waveforms in both the domains have the same energy or power. A compressed signal varies more rapidly and, hence, requires higher-frequency components to synthesize. Therefore, the spectrum is expanded. The reverse is the case for signal expansion.

Consider the transform pair $\sin(2t) \Longleftrightarrow (j\pi)(\delta(\omega+2) - \delta(\omega-2))$. $\sin(6t)$ is a time-compressed version of $\sin(2t)$ with $a = 3$. Using the property, the transform of $\sin(6t)$ is obtained from that of $\sin(2t)$ as follows.

$$\frac{1}{3}(j\pi)\left(\delta\left(\frac{\omega}{3}+2\right) - \delta\left(\frac{\omega}{3}-2\right)\right) = \frac{1}{3}(j\pi)\left(\delta\left(\frac{\omega+6}{3}\right) - \delta\left(\frac{\omega-6}{3}\right)\right)$$
$$= (j\pi)(\delta(\omega+6) - \delta(\omega-6))$$

Note that $\delta(a\omega) = (1/|a|)\delta(\omega)$.

9.2.11 Time-differentiation

The derivative of a typical spectral component $X(j\omega_a)e^{j\omega_a t}$ is $j\omega_a X(j\omega_a)e^{j\omega_a t}$. Therefore, if the transform of a time-domain function $x(t)$ is $X(j\omega)$, then the transform of its derivative is given by $j\omega X(j\omega)$. That is,

$$\frac{dx(t)}{dt} \Longleftrightarrow j\omega X(j\omega)$$

Note that, the spectral value with $\omega = 0$ is zero, as the dc component is lost in differentiating a signal. The factor ω implies that the magnitude of the high-frequency components is enhanced more and, hence, rapid time variations of the signal are accentuated. The property is valid only if the derivative function is Fourier transformable. For example,

$$e^{-t}u(t) \Longleftrightarrow \frac{1}{j\omega + 1} \quad \text{and} \quad \frac{d(e^{-t}u(t))}{dt} = (\delta(t) - e^{-t}u(t)) \Longleftrightarrow \frac{j\omega}{j\omega + 1}$$

In general,

$$\frac{d^n x(t)}{dt^n} \Longleftrightarrow (j\omega)^n X(j\omega)$$

Consider finding the FT, shown in Figure 9.5(b), of the triangular waveform $x(t) = 0.5(t + 2)u(t + 2) - tu(t) + 0.5(t - 2)u(t - 2)$, shown in Figure 9.5(a). This problem is solved using the convolution property. Now, we use the differentiation property. The FT of scaled and shifted impulse function can be found easily. Therefore, the idea is to reduce the given function to a set of impulses by differentiating it successively. (This method is applicable to signals those are characterized or approximated by any piecewise polynomial function with finite energy.) Then, the FT of the impulses can be related to the FT of the given function by the differentiation property. The first and second derivatives of the triangular waveform, $dx(t)/dt = 0.5u(t + 2) - u(t) + 0.5u(t - 2)$ and $d^2 x(t)/dt^2 = 0.5\delta(t + 2) - \delta(t) + 0.5\delta(t - 2)$, are shown, respectively, in Figures 9.5(c) and (d). Let the FT of the triangular waveform be $X(j\omega)$. Then, the FT of the impulses of $d^2 x(t)/dt^2$ shown in Figure 9.5(d), $(0.5e^{j2\omega} - 1 + 0.5e^{-j2\omega})$, must

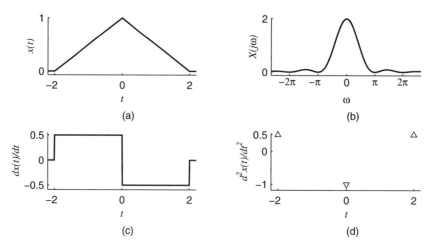

Figure 9.5 (a) The triangular waveform; (b) its spectrum; (c) the first derivative of the triangular waveform; (d) the second derivative of the triangular waveform

be equal to $(j\omega)^2 X(j\omega)$. That is,

$$(0.5e^{j2\omega} - 1 + 0.5e^{-j2\omega}) = \cos(2\omega) - 1 = -2\sin^2(\omega) = -\omega^2 X(j\omega)$$

Solving for $X(j\omega)$, we get the FT of the triangular waveform as

$$X(j\omega) = 2\left(\frac{\sin(\omega)}{\omega}\right)^2$$

9.2.12 Time-integration

The indefinite integral, $y(t)$, of a time-domain signal, $x(t)$, can be expressed as the convolution of $x(t)$ and the unit-step signal, $u(t)$, as

$$y(t) = \int_{-\infty}^{t} x(\tau)d\tau = \int_{-\infty}^{\infty} x(\tau)u(t-\tau)d\tau = x(t) * u(t)$$

As convolution in the time-domain corresponds to multiplication in the frequency-domain, with $x(t) \Longleftrightarrow X(j\omega)$ and $u(t) \Longleftrightarrow \frac{1}{j\omega} + \pi\delta(\omega)$, we get

$$\int_{-\infty}^{t} x(\tau)d\tau \Longleftrightarrow X(j\omega)\left(\frac{1}{j\omega} + \pi\delta(\omega)\right) = \frac{X(j\omega)}{j\omega} + \pi X(j0)\delta(\omega)$$

Note that, if $X(j0) = 0$, the integration operation can be considered as the inverse of the differentiation operation. The property is valid only if $y(t)$ is Fourier transformable. The factor ω in the denominator implies that the magnitude of the high-frequency components is reduced more and, hence, rapid time variations of the signal are reduced, resulting in a smoother signal.

Consider the signal $x(t) = u(t) - u(t-2)$, shown in Figure 9.6(a), with the FT $X(j\omega) = \frac{1}{j\omega}(1 - e^{-j2\omega})$ and $X(j0) = 2$. Now, using the property,

$$y(t) = \int_{-\infty}^{t} x(\tau)d\tau \Longleftrightarrow Y(j\omega) = \frac{X(j\omega)}{j\omega} + 2\pi\delta(\omega) = 2\pi\delta(\omega) + \frac{(e^{-j2\omega} - 1)}{\omega^2}$$

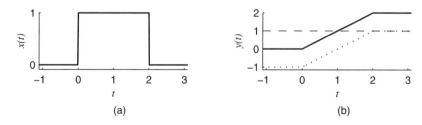

Figure 9.6   (a) Signal $x(t)$; (b) the integral of $x(t)$, $y(t)$ (solid line), and its two components

The integral of $x(t)$ is $y(t) = tu(t) - (t - 2)u(t - 2)$, shown in Figure 9.6(b) along with its two components corresponding to the two terms of its transform.

9.2.13 Frequency-differentiation

Differentiating both sides of the FT definition with respect to ω yields

$$(-jt)x(t) \Longleftrightarrow \frac{dX(j\omega)}{d\omega} \quad \text{or} \quad tx(t) \Longleftrightarrow j\frac{dX(j\omega)}{d\omega}$$

The property is valid only if the resulting function is Fourier transformable. In general,

$$(-jt)^n x(t) \Longleftrightarrow \frac{d^n X(j\omega)}{d\omega^n} \quad \text{or} \quad (t)^n x(t) \Longleftrightarrow (j)^n \frac{d^n X(j\omega)}{d\omega^n}$$

For example,

$$e^{-2t}u(t) \Longleftrightarrow \frac{1}{j\omega + 2} \quad \text{and} \quad te^{-2t}u(t) \Longleftrightarrow \frac{1}{(j\omega + 2)^2}$$

9.2.14 Parseval's Theorem and the Energy Transfer Function

As the frequency-domain representation of a signal is an equivalent representation, energy E of a signal can also be expressed in terms of its spectrum. Note that this theorem is applicable only to the FT of energy signals. From the frequency-domain convolution property, we get

$$\int_{-\infty}^{\infty} x(t)y(t)e^{-j\omega t}dt = \frac{1}{2\pi} \int_{-\infty}^{\infty} X(jv)Y(j(\omega - v))dv$$

Letting $\omega = 0$ and then replacing v by ω, we get

$$\int_{-\infty}^{\infty} x(t)y(t)dt = \frac{1}{2\pi} \int_{-\infty}^{\infty} X(j\omega)Y(-j\omega)d\omega$$

Assuming $x^*(t) = y(t)$, $X^*(-j\omega) = Y(j\omega)$ and $X^*(j\omega) = Y(-j\omega)$. Therefore, we get

$$\int_{-\infty}^{\infty} x(t)x^*(t)dt = \frac{1}{2\pi} \int_{-\infty}^{\infty} X(j\omega)X^*(j\omega)d\omega$$

$$E = \int_{-\infty}^{\infty} |x(t)|^2 dt = \frac{1}{2\pi} \int_{-\infty}^{\infty} |X(j\omega)|^2 d\omega$$

This relationship is called Parseval's theorem. This expression is the limiting form of the corresponding expression for DTFT as the sampling interval of the time-domain signal tends to zero. Alternately, this expression can also be considered as the limiting form of the corresponding expression for the FS as the period of the signal tends to infinity. For real signals, as $|X(j\omega)|$ is even, we get

$$E = \int_{-\infty}^{\infty} |x(t)|^2 dt = \frac{1}{\pi} \int_{0}^{\infty} |X(j\omega)|^2 d\omega$$

The quantity $|X(j\omega)|^2$ is called the energy spectral density of the signal, since $\frac{1}{2\pi}|X(j\omega)|^2 d\omega$ is the signal energy over the infinitesimal frequency band ω to $\omega + d\omega$.

Example 9.7. Find the energy of the signal $x(t) = e^{-t}u(t)$. Find the value of T such that 99% of the signal energy lies in the range $0 \le t \le T$. What is the corresponding signal bandwidth B, where B is such that 99% of the spectral energy lies in the range $0 \le \omega \le B$.

Solution
From the transform pair of Example 9.2, we get

$$e^{-t}u(t) \Longleftrightarrow \frac{1}{1 + j\omega}$$

The energy E of the signal is

$$E = \int_{-\infty}^{\infty} |x(t)|^2 dt = \int_{0}^{\infty} e^{-2t} dt = \frac{1}{2}$$

By changing the upper limit to T, we get

$$\int_{0}^{T} e^{-2t} dt = -\frac{1}{2}(e^{-2T} - 1) = \frac{0.99}{2} = 0.495$$

Solving for T, we get $T = 2.3026$ s. This value is required in order to truncate the signal for numerical analysis.
Using the spectrum,

$$\frac{1}{\pi} \int_{0}^{B} \frac{d\omega}{1 + \omega^2} = \frac{1}{\pi} \tan^{-1}(B) = 0.495 \quad \text{or} \quad B = \tan(0.495\pi) = 63.6567$$

Using this value, we can determine the sampling interval required to sample this signal. As the sampling frequency must be greater than twice of that of the highest frequency component, the sampling frequency must be greater than (2)(63.6567) radians per second. Therefore, the sampling interval must be smaller than $2\pi/(2)(63.6567) = 0.0494$ s. $\quad\square$

Since $|X(j\omega)|^2 = X(j\omega)X^*(j\omega) = X(j\omega)X(-j\omega)$ for real signals, $x(t) * x(-t) \Longleftrightarrow |X(j\omega)|^2$. The convolution $x(t) * x(-t)$, called the autocorrelation of $x(t)$, is defined as

$$x(t) * x(-t) = \int_{-\infty}^{\infty} x(\tau)x(\tau - t)d\tau \Longleftrightarrow |X(j\omega)|^2$$

The input and output of a LTI system, in the frequency-domain, are related by the transfer function $H(j\omega)$ as $Y(j\omega) = H(j\omega)X(j\omega)$, where $X(j\omega)$, $Y(j\omega)$, and $H(j\omega)$ are the FT of the input, output, and impulse response of the system. The output energy spectrum is given by

$$|Y(j\omega)|^2 = Y(j\omega)Y^*(j\omega)$$
$$= H(j\omega)X(j\omega)H^*(j\omega)X^*(j\omega) = |H(j\omega)|^2|X(j\omega)|^2$$

The quantity $|H(j\omega)|^2$ is called the energy transfer function, as it relates the input and output energy spectral densities of a system.

9.3 Fourier Transform of Mixed Classes of Signals

As the most general version of Fourier analysis, the FT is capable of representing all types of signals. Therefore, the relation between the FT and other versions of Fourier analysis is important in dealing with mixed class of signals. The signal $x(t)$ and its FT $X(j\omega)$ are, in general, continuous and aperiodic. The inverse FT of a sampled spectrum $X_s(j\omega)$ yields a periodic signal, which is the sum of a periodic repetition of $x(t)$. This version corresponds to the FS. On the other hand, the FT of a sampled signal $x_s(t)$ yields a periodic spectrum, which is the sum of a periodic repetition of $X(j\omega)$. This version corresponds to the DTFT. Sampling in both the domains corresponds to the DFT with both the signal and its spectrum sampled and periodic.

9.3.1 The FT of a Continuous Periodic Signal

A periodic signal $x(t)$ is reconstructed using its FS coefficients $X_{cs}(k)$ as

$$x(t) = \sum_{k=-\infty}^{\infty} X_{cs}(k)e^{jk\omega_0 t},$$

where ω_0 is the fundamental frequency. Since the FT of $e^{jk\omega_0 t}$ is $2\pi\delta(\omega - k\omega_0)$, we get, from the linearity property of the FT,

$$x(t) = \sum_{k=-\infty}^{\infty} X_{cs}(k)e^{jk\omega_0 t} \Longleftrightarrow X(j\omega) = 2\pi \sum_{k=-\infty}^{\infty} X_{cs}(k)\delta(\omega - k\omega_0)$$

Therefore, the FT of a periodic signal is a sum of impulses with strength $2\pi X_{cs}(k)$ occurring at intervals of ω_0.

Example 9.8. Find the FT of the signal $x(t) = \cos(\omega_0 t)$.

Solution
The FS spectrum for $\cos(\omega_0 t)$ is $\frac{1}{2}(\delta(k-1) + \delta(k+1))$. Multiplying this result by 2π and with the discrete impulse $\delta(k-1)$ corresponding to the continuous impulse $\delta(\omega - \omega_0)$, we get the FT as $\pi(\delta(\omega - \omega_0) + \delta(\omega + \omega_0))$. Hence, $\cos(\omega_0 t) \Longleftrightarrow \pi(\delta(\omega - \omega_0) + \delta(\omega + \omega_0))$. Similarly, $\sin(\omega_0 t) \Longleftrightarrow (j\pi)(\delta(\omega + \omega_0) - \delta(\omega - \omega_0))$. For example, the FS and FT spectra of $\cos(3t)$, shown in Figure 9.7(a), are shown in Figures 9.7(b) and (c), respectively. □

The spectra in Figures 9.7(b) and (c) are the equivalent representations of a single sinusoid by the FS and the FT. In Figure 9.7(b), the discrete spectrum $X_{cs}(k)$ consists of two nonzero discrete impulses of value 0.5. In Figure 9.7(c), the continuous spectrum $X(j\omega)$ consists of two continuous impulses with the value of their integrals being π, which, after dividing by the scale factor 2π, becomes 0.5. The amplitude of a constituent complex exponential of a signal $x(t)$ is $X_{cs}(k)$ in the case of the FS and $(1/2\pi)X(j\omega)d\omega$ in the case of the FT. Note that $(\delta(\omega - \omega_0)d\omega)|_{\omega=\omega_0} = 1$. Remember that both the spectra in Figures 9.7(b) and (c) represent the same waveform and, from either spectra, we get $0.5(e^{j3t} + e^{-j3t}) = \cos(3t)$.

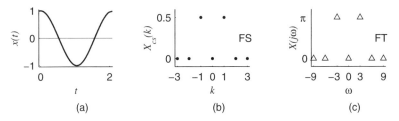

Figure 9.7   (a) The sinusoid $\cos(3t)$; (b) the FS spectrum, $X_{cs}(k)$, of the sinusoid; (c) the FT, $X(j\omega)$, of the sinusoid

9.3.2 Determination of the FS from the FT

Let $x(t)$ be a periodic signal of period T. Let us define an aperiodic signal $x_p(t)$ that is identical with $x(t)$ over one period from t_1 to $t_1 + T$ and is zero otherwise, where t_1 is arbitrary. The FT of this signal is

$$X_p(j\omega) = \int_{-\infty}^{\infty} x_p(t)e^{-j\omega t} dt = \int_{t_1}^{t_1+T} x(t)e^{-j\omega t} dt$$

The FS spectrum for $x(t)$ is

$$X_{cs}(k) = \frac{1}{T}\int_{t_1}^{t_1+T} x(t)e^{-jk\omega_0 t} dt \qquad \omega_0 = \frac{2\pi}{T}$$

Comparing the FS and FT definitions of the signals, we get

$$X_{cs}(k) = \frac{1}{T}X_p(j\omega)|_{\omega=k\omega_0} = \frac{1}{T}X_p(jk\omega_0)$$

The discrete samples of $\frac{1}{T}X_p(j\omega)$, at intervals of ω_0, constitute the FS spectrum for the periodic signal $x(t)$. While the spectral values at discrete frequencies are adequate to reconstruct one period of the periodic waveform using the inverse FS, spectral values at continuum of frequencies are required to reconstruct one period of the periodic waveform and the infinite extent zero values of the aperiodic waveform using the inverse FT. Note the similarity of this relationship to that between the DTFT and the DFT.

Example 9.9. Find the FS spectrum for the periodic signal $x(t)$, one period of which is defined as

$$x(t) = \begin{cases} 1 & \text{for } |t| < 1 \\ 0 & \text{for } 1 < |t| < 2 \end{cases}$$

Solution
Using the derivative method, the FT of $x_p(t)$ is obtained as follows:

$$j\omega X_p(j\omega) = e^{j\omega} - e^{-j\omega} \quad \text{and} \quad X_p(j\omega) = 2\frac{\sin(\omega)}{\omega}$$

Since $X_{cs}(k) = \frac{1}{T}X_p(jk\omega_0)$, with $T = 4$ and $\omega = k\omega_0 = k\frac{2\pi}{4} = \frac{\pi}{2}k$, we get

$$X_{cs}(k) = \frac{2}{4}\frac{\sin\left(\frac{\pi}{2}k\right)}{\frac{\pi}{2}k} = \frac{\sin\left(\frac{\pi}{2}k\right)}{\pi k} \qquad\qquad \square$$

9.3.3 The FT of a Sampled Signal and the Aliasing Effect

Let the FT of a signal $x(t)$ be $X(j\omega)$. The sampled version of this signal, $x_s(t)$, is obtained by multiplying it with an impulse train, $s(t) = \sum_{n=-\infty}^{\infty} \delta(t - nT_s)$, where T_s is the period and n is an integer. That is,

$$x_s(t) = x(t)s(t) = x(t) \sum_{n=-\infty}^{\infty} \delta(t - nT_s) = \sum_{n=-\infty}^{\infty} x(nT_s)\delta(t - nT_s)$$

The FS representation of the impulse train, from Chapter 7, is given as

$$s(t) = \frac{1}{T_s} \sum_{k=-\infty}^{\infty} e^{jk\omega_s t}$$

where $\omega_s = 2\pi/T_s$. Therefore, the sampled signal $x_s(t)$ is given by

$$x_s(t) = \frac{1}{T_s} \sum_{k=-\infty}^{\infty} x(t)e^{jk\omega_s t}$$

$$= \frac{1}{T_s}(\cdots + x(t)e^{-j\omega_s t} + x(t) + x(t)e^{j\omega_s t} + \cdots)$$

Let the FT of $x_s(t)$ be $X_s(j\omega)$. Then, using the linearity and frequency shift properties of the Fourier transform, we get

$$X_s(j\omega) = \frac{1}{T_s}(\cdots + X(j(\omega + \omega_s)) + X(j\omega) + X(j(\omega - \omega_s)) + \cdots)$$

$$= \frac{1}{T_s} \sum_{k=-\infty}^{\infty} X(j(\omega - k\omega_s))$$

This expression represents the convolution of the spectra of $x(t)$ and $s(t)$ (since it is the FT of their product) and we could just as well have obtained the result through the frequency-domain convolution property, as we shall see later. As the FT of the sampled signal is expressed as a sum of the shifted versions of that of the corresponding continuous signal, it is easy to visualize the form of $X_s(j\omega)$ if we know $X(j\omega)$. The sampling of a signal has made the resulting spectrum periodic with period ω_s, the sampling frequency, in addition to scaling the amplitude by the factor $1/T_s$, where T_s is the sampling interval. The periodicity is the result of the reduction of the range of frequencies, due to sampling, over which sinusoids can be distinguished. The factor

$1/T_s$ arises from the fact that

$$x(t) = \int_{-\infty}^{\infty} x(\tau)\delta(t-\tau)d\tau = \lim_{T_s \to 0} \sum_{n=-\infty}^{\infty} x(nT_s)T_s\delta(t-nT_s) = \lim_{T_s \to 0} T_s x_s(t)$$

Figures 9.8(a) and (b) show, respectively, the continuous sinc function and its aperiodic FT spectrum.

$$x(t) = \frac{\sin\left(\frac{2\pi}{3}t\right)}{2\pi t} \iff X(j\omega) = 0.5\left(u\left(\omega+\frac{2\pi}{3}\right)-u\left(\omega-\frac{2\pi}{3}\right)\right)$$

Figures 9.8(c) and (d) show the sampled sinc function, with $T_s = 0.5$, and its periodic FT spectrum with period $2\pi/0.5 = 4\pi$ radians and amplitude $0.5/0.5 = 1$.

$$x_s(t) = \sum_{n=-\infty}^{\infty} \frac{\sin\left(\frac{2\pi}{3}(0.5n)\right)}{2\pi(0.5n)}\delta(t-0.5n) \iff$$

$$X_s(j\omega) = \sum_{k=-\infty}^{\infty}\left(u\left(\omega+\frac{2\pi}{3}-4k\pi\right)-u\left(\omega-\frac{2\pi}{3}-4k\pi\right)\right)$$

At any discontinuity of the time-domain function, the strength of the sample should be equal to the average value of the right- and left-hand limits.

While the FT $X(j\omega)$ of $x(t)$ uniquely determines the FT $X_s(j\omega)$ of $x_s(t)$, the converse is not necessarily true. By sampling the signal, we simultaneously reduce the effective frequency range of the sinusoids available to represent the signal and, hence, the FT of the sampled signal becomes periodic due to aliasing effect. Therefore, if the signal is band-limited and the sampling frequency is greater than twice the highest frequency component of the signal, we can recover its exact FT from that of its sampled version by lowpass filtering (since the periodic repetition of $X(j\omega)$, yielding $X_s(j\omega)$, does not result in the overlapping of its nonzero portions). If the sampling frequency is not sufficiently high, we can only recover a corrupted version of its FT spectrum, as shown in Figures 9.8(i) and (j), since the periodic repetition of $X(j\omega)$ results in the overlapping of its nonzero portions.

$$x_s(t) = \sum_{n=-\infty}^{\infty} \frac{\sin\left(\frac{2\pi}{3}(2n)\right)}{2\pi(2n)}\delta(t-2n) \iff$$

$$X_s(j\omega) = \sum_{k=-\infty}^{\infty} 0.25\left(u\left(\omega+\frac{2\pi}{3}-k\pi\right)-u\left(\omega-\frac{2\pi}{3}-k\pi\right)\right)$$

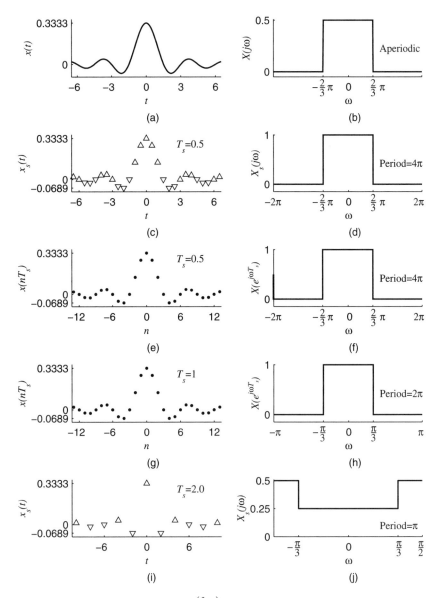

Figure 9.8 (a) The sinc function $\sin\left(\frac{2\pi}{3}t\right)/2\pi t$; (b) its FT spectrum; (c) samples of (a) with $T_s = 0.5$s; (d) its periodic FT spectrum; (e) discrete samples of (a) with $T_s = 0.5$s; (f) its DTFT spectrum with period 4π radians, which is the same as in (d); (g) the same samples as in (e) with $T_s = 1$; (h) its DTFT spectrum with period 2π radians; (i) samples of (a) with $T_s = 2$ s; (j) its periodic FT spectrum

As the FT of $\delta(t - nT_s)$ is $e^{-jn\omega T_s}$ and due to the linearity property of the FT, the FT of the sampled signal $x_s(t) = \sum_{n=-\infty}^{\infty} x(nT_s)\delta(t - nT_s)$ is also given by

$$X_s(j\omega) = \sum_{n=-\infty}^{\infty} x(nT_s)e^{-jn\omega T_s}$$

This expression, which, of course, is completely equivalent to that derived earlier for $X_s(j\omega)$, reminds us that the relation is a FS with the roles of the domains interchanged and corresponds to the DTFT. The time-domain samples $x(nT_s)$ are the FS coefficients of the corresponding continuous periodic spectrum $X_s(j\omega)$.

9.3.4 The FT of a Sampled Aperiodic Signal and the DTFT

Let us construct a sequence with the discrete sample values, at intervals of T_s, of the signal $x(t)$. These sample values are the same as the strengths of the corresponding impulses $x(nT_s)\delta(t - nT_s)$ of the sampled signal. The DTFT of $x(nT_s)$ is defined as

$$X(e^{j\omega T_s}) = \sum_{n=-\infty}^{\infty} x(nT_s)e^{-jn\omega T_s}$$

That is, the DTFT of a sequence $x(nT_s)$ and the FT of the corresponding sampled signal, $\sum_{n=-\infty}^{\infty} x(nT_s)\delta(t - nT_s)$, are the same when the DTFT version includes the sampling interval, T_s. Figures 9.8(e) and (f) show, respectively, the discrete samples of the sinc function

$$x(0.5n) = \frac{\sin\left(\frac{2\pi}{3}(0.5n)\right)}{2\pi(0.5n)}$$

with $T_s = 0.5$ s and its DTFT spectrum with period 4π radians, which is the same as in (d).

Usually, the DTFT spectrum is computed with the assumption of $T_s = 1$ s. Figures 9.8(g) and (h) show, respectively, the samples as in (e) with $T_s = 1$ s and its DTFT spectrum with period 2π radians. The FT of the corresponding sampled continuous signal $x_s(t)$ is obtained by scaling the frequency axis of this DTFT spectrum so that the period of the spectrum becomes $2\pi/T_s$, as can be seen from Figs 9.8(g), (h), (c), and (d).

9.3.5 The FT of a Sampled Periodic Signal and the DFT

The FT of a band-limited periodic signal $x(t)$, from earlier results, is

$$x(t) = \sum_{k=-N}^{N} X_{cs}(k)e^{jk\omega_0 t} \iff X(j\omega) = 2\pi \sum_{k=-N}^{N} X_{cs}(k)\delta(\omega - k\omega_0),$$

where $\omega_0 = 2\pi/T$, the fundamental frequency of $x(t)$. Let us sample the periodic signal by multiplying it with an impulse train

$$s(t) = \sum_{n=-\infty}^{\infty} \delta(t - nT_s) \iff S(j\omega) = \frac{2\pi}{T_s} \sum_{m=-\infty}^{\infty} \delta(\omega - m\omega_s)$$

with the interval between impulses being $T_s = 2\pi/\omega_s$. Then, as multiplication in the time-domain corresponds to convolution in the frequency domain, the FT $X_s(j\omega)$ of the sampled signal $x_s(t) = x(t)s(t)$ is $(1/2\pi)X(j\omega) * S(j\omega)$. The FT of the sampled signal, as convolution of a signal with an impulse is the relocation of the origin of the signal at the location of the impulse, is

$$X_s(j\omega) = \frac{2\pi}{T_s} \sum_{m=-\infty}^{\infty} \sum_{k=-N}^{N} X_{cs}(k)\delta(\omega - k\omega_0 - m\omega_s)$$

where $\omega_s = 2\pi/T_s$. As $X(k) = (2N + 1)X_{cs}(k)$, where $X(k)$ is the DFT of the $2N + 1$ discrete samples of $x(t)$ over one period, we get

$$X_s(j\omega) = \frac{2\pi}{(2N + 1)T_s} \sum_{m=-\infty}^{\infty} \sum_{k=-N}^{N} X(k)\delta(\omega - k\omega_0 - m\omega_s)$$

This FT corresponds to the sampled periodic time-domain signal

$$x_s(t) = \sum_{n=-\infty}^{\infty} x(nT_s)\delta(t - nT_s)$$

The period of the time-domain signal $x(n)$ of the DFT is $2N + 1$ samples and that of corresponding sampled continuous signal $x_s(t)$ is $(2N + 1)T_s = T$ s. The period of the FT spectrum is $\omega_s = (2\pi/T_s)$ radians and the spectral samples are placed at intervals of $\omega_0 = [2\pi/(2N + 1)T_s] = 2\pi/T$ radians.

Consider the discrete samples, over two periods, of the continuous cosine wave $\cos(2\frac{2\pi}{48}t)$ with sampling interval $T_s = 3$ s and its DFT spectrum shown, respectively, in Figures 9.9(a) and (b). Both the waveform and its spectrum are periodic with period $N = 16$ samples. The sampled version of the cosine wave is shown in Figure 9.9(c). The waveform is periodic with period $NT_s = T = 48$ s. The

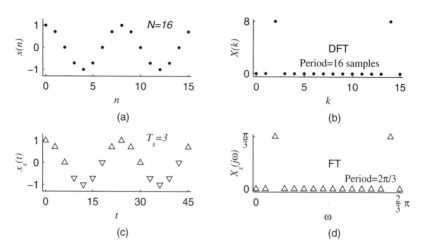

Figure 9.9 (a) The discrete samples, over two periods, of the continuous cosine wave $\cos(2\frac{2\pi}{48}t)$ with sampling interval $T_s = 3$ s; (b) its DFT spectrum; (c) the sampled version of the cosine wave $\cos(2\frac{2\pi}{48}t)$; (d) its periodic FT spectrum

FT spectrum of the waveform in Figure 9.9(c) is shown in Figure 9.9(d). The spectrum is periodic with period $2\pi/T_s = 2\pi/3$ radians. The spectral samples are placed at intervals of $2\pi/NT_s = 2\pi/48 = 0.1309$ radians.

$$x_s(t) = \sum_{n=-\infty}^{\infty} \cos\left(2\frac{2\pi}{48}n(3)\right)\delta(t-3n) \Longleftrightarrow$$

$$X_s(j\omega) = \frac{\pi}{3}\sum_{m=-\infty}^{\infty}\left(\delta\left(\omega - 2\frac{2\pi}{48} - \frac{2m\pi}{3}\right) + \delta\left(\omega + 2\frac{2\pi}{48} - \frac{2m\pi}{3}\right)\right)$$

The point is that we should mean the same waveform by looking at DFT and FT spectra. The term $2m\pi/3$ indicates that the spectrum is periodic with period $2\pi/3$ radians and, hence, the time-domain waveform is sampled with a sampling interval of 3 s. The two impulse terms, with strength $\pi/3$, indicate a cosine waveform with frequency $2(2\pi/48)$ radians and amplitude one. The DFT spectrum indicates a cosine waveform $\cos(2\frac{2\pi}{16}n)$. With a sampling interval of three seconds, this waveform corresponds to $\cos(2\frac{2\pi}{48}t)$.

Consider the differences between the cosine waveforms with amplitude one and their spectra in Figures 9.7 and 9.9. The waveform is continuous in Figure 9.7(a) and makes one cycle in the fundamental period. The FS and FT spectra in Figure 9.7(b) and (c) are aperiodic. The waveforms in Figures 9.9(a) and (c), with two cycles in the fundamental period, are sampled, and, hence, their spectra are periodic with the spectral values of the second harmonic in one period the same as those shown in Figures 9.7(b) and (c) except for scale factors. As the DFT coefficients $X(k)$ are equal to N times the FS coefficients $X_{cs}(k)$, the spectral values in Figure 9.9(b), with the number of samples $N = 16$, are $(0.5)16 = 8$. As the FT spectrum of a sampled signal $X_s(j\omega)$ is $(1/T_s)X(j\omega)$, the spectral values in Figure 9.9(d) are $\pi/3$.

9.3.6 Approximation of a Continuous Signal from its Sampled Version

The zero-order hold filter is commonly used to approximate a continuous signal $x(t)$ from its sampled version $x_s(t)$. The impulse response of this filter is a rectangular pulse of unit height and width T_s, $h(t) = u(t) - u(t - T_s)$, where T_s is the sampling interval of $x_s(t)$. By passing $x_s(t)$ through this filter, we get an output signal, which is the convolution of $x_s(t)$ and $h(t)$, that is a staircase approximation of $x(t)$. The convolution of $x_s(t)$, which is a sum of impulses, with $h(t)$ results in replacing each impulse of $x_s(t)$ by a pulse of width T_s and height equal to its strength (holding the current sample value until the next sample arrives).

9.4 Approximation of the Fourier Transform

In approximating the FS by the DFT, we determine the appropriate sampling interval and take samples over one period. In approximating the FT by the DFT, we have to fix the record length as well. These two parameters have to be fixed so that most of the energy of the signal is included in the selected record length and the continuous spectrum of the FT is represented by a sufficiently accurate and dense set of spectral samples.

The integral in Equation (9.1) is approximated by the rectangular rule of numerical integration. The summation interval can start from zero, since the truncated signal, of length T, is assumed periodic by the DFT, although the input signal can be nonzero in any interval. We divide the period T into N intervals of width $T_s = T/N$ and represent the signal at N points as $x(0), x(T/N), x(2T/N), \ldots, x((N-1)T/N)$. The sampling interval in the time-domain is T_s seconds and that in the frequency-domain is $2\pi/NT_s = 2\pi/T$ radians per second. Now, Equation (9.1) is approximated as

$$X\left(j\frac{2\pi k}{NT_s}\right) = T_s \sum_{n=0}^{N-1} x(nT_s)e^{-j\frac{2\pi}{N}nk} \qquad k = 0, 1, \ldots, N-1 \qquad (9.3)$$

Equation (9.2) is approximated as

$$x(nT_s) = \frac{1}{NT_s} \sum_{k=0}^{N-1} X\left(j\frac{2\pi k}{NT_s}\right) e^{j\frac{2\pi}{N}nk} \qquad n = 0, 1, \ldots, N-1 \qquad (9.4)$$

Except for the scale factors, Equations (9.3) and (9.4) are, respectively, the DFT and the IDFT of N samples. By multiplying the DFT coefficients by the sampling interval

T_s, we get the approximate samples of the FT. By multiplying the IDFT values by $1/T_s$, we get the approximate samples of the time-domain signal.

Example 9.10. Approximate the magnitude of the FT of the signal $x(t) = e^{-t}u(t)$ using the DFT.

Solution
From the transform pair of Example 9.7, we get

$$X(j\omega) = \frac{1}{1 + j\omega} \quad \text{and} \quad |X(j\omega)| = \frac{1}{\sqrt{1 + \omega^2}}$$

Figure 9.10(a) shows the exponential signal $e^{-t}u(t)$ with four samples over a period of $T = 8$ s. Figure 9.10(b) shows the magnitude of the FT and the samples of the FT obtained through the DFT with $N = 4$ and $N = 1024$ samples. While the signal is of infinite duration, we have truncated it to 8 s duration. The truncated signal has most of the energy of the untruncated signal. This signal has also an infinite bandwidth. Here again, the spectral values become negligible beyond some finite range. The sample values of the signal in Figure 9.10(a) are $\{0.5, 0.1353, 0.0183, 0.0025\}$. The first sample value occurs at a discontinuity and it is the average of the left- and right-hand limits at the discontinuity. The magnitude of the DFT of these values, after scaling by $T_s = 2$ s, is $\{1.3123, 0.9993, 0.7610, 0.9993\}$. As the second half of the DFT spectrum is redundant, only the first three values are useful. The corresponding samples of the FT are $\{1, 0.7864, 0.5370\}$.

The spectral samples obtained by the DFT are very inaccurate since the number of samples $N = 4$ is very small and the frequency range is inadequate thereby resulting in large amount of aliasing. The magnitude of the first five samples of the FT obtained through the DFT with $N = 1024$ are $\{0.9997, 0.7862, 0.5368, 0.3905, 0.3032\}$. The corresponding samples of the FT are $\{1, 0.7864, 0.5370, 0.3907, 0.3033\}$. This time, while the DFT values are still inaccurate, the error is much less compared with the last case. In general, the DFT values will never be exactly equal to the analytical values, but can be made sufficiently accurate by increasing the number of samples. ☐

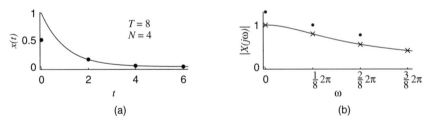

(a)   (b)

Figure 9.10   (a) The exponential waveform $x(t) = e^{-t}u(t)$, with four samples over the range $0 \le t < 8$; (b) the magnitude of the FT (solid line) and the samples of the FT obtained through the DFT with $N = 4$ (dots) and $N = 1024$ (crosses) samples

In order to approximate the FT of an arbitrary signal by the DFT, a trial and error procedure is used. A set of samples over a reasonable record length of the signal with an initial sampling interval is taken and the DFT is computed. Then, keeping the record length the same, we double the number of samples. That is, we reduce the sampling interval by one-half and the DFT is computed. This process is repeated until the spectral values near the middle of the spectrum for real signals (at the end of the spectrum for complex signals) become negligibly small, which ensures very little aliasing. Now, the sampling interval is fixed. Truncation of a signal is multiplying it with a rectangular window. As the window becomes longer, the truncation becomes less. In the frequency domain, the spectrum of the window becomes more closer to an impulse from that of a sinc function. The convolution of the spectra of the untruncated signal and the window distorts the spectrum of the signal to a lesser extent. Therefore, keeping the sampling interval the same, we keep doubling the record length and use the DFT to compute the spectral samples. When truncation becomes negligible, the spectral values with two successive lengths will be almost the same. Now, the record length is fixed.

9.5 Applications of the Fourier Transform

9.5.1 Transfer Function and System Response

The input–output relationship of a LTI system is given by the convolution operation in the time-domain. Since convolution corresponds to multiplication in the frequency-domain, we get

$$y(t) = \int_{-\infty}^{\infty} x(\tau)h(t - \tau)\mathrm{d}\tau \Longleftrightarrow Y(j\omega) = X(j\omega)H(j\omega),$$

where $x(t)$, $h(t)$, and $y(t)$ are, respectively the system input, impulse response, and output, and $X(j\omega)$, $H(j\omega)$, and $Y(j\omega)$ are their respective transforms. As input is transferred to output by multiplication with $H(j\omega)$, $H(j\omega)$ is called the transfer function of the system. The transfer function, which is the transform of the impulse response, characterizes a system in the frequency-domain just as the impulse response does in the time-domain.

Since the impulse function, whose FT is one (a uniform spectrum), is composed of complex exponentials $e^{j\omega t}$ of all frequencies with equal magnitude and zero-phase, the transform of the impulse response, the transfer function, is also called the frequency response of the system. Therefore, an exponential $Ae^{j(\omega_a t+\theta)}$ is changed to $(|H(j\omega_a)|A)e^{j(\omega_a t+(\theta+\angle(H(j\omega_a))))}$ at the output. A real sinusoidal input signal $A\cos(\omega_a t + \theta)$ is also changed at the output by the same amount of amplitude and phase of the complex scale factor $H(j\omega_a)$. That is, $A\cos(\omega_a t + \theta)$ is changed to $(|H(j\omega_a)|A)\cos(\omega_a t + (\theta + \angle(H(j\omega_a))))$. The steady-state response of a stable system to the input $Ae^{j(\omega_a t+\theta)}u(t)$ is also the same.

As $H(j\omega) = Y(j\omega)/X(j\omega)$, the transfer function can also be described as the ratio of the transform $Y(j\omega)$ of the response $y(t)$ to an arbitrary signal $x(t)$ to that of its transform $X(j\omega)$, provided $|X(j\omega)| \neq 0$ for all frequencies and the system is initially relaxed.

As the transform of the derivative of a signal is its transform multiplied by a factor, we can readily find the transfer function from the differential equation. Consider the second-order differential equation of a stable and initially relaxed LTI continuous system.

$$\frac{d^2y(t)}{dt^2} + a_1\frac{dy(t)}{dt} + a_0y(t) = b_2\frac{d^2x(t)}{dt^2} + b_1\frac{dx(t)}{dt} + b_0x(t)$$

Taking the FT of both sides, we get

$$(j\omega)^2Y(j\omega) + a_1(j\omega)Y(j\omega) + a_0Y(j\omega) =$$
$$(j\omega)^2b_2X(j\omega) + b_1(j\omega)X(j\omega) + b_0X(j\omega)$$

The transfer function $H(j\omega)$ is obtained as

$$H(j\omega) = \frac{Y(j\omega)}{X(j\omega)} = \frac{(j\omega)^2b_2 + (j\omega)b_1 + b_0}{(j\omega)^2 + a_1(j\omega) + a_0}$$

Example 9.11.   Find the response, using the FT, of the system governed by the differential equation

$$\frac{dy(t)}{dt} + y(t) = x(t)$$

to the input $x(t) = 2\cos(t + \frac{\pi}{4})$.

Solution

$$H(j\omega) = \frac{1}{1 + j\omega}$$

Substituting $\omega = 1$, we get

$$H(j1) = \frac{1}{1 + j1} = \frac{1}{\sqrt{2}}\angle\left(-\frac{\pi}{4}\right)$$

The response of the system to the input $x(t) = 2\cos\left(t + \frac{\pi}{4}\right)$ is $y(t) = \frac{2}{\sqrt{2}}\cos\left(t + \frac{\pi}{4} - \frac{\pi}{4}\right) = \sqrt{2}\cos(t)$.   □

Example 9.12. Find the impulse response, using the FT, of the system governed by the differential equation

$$\frac{dy(t)}{dt} + 3y(t) = x(t)$$

Solution

$$H(j\omega) = \frac{1}{3 + j\omega}$$

The impulse response of the system, which is the inverse FT of $H(j\omega)$, is $h(t) = e^{-3t}u(t)$. □

Example 9.13. Find the zero-state response of the system governed by the differential equation

$$\frac{d^2y(t)}{dt^2} + 4\frac{dy(t)}{dt} + 4y(t) = \frac{d^2x(t)}{dt^2} + \frac{dx(t)}{dt} + 2x(t)$$

with the input $x(t) = u(t)$, the unit-step function.

Solution

$$H(j\omega) = \frac{(j\omega)^2 + (j\omega) + 2}{(j\omega)^2 + 4(j\omega) + 4}$$

With $X(j\omega) = \pi\delta(\omega) + \frac{1}{j\omega}$,

$$Y(j\omega) = H(j\omega)X(j\omega) = \frac{(j\omega)^2 + (j\omega) + 2}{(j\omega)((j\omega)^2 + 4(j\omega) + 4)} + \frac{\pi\delta(\omega)((j\omega)^2 + (j\omega) + 2)}{(j\omega)^2 + 4(j\omega) + 4}$$

Expanding into partial fractions, we get

$$Y(j\omega) = \frac{0.5}{j\omega} + \frac{0.5}{j\omega + 2} - \frac{2}{(j\omega + 2)^2} + \frac{\pi}{2}\delta(\omega)$$

Taking the inverse FT, we get the zero-state response

$$y(t) = (0.5 + 0.5e^{-2t} - 2te^{-2t})u(t)$$

The steady-state response is $0.5u(t)$ and the transient response due to the input is $(0.5e^{-2t} - 2te^{-2t})u(t)$. □

Systems with nonzero initial conditions cannot be directly analyzed with FT. Further, handling of the frequency variable $j\omega$ is relatively more difficult. For these reasons, the Laplace transform is preferable for system analysis. However, the FT, wherever it is more suitable, is efficient, as it can be approximated by the DFT using fast algorithms.

9.5.2 Ideal Filters and their Unrealizability

Filters are prominent examples of LTI systems for signal analysis, manipulation, and processing. Common applications of filters include removing noise from signals and selection of individual channels in radio or television receivers. We present, in this subsection, the constraints involved in the realization of practical filters.

The frequency response of an ideal lowpass filter is shown in Figure 9.11. As it is even-symmetric, the specification of the response over the interval from $\omega = 0$ to $\omega = \infty$, shown in thick lines, characterizes a filter.

$$H(j\omega) = \begin{cases} 1 & \text{for } 0 \leq \omega < \omega_c \\ 0 & \text{for } \omega > \omega_c \end{cases}$$

From $\omega = 0$ to $\omega = \omega_c$, the filter passes frequency components of a signal with a gain of 1 and rejects the other frequency components, since the output of the filter, in the frequency-domain, is given by $Y(j\omega) = H(j\omega)X(j\omega)$. The magnitudes of the frequency components of the signal, $X(j\omega)$, with frequencies up to ω_c are multiplied by 1 and the rest by 0. The range of frequencies from 0 to ω_c is called the passband and the range from ω_c to ∞ is called the stopband. This ideal filter model is practically unrealizable since its impulse response (inverse of $H(j\omega)$) extends from $t = -\infty$ to $t = \infty$, which requires a noncausal system. Practical filters approximate this model.

The impulse response of practical systems must be causal. The even and odd components, for $t > 0$, of a causal time function $x(t)$ are given as

$$x_e(t) = \frac{x(t) + x(-t)}{2} = \frac{x(t)}{2} \quad \text{and} \quad x_o(t) = \frac{x(t) - x(-t)}{2} = \frac{x(t)}{2}$$

That is, $x(t) = 2x_e(t) = 2x_o(t)$, $t > 0$ and $x_e(t) = -x_o(t)$, $t < 0$. As the FT of an even signal is real and that of an odd signal is imaginary, $x(t)$ can be obtained by

Figure 9.11   Frequency response of an ideal lowpass filter

finding the inverse FT of either the real part or the imaginary part of its spectrum $X(j\omega)$. That is,

$$x(t) = \frac{2}{\pi} \int_0^\infty \text{Re}(X(j\omega))\cos(\omega t)d\omega = -\frac{2}{\pi} \int_0^\infty \text{Im}(X(j\omega))\sin(\omega t)d\omega \qquad t > 0$$

The point is that the real and imaginary parts or, equivalently, the magnitude and the phase of the FT of a causal signal are related. This implies that there are constraints, for the realizability, on the magnitude of the frequency response, $H(j\omega)$, of a practical filter. These constraints are given by the Paley–Wiener criterion as

$$\int_{-\infty}^{\infty} \frac{|\log_e |H(j\omega)||}{1 + \omega^2} d\omega < \infty$$

To satisfy this criterion, the magnitude of the frequency response $|H(j\omega)|$ can be zero at discrete points but not over any continuous band of frequencies. If $H(j\omega)$ is zero over a band of frequencies, $|\log_e |H(j\omega)|| = \infty$ and the condition is violated. On the other hand, if $H(j\omega)$ is zero at a finite set of discrete frequencies, the value of the integral may still be finite, although the integrand is infinite at these frequencies. In addition, any transition of this function cannot vary more rapidly than by exponential order. The $H(j\omega)$ of the ideal filter shown in Figure 9.11 does not meet the Paley–Wiener criterion. Further, the order of the filter must be infinite to have a constant gain all over the passband. Therefore, neither the flatness of the bands nor the sharpness of the transition between the bands of ideal filters is realizable by practical filters.

9.5.3 Modulation and Demodulation

Modulation and demodulation operations are fundamental to communication applications. These operations are required in signal communication because of different frequency ranges required for the signals to be communicated and for efficient transmission of signals. As the antenna size is inversely proportional to the frequency of the signal, the lower the frequency of the signal the larger is the required antenna size. For example, an antenna of size about 30 km is required to transmit the audio signal efficiently. Therefore, it is a necessity to embed the audio signal, called the message signal, in a much higher-frequency signal, called the carrier signal, which can be transmitted more efficiently. The operation of embedding the message signal in a carrier signal is called modulation. The embedding involves the variation of some property of the carrier signal in accordance with the message signal. At the receiving end, the message signal has to extracted from the modulated carrier signal. This operation is called demodulation. There are different types of these operations with distinct characteristics. We understand these operations using the property of the FT that the multiplication of two signals in the time-domain corresponds to convolution in the frequency-domain.

9.5.3.1 Double Sideband, Suppressed Carrier (DSB-SC), Amplitude Modulation

In this type of modulation, the amplitude A of the carrier signal, $A\cos(\omega_c(t) + \theta_c)$, is varied in some manner with the message signal, $m(t)$, where ω_c and θ_c are constants. Let the FT of $m(t)$ be $M(j\omega)$. Then, the FT of the product of the message and carrier signals, with $A = 1$ and $\theta_c = 0$, is given as

$$m(t)\cos(\omega_c(t)) \quad \Longleftrightarrow \quad \frac{1}{2}(M(j(\omega + \omega_c)) + M(j(\omega - \omega_c)))$$

After modulation, a copy of the spectrum of the message signal is placed at ω_c and another copy is placed at $-\omega_c$. Each copy of the spectrum of the message signal has the upper sideband (USB) portion (the right half of the spectrum centered at ω_c and the left half of the spectrum centered at $-\omega_c$) and the lower sideband (LSB) portion (the left half of the spectrum centered at ω_c and the right half of the spectrum centered at $-\omega_c$). As there are two sidebands and no carrier in the spectrum, this form of modulation is called double sideband, suppressed carrier, amplitude modulation. Note that, the message signal can be recovered from either sideband.

Let the message signal be $m(t) = \sin(2\pi t)$ and the carrier signal be $c(t) = \cos(20\pi t)$, as shown in Figure 9.12(a). For illustration, we are using a sine wave of 1 Hz as the message signal and cosine wave of 10 Hz as the carrier signal. However, it should be noted that, in practice, the message signal will have a finite bandwidth and the carrier frequency will be much higher. For example, the bandwidth of a message signal could be 3 kHz with a carrier frequency 3000 kHz. The product $m(t)c(t)$ is shown in

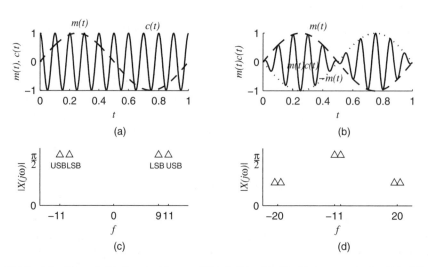

Figure 9.12   (a) $m(t) = \sin(2\pi t)$ and $c(t) = \cos(20\pi t)$; (b) $m(t)c(t)$; (c) the magnitude of the FT spectrum of $m(t)c(t)$; (d) the magnitude of the FT spectrum of $(m(t)c(t))c(t)$

Figure 9.12(b). The envelopes of $m(t)c(t)$ are $m(t)$ and $-m(t)$, since $m(t)\cos(20\pi t) = m(t)$ when $\cos(20\pi t) = 1$ and $m(t)\cos(20\pi t) = -m(t)$ when $\cos(20\pi t) = -1$. For this specific example, the FT $X(j\omega)$ of $m(t)c(t)$ is

$$\frac{j\pi}{2}(\underbrace{-\delta(f + (10 + 1)) + \delta(f + (-10 - 1))}_{\text{USB}} + \underbrace{\delta(f + (10 - 1)) - \delta(f + (-10 + 1))}_{\text{LSB}})$$

Therefore, the magnitude of the FT spectrum of $m(t)c(t)$, shown in Figure 9.12(c), has impulses with strength $\pi/2$ at -11 Hz, -9 Hz, 9Hz, and 11 Hz.

Demodulation. To demodulate the modulated signal, $m(t)\cos(\omega_c(t))$, at the receiver, we simply multiply the signal by the carrier, $\cos(\omega_c(t))$. This results in

$$m(t)\cos^2(\omega_c(t)) = \frac{1}{2}(m(t) + m(t)\cos(2\omega_c(t))$$

The FT of this signal is

$$\frac{1}{2}M(j\omega) + \frac{1}{4}(M(j(\omega + 2\omega_c)) + M(j(\omega - 2\omega_c)))$$

The spectrum of the message signal is centered at $\omega = 0$ and can be recovered by lowpass filtering. The other two spectra are the transform of $m(t)$ modulated by a carrier with frequency $2\omega_c$. For the specific example, the magnitude of the spectrum of $(m(t)c(t))c(t)$ is shown in Figure 9.12(d).

To use this type of demodulation, we have to generate the carrier signal with the same frequency and phase. This requires a complex receiver. While this form is used in certain applications, for commercial radio broadcasting, another type of modulation and demodulation, described next, is most commonly used.

9.5.3.2 Double Sideband, with Carrier (DSB-WC), Amplitude Modulation

In this type of modulation, the amplitude of the carrier signal, $\cos(\omega_c(t))$, is varied in some manner with the modulating signal, $(1 + km(t))$, where ω_c and k are constants. Let the FT of $m(t)$ be $M(j\omega)$. Then, the FT of the product of the message and carrier signal is given as

$$(1 + km(t))\cos(\omega_c t) \Longleftrightarrow$$
$$\frac{k}{2}(M(j(\omega + \omega_c)) + M(j(\omega - \omega_c))) + \pi(\delta(\omega + \omega_c) + \delta(\omega - \omega_c))$$

After modulation, a copy of the spectrum of the message signal is placed at ω_c and another copy is placed at $-\omega_c$. As there are two sidebands and the carrier in the spectrum of the transmitted signal, this form of modulation is called double sideband,

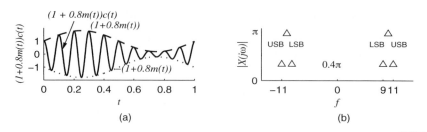

Figure 9.13 (a) $(1 + 0.8m(t))c(t)$; (b) the magnitude of its FT spectrum

with carrier, amplitude modulation. This form of modulation is intended for simple receivers with out the need for generating the carrier signal. For example, let the signals $m(t)$ and $c(t)$ be the same as shown in Figure 9.12(a). Figures 9.13(a) and (b) show, respectively, $(1 + 0.8m(t))c(t)$, and the magnitude of the FT spectrum of $(1 + 0.8m(t))c(t)$. The signal can be demodulated by a simple envelope detector circuit or a rectifier followed by a lowpass filter, if the message signal rides on the carrier signal. That is, $(1 + km(t)) \geq 0$ for all values of t.

The basis of modulation in the two cases studied is frequency shifting. One advantage of this type of modulation is the transmission of several signals over the same channel using the frequency-division multiplexing method. The signals share portions of the bandwidth of the channel with adequate separation between them.

9.5.3.3 Pulse Amplitude Modulation (PAM)

In the modulation types so far presented, the carrier is a sinusoid. The use of a pulse train as the carrier and modulating its amplitude in accordance with the message signal is called pulse amplitude modulation (PAM). The pulse train consists of constant width and amplitude pulses with uniform spacing between them. The message signal modulates the amplitude of the pulses. This is essentially the same as that of sampling of continuous signals using an impulse train, presented earlier. The difference is that the sampling pulse, unlike the impulse, has a finite width. The FS spectrum of this signal is $\sin(a k\omega_s)/k\pi$, where a is half the width of the pulse, $\omega_s = 2\pi/T_s$, and T_s is the sampling interval. Proceeding as in the case of the impulse sampling, we get the FT of the modulated signal as

$$X_s(j\omega) = \left(\cdots + \frac{\sin(a\omega_s)}{\pi} X(j(\omega + \omega_s)) + \frac{2a}{T_s} X(j\omega) + \cdots \right)$$

The spectrum, centered at $\omega = 0$, is unaltered except for a scale factor compared with that of the signal. Therefore, we can recover the original spectrum using a lowpass filter. Using this type of modulation, several message signals can be transmitted over the same channel by the method called time-division multiplexing. The time between two pulses of a modulated signal can be used by other modulated signals.

9.6 Summary

- In this chapter, we have studied the FT, its properties, its approximation by the DFT, and some of its applications.
- The FT is primarily intended to analyze continuous aperiodic signals in terms of a continuum of complex exponentials over an infinite frequency range.
- The FT is the most general type of Fourier analysis and, hence, it can be used to analyze a mixed class of signals.
- The FT can be approximated by the DFT to a desired accuracy with proper choice of the record length and the number of samples.
- The FT has wide applications in signal and system analysis.

Further Reading

1. Guillemin, E. A., *Theory of Linear Physical Systems*, John Wiley, New York, 1963.
2. Lathi, B. P., *Linear Systems and Signals*, Oxford University Press, New York, 2004.
3. Sundararajan, D., *Discrete Fourier Transform, Theory, Algorithms, and Applications*, World Scientific, Singapore, 2001.

Exercises

9.1   Starting from the defining equations of the exponential form of the FS and the inverse FS, derive the defining equations of the FT and the inverse FT as the period of the time-domain waveform tends to infinity.

9.2   Derive an expression, using the defining integral, for the FT of the signal $x(t) = e^{-at}\cos(\omega_0 t)u(t)$, $a > 0$. With $a = 0.4$ and $\omega_0 = 3$, compute $X(j0)$.

9.3   Derive an expression, using the defining integral, for the FT of the signal $x(t) = e^{-at}\sin(\omega_0 t)u(t)$, $a > 0$. With $a = 0.1$ and $\omega_0 = 2$, compute $X(j0)$.

9.4   Derive an expression, using the defining integral, for the FT of the signal

$$x(t) = \begin{cases} 1 - |t| & \text{for } |t| < 1 \\ 0 & \text{elsewhere} \end{cases}$$

Compute $X(j0)$ and $X(j(2\pi))$.

9.5   Derive an expression, using the defining integral, for the FT of the signal $x(t) = e^{-3|t|}$.

Compute $X(j0)$.

9.6   Derive an expression, using the defining integral, for the FT of the signal

$$x(t) = \begin{cases} \cos(10t) & \text{for } |t| < 1 \\ 0 & \text{for } |t| > 1 \end{cases}$$

Compute $X(j0)$.

*9.7 Derive an expression, using the defining integral, for the FT of the signal $x(t) = te^{-2t}u(t)$.

Compute $X(j0)$.

9.8 Apply a limiting process so that $x(t)$ degenerates into the impulse function and, hence, derive the FT of the impulse function $\delta(t)$.

9.8.1 $x(t) = ae^{-at}u(t)$, $a > 0$, as $a \to \infty$.

9.8.2 $x(t) = \begin{cases} \frac{1}{2a} & \text{for } |t| < a \\ 0 & \text{for } |t| > a \end{cases}$, $a > 0$, as $a \to 0$.

9.8.3 $x(t) = \begin{cases} \frac{1}{a} & \text{for } 0 < t < a \\ 0 & \text{elsewhere} \end{cases}$, $a > 0$, as $a \to 0$.

9.8.4 $x(t) = \begin{cases} \frac{1}{a} & \text{for } -a < t < 0 \\ 0 & \text{elsewhere} \end{cases}$, $a > 0$, as $a \to 0$.

9.8.5 $x(t) = \begin{cases} \frac{1}{a^2}(a - |t|) & \text{for } |t| < a \\ 0 & \text{for } |t| > a \end{cases}$, $a > 0$, as $a \to 0$.

9.8.6 $x(t) = \begin{cases} \frac{1}{2a^2}(2a + t) & \text{for } -2a < t < 0 \\ 0 & \text{elsewhere} \end{cases}$, $a > 0$, as $a \to 0$.

9.9 Apply a limiting process so that $x(t)$ degenerates into the dc function and, hence, derive the FT of the dc function, $x(t) = 1$.

$x(t) = \begin{cases} 1 & \text{for } |t| < a \\ 0 & \text{for } |t| > a \end{cases}$, $a > 0$, as $a \to \infty$.

9.10 Derive the FT of the function $y(t) = \begin{cases} -1 & \text{for } t < 0 \\ 1 & \text{for } t > 0 \end{cases}$ by applying a limiting process to the signal $x(t) = e^{-at}u(t) - e^{at}u(-t)$, as $a \to 0$.

9.11 Apply a limiting process so that $x(t)$ degenerates into the cosine function and, hence, derive the FT of the cosine function, $\cos(t)$.

$x(t) = \begin{cases} \cos(t) & \text{for } |t| < a \\ 0 & \text{for } |t| > a \end{cases}$, $a > 0$, as $a \to \infty$

9.12 Apply a limiting process so that $e^{-a|t|}\sin(t)$, $a > 0$ degenerates into $\sin(t)$, as $a \to 0$, and, hence, derive the FT of $\sin(t)$.

9.13 Derive the FT of the unit-step function $u(t)$ using the FT of the functions

$$x(t) = \begin{cases} 1 & \text{for } t > 0 \\ -1 & \text{for } t < 0 \end{cases} \qquad \text{and} \qquad y(t) = 1$$

9.14 Using the duality property, find the FT of the signal $x(t)$.

9.14.1 $x(t) = \frac{1}{2+jt}$.

*9.14.2 $x(t) = 2\frac{\sin(3t)}{t}$.

9.14.3 $x(t) = \pi\delta(t) + \frac{1}{jt}$.

9.15 Using the linearity and frequency-shifting properties, find the FT of $x(t)$.

9.15.1 $x(t) = \cos(\omega_0 t)u(t)$.

9.15.2 $x(t) = \sin(\omega_0 t)u(t)$.

*9.15.3 $x(t) = \begin{cases} \cos(\omega_0 t) & \text{for } |t| < a \\ 0 & \text{for } |t| > a \end{cases}$, $a > 0$.

9.15.4 $x(t) = \begin{cases} \sin(\omega_0 t) & \text{for } |t| < a \\ 0 & \text{for } |t| > a \end{cases}$, $a > 0$.

*9.16 Derive the inverse FT of the function

$$X(j\omega) = \frac{1}{\omega^2}\left(e^{-j4\omega} - 1\right)$$

using the time-domain convolution property.

9.17 Using the time-domain convolution property, find the FT of the convolution of $x(t)$ and $h(t)$.

9.17.1 $x(t) = \begin{cases} -2 & \text{for } 0 < t < 4 \\ 0 & \text{for } t < 0 \text{ and } t > 4 \end{cases}$ and

$h(t) = \begin{cases} 3 & \text{for } 0 < t < 5 \\ 0 & \text{for } t < 0 \text{ and } t > 5 \end{cases}$

9.17.2 $x(t) = e^{-2t}u(t)$ and $h(t) = e^{-3t}u(t)$.

9.17.3 $x(t) = e^{-t}u(t)$ and $h(t) = \begin{cases} 1 & \text{for } 0 < t < 1 \\ 0 & \text{for } t < 0 \text{ and } t > 1 \end{cases}$

9.17.4 $x(t) = \begin{cases} (1 - |t|) & \text{for } |t| < 1 \\ 0 & \text{otherwise} \end{cases}$ and $h(t) = \begin{cases} 1 & \text{for } 0 < t < 1 \\ 0 & \text{for } t < 0 \text{ and } t > 1 \end{cases}$

9.17.5 $x(t) = e^{-at}u(t)$, $a > 0$ and $h(t) = x(t)$.

9.18 Using the frequency-domain convolution property, find the FT of the product of $x(t)$ and $h(t)$.

9.18.1 $x(t) = \cos(\omega_0 t)$ and $h(t) = u(t)$.

*9.18.2 $x(t) = \sin(\omega_0 t)$ and $h(t) = u(t)$.

9.18.3 $x(t) = \cos(\omega_0 t)$ and $h(t) = \begin{cases} 1 & \text{for } |t| < a \\ 0 & \text{for } |t| > a \end{cases}$, $a > 0$.

9.18.4 $x(t) = \sin(\omega_0 t)$ and $h(t) = \begin{cases} 1 & \text{for } |t| < a \\ 0 & \text{for } |t| > a \end{cases}$, $a > 0$.

9.19 Derive the FT of the function $x(t) = e^{-a|t|}$, $a > 0$ using the linearity and time-reversal properties.

9.20 Using the time-scaling property, find the FT of the signal $x(at)$.

9.20.1 $x(t) = \cos(t)$ and $a = -2$.

9.20.2 $x(t) = e^{-2t}u(t)$ and $a = 2$.

9.20.3 $x(t) = e^{-2t}u(t)$ and $a = \frac{1}{2}$.

9.20.4 $x(t) = e^{-2t}u(t)$ and $a = -\frac{1}{2}$.

9.20.5 $x(t) = \begin{cases} 1 & \text{for } |t| < 2 \\ 0 & \text{for } |t| > 2 \end{cases}$ and $a = 2$.

9.20.6 $x(t) = \begin{cases} 1 & \text{for } |t| < 2 \\ 0 & \text{for } |t| > 2 \end{cases}$ and $a = -2$.

9.20.7 $x(t) = \begin{cases} 1 & \text{for } |t| < 2 \\ 0 & \text{for } |t| > 2 \end{cases}$ and $a = \frac{1}{2}$.

9.20.8 $x(t) = u(t)$ and $a = 3$.

*9.20.9 $x(t) = u(t)$ and $a = -2$.

9.20.10 $x(t) = u(t - 4)$ and $a = 2$.

9.21   Using the time-differentiation property, find the FT of the derivative of the signal $x(t) = \sin(4t)$.

9.22   Using the time-differentiation property, find the FT of the signal $x(t)$.

9.22.1 $x(t) = \begin{cases} (1 - t) & \text{for } 0 < t < 1 \\ 0 & \text{for } t < 0 \text{ and } t \geq 1 \end{cases}$

9.22.2 $x(t) = \begin{cases} t & \text{for } 0 < t < 1 \\ 0 & \text{for } t \leq 0 \text{ and } t > 1 \end{cases}$

9.22.3 $x(t) = \begin{cases} 1 & \text{for } -1 < t < 0 \\ -1 & \text{for } 0 < t < 1 \\ 0 & \text{for } t < -1 \text{ and } t > 1 \end{cases}$

*9.22.4 $x(t) = \begin{cases} t & \text{for } 0 \leq t < 1 \\ 1 & \text{for } 1 \leq t < 2 \\ (3 - t) & \text{for } 2 \leq t < 3 \\ 0 & \text{for } t < 0 \text{ and } t > 3 \end{cases}$

9.22.5 $x(t) = e^{-2|t|}$.

9.23   Using the time-integration property, find the FT of $y(t)$, where

$$y(t) = \int_{-\infty}^{t} x(\tau)d\tau$$

9.23.1 $x(t) = \delta(t - 3)$.

9.23.2 $x(t) = \begin{cases} 2 & \text{for } -1 < t < 0 \\ -2 & \text{for } 0 < t < 1 \\ 0 & \text{for } t < -1 \text{ and } t > 1 \end{cases}$

9.23.3 $x(t) = \cos(3t)$.

9.23.4 $x(t) = \begin{cases} \sin(t) & \text{for } 0 \leq t < \frac{\pi}{2} \\ 0 & \text{for } t < 0 \text{ and } t > \frac{\pi}{2} \end{cases}$

9.23.5 $x(t) = \begin{cases} \cos(t) & \text{for } 0 < t < \pi \\ 0 & \text{for } t < 0 \text{ and } t > \pi \end{cases}$

*9.23.6 $x(t) = e^{-t}u(t)$.

9.23.7 $x(t) = u(t)$.

9.24 Using the frequency-differentiation property, find the FT of the signal $x(t)$.

9.24.1 $x(t) = t^2 e^{-t}u(t)$.

9.24.2 $x(t) = tu(t)$.

9.24.3 $x(t) = te^{-2|t|}$.

*9.24.4 $x(t) = \begin{cases} t & \text{for } 0 < t < 1 \\ 0 & \text{for } t < 0 \text{ and } t > 1 \end{cases}$

9.24.5 $x(t) = \begin{cases} t & \text{for } -1 < t < 1 \\ 0 & \text{for } t < -1 \text{ and } t > 1 \end{cases}$

9.24.6 $x(t) = \begin{cases} t\sin(t) & \text{for } 0 < t < \pi \\ 0 & \text{for } t < 0 \text{ and } t > \pi \end{cases}$

9.25 Using the linearity, time-shifting, frequency differentiation properties and the FT of $u(t)$, find the FT of the signal

$$x(t) = \begin{cases} 0 & \text{for } t < 0 \\ t & \text{for } 0 \leq t \leq 3 \\ 3 & \text{for } t > 3 \end{cases}$$

9.26 Find the energy of the signal $x(t) = e^{-2t}u(t)$. Find the value of T such that 90% of the signal energy lies in the range $0 \leq t \leq T$. What is the corresponding signal bandwidth.

9.27 Derive the Parseval's theorem for aperiodic signals from that for the Fourier series of periodic signals, as the period tends to infinity.

9.28 Using the complex FS coefficients of the periodic signal $x(t)$, find its FT.

9.28.1 $x(t) = \sum_{n=-\infty}^{\infty} \delta(t - nT)$.

9.28.2 $x(t) = 2 + 3\cos(2t) + 4\sin(4t) - 5e^{-j6t} + 6e^{j10t}$.

*9.28.3 $x(t) = -1 - 3\sin(3t) + 2\cos(5t) + 6e^{-j7t}$.

9.28.4 $x(t) = 3 - 2\cos(10t) + 3\sin(15t) - e^{j25t}$.

9.29 Using the FT, find the complex FS coefficients of the periodic signal $x(t)$.

9.29.1 $x(t) = 5e^{-j(t+\frac{\pi}{3})}$.

9.29.2 $x(t) = 2\cos(2t - \frac{\pi}{4})$.

9.29.3 $x(t) = 3\sin(3t - \frac{\pi}{6})$.

*9.29.4 $x(t) = \sum_{n=-\infty}^{\infty}(t - 2n)(u(t - 2n) - u(t - 2 - 2n))$.

9.30 Find the inverse FT, $x(t)$, of $X(j\omega)$. Find the sampled signal $x_s(t)$ and its transform $X_s(j\omega)$ for the sampling interval $T_s = 0.25, 0.5, 1, 2$ and 3 s.

9.30.1

$$X(j\omega) = \begin{cases} \cos(\omega) & \text{for } |\omega| < \pi \\ 0 & \text{elsewhere} \end{cases}$$

9.30.2

$$X(j\omega) = \begin{cases} \sin(2\omega) & \text{for } |\omega| < \pi \\ 0 & \text{elsewhere} \end{cases}$$

9.31 Find the FT of $x(t)$ and its sampled versions with the sampling interval $T_s = 0.01, 0.1, 1$ and 10 s. What are the spectral values of $x(t)$ and its sampled versions at $\omega = 0$.
   9.31.1 $x(t) = e^{-t}u(t)$.
   *9.31.2 $x(t) = e^{-|t|}$.
9.32 Find the FT of $x(t)$ and its sampled versions with the sampling interval $T_s = 0.1, 0.5, 1$ and 2 s.
   9.32.1 $x(t) = 2\cos\left(\frac{2\pi}{32}t\right) + \sin\left(3\frac{2\pi}{32}t\right)$.
   9.32.2 $x(t) = 4\sin\left(\frac{2\pi}{24}t\right) + \cos\left(5\frac{2\pi}{24}t\right)$.
*9.33 Approximate the samples of the FT of the signal

$$x(t) = \begin{cases} 1 - |t| & \text{for } |t| < 1 \\ 0 & \text{elsewhere} \end{cases}$$

using the DFT with $N = 4$ samples. Assume that the signal is periodically extended with period $T = 2$ s. Compare the first two samples of the FT obtained using the DFT with that of the exact values.
9.34 Approximate the samples of the FT of the signal

$$x(t) = \begin{cases} 1 & \text{for } |t| < 2 \\ 0 & \text{for } |t| > 2 \end{cases}$$

using the DFT with $N = 4$ samples. Assume that the signal is periodically extended with period $T = 8$ s. Compare the first two samples of the FT obtained using the DFT with that of the exact values.
9.35 Find the response $y(t)$, using the FT, of the system governed by the differential equation

$$\frac{dy(t)}{dt} + y(t) = e^{jt}$$

Verify your solution by substituting it into the differential equation.

9.36 Using the FT, find the zero-state response $y(t)$ of the system governed by the differential equation

$$2\frac{dy(t)}{dt} + 3y(t) = \delta(t)$$

Verify your solution by substituting it into the differential equation.

9.37 Using the FT, find the zero-state response $y(t)$ of the system governed by the differential equation

$$\frac{dy(t)}{dt} + 2y(t) = u(t)$$

Verify your solution by substituting it into the differential equation.

9.38 Using the FT, find the zero-state response $y(t)$ of the system governed by the differential equation

$$3\frac{dy(t)}{dt} + 2y(t) = 4e^{-2t}u(t)$$

Verify your solution by substituting it into the differential equation.

10

The z-Transform

In Fourier analysis, we decompose a signal in terms of its constituent constant-amplitude sinusoids. Systems are modeled in terms of their responses to sinusoids. This representation provides an insight into the signal and system characteristics and makes the evaluation of important operations, such as convolution, easier. The general constraint on the signal to be analyzed is that it is absolutely or square integrable/summable. Even with this constraint, the use of Fourier analysis is extensive in signal and system analysis. However, we still need the generalization of Fourier analysis so that a larger class of signals and systems can be analyzed in the frequency-domain, retaining all the advantages of the frequency-domain methods. The generalization of the Fourier analysis for discrete signals, called the z-transform, is described in this chapter.

The differences between the z-transform and Fourier analysis are presented in Section 10.1. In Section 10.2, the z-transform is derived starting from the DTFT definition. In Section 10.3, the properties of the z-transform are described. In Section 10.4, the inverse z-transform is derived and two frequently used methods to find the inverse z-transform are presented. Typical applications of the z-transform are described in Section 10.5.

10.1 Fourier Analysis and the z-Transform

In Fourier analysis, we analyze a waveform in terms of constant-amplitude sinusoids $A\cos(\omega n + \theta)$, shown in Figure 2.1. Fourier analysis is generalized by making the basis signals a larger set of sinusoids, by including sinusoids with exponentially varying amplitudes $Ar^n\cos(\omega n + \theta)$, shown in Figure 2.5. This extension enables us to analyze a larger set of signals and systems than is possible with Fourier analysis. The sinusoids, whether they have constant-amplitude or varying amplitude, have the key advantages of being the basis signals in terms of ease of signal decomposition and

A Practical Approach to Signals and Systems D. Sundararajan
© 2008 John Wiley & Sons (Asia) Pte Ltd

efficient signal and system analysis. In Fourier analysis, we use fast algorithms to obtain the frequency-domain representation of signals. In the case of the transforms that use sinusoids with exponentially varying amplitudes, it is found that a short table of transform pairs is adequate for most practical purposes.

In the Fourier representation, the spectrum of a one-dimensional signal is also one-dimensional, the spectral coordinates being the frequency ω and the complex amplitude of the complex sinusoids. In the case of the generalized transforms, the rate of change of the amplitude of the exponentially varying amplitude sinusoids is also a parameter. This makes the spectrum of a one-dimensional signal two-dimensional, a surface. The spectrum provides infinite spectral representations of the signal, that is the spectral values along any appropriate closed contour of the two-dimensional spectrum could be used to reconstruct the signal. Therefore, a signal may be reconstructed using constant-amplitude sinusoids or exponentially decaying sinusoids or exponentially growing sinusoids, or an infinite combination of these types of sinusoids.

The advantages of the z-transform include: the pictorial description of the behavior of the system obtained by the use of the complex frequency; the ability to analyze unstable systems or systems with exponentially growing inputs; automatic inclusion of the initial conditions of the system in finding the output; and easier manipulation of the expressions involving the variable z than those with $e^{j\omega}$.

10.2 The z-Transform

We assume, in this chapter, that all signals are causal, that is $x(n) = 0$, $n < 0$, unless otherwise specified. This leads to the one-sided or unilateral version of the z- transform, which is mostly used for practical system analysis. If a signal $x(n)u(n)$ is not Fourier transformable, then its exponentially weighted version $(x(n)r^{-n})$ may be Fourier transformable for the positive real quantity $r > 1$. If $x(n)u(n)$ is Fourier transformable, $(x(n)r^{-n})$ may still be transformable for some values of $r < 1$. The DTFT of this signal is

$$\sum_{n=0}^{\infty}(x(n)r^{-n})e^{-j\omega n}$$

By combining the exponential factors, we get

$$X(re^{j\omega}) = \sum_{n=0}^{\infty}x(n)(re^{j\omega})^{-n}$$

This equation can be interpreted as the generalized Fourier analysis of the signal $x(n)$ using exponentials with complex exponents or sinusoids with varying amplitudes as the basis signals. By substituting $z = re^{j\omega}$, we get the defining equation of the one-sided or unilateral z-transform of $x(n)$ as

$$X(z) = \sum_{n=0}^{\infty} x(n)z^{-n} \tag{10.1}$$

Expanding the summation, we get

$$X(z) = x(0) + x(1)z^{-1} + x(2)z^{-2} + x(3)z^{-3} + \cdots$$

where z is a complex variable. Therefore, the basis functions used in the z-transform are of the form $z^n = e^{(\sigma + j\omega)n} = r^n e^{j\omega n} = r^n(\cos(\omega n) + j\sin(\omega n)) = (a + jb)^n$. While $X(e^{j\omega})$ is the DTFT of $x(n)$, $X(z) = X(re^{j\omega})$ is the DTFT of $x(n)r^{-n}$ for all values of r for which $\sum_{n=0}^{\infty} |x(n)r^{-n}| < \infty$. If the value one is included in these values of r, then $X(e^{j\omega})$ can be obtained from $X(z)$ by the substitution $z = e^{j\omega}$. The z-transform of a signal $x(n)$, $X(z)$, exists for $|z| > r_0$ if $|x(n)| \le r_0^n$ for some constant r_0. For example, $x(n) = a^{n^2}$ does not have a z-transform. In essence, the z-transform of a signal, whether it is converging or not, is the DTFT of all its versions, obtained by multiplying it by a real exponential of the form r^{-n}, so that the modified signal is guaranteed to converge.

The z-transform, $X(z)$, represents a sequence only for the set of values of z for which it converges, that is the magnitude of $X(z)$ is not infinite. The region that comprises this set of values in the z-plane (a complex plane used for displaying the z-transform) is called the region of convergence (ROC). For a given positive number c, the equation $|z| = |a + jb| = c$ or $a^2 + b^2 = c^2$ describes a circle in the z-plane with center at the origin and radius c. Consequently, the condition $|z| > c$ for ROC specifies the region outside this circle. If the ROC of the z-transform of a sequence includes the unit-circle, then its DTFT can be obtained from $X(z)$ by replacing z with $e^{j\omega}$.

Example 10.1. Find the z-transform of the unit-impulse signal, $\delta(n)$.

Solution
Using the definition, we get

$$X(z) = 1, \text{ for all } z \qquad \text{and} \qquad \delta(n) \Longleftrightarrow 1, \text{ for all } z$$

The transform pair for a delayed impulse $\delta(n - m)$ is

$$\delta(n - m) \Longleftrightarrow z^{-m} \qquad |z| > 0,$$

where m is positive.   □

Example 10.2. Find the z-transform of the finite sequence with its only nonzero samples specified as $\{x(0) = 5, x(2) = 4, x(5) = -2\}$.

Solution
Using the definition, we get

$$X(z) = 5 + 4z^{-2} - 2z^{-5} = \frac{5z^5 + 4z^3 - 2}{z^5} \qquad |z| > 0$$

□

The geometric sequence, $a^n u(n)$, is fundamental to the study of linear discrete systems, as it is more convenient to express the natural response of systems in that form.

Example 10.3. Find the z-transform of the geometric sequence, $a^n u(n)$.

Solution
Substituting $x(n) = a^n$ in the defining equation of the z-transform, we get

$$X(z) = \sum_{n=0}^{\infty} (a^{-1}z)^{-n} = 1 + (a^{-1}z)^{-1} + (a^{-1}z)^{-2} + (a^{-1}z)^{-3} + \cdots$$

$$= \frac{1}{1 - (a^{-1}z)^{-1}} = \frac{z}{z - a} \qquad |z| > |a|$$

It is known that the geometric series $1 + r + r^2 + \cdots$ converges to $1/(1-r)$, if $|r| < 1$. If $|z| > |a|$, the common ratio of the series $r = a/z$ has magnitude that is less than one. Therefore, the ROC of the z-transform is given as $|z| > |a|$ and we get the transform pair

$$a^n u(n) \Longleftrightarrow \frac{z}{z - a} \qquad |z| > |a|$$

□

Note that the DTFT of $a^n u(n)$ does not exist for $a > 1$, whereas the z-transform exists for all values of a as long as $|z| > |a|$. The z-transform spectrum of a sequence is usually displayed by the locations of zeros and poles of the z-transform and its magnitude.

The pole-zero plot and the magnitude of the z-transform $z/(z - 0.8)$ of the signal $a^n u(n)$ with $a = 0.8$ are shown, respectively, in Figures 10.1(a) and (b). When $z = 0.8$, $|X(z)| = \infty$.

This point, marked by the cross in Figure 10.1(a) is called a pole of $X(z)$ (the peak in Figure 10.1(b)). When $z = 0$, $X(z) = 0$. This point marked by the open circle in Figure 10.1(a) is called a zero of $X(z)$ (the valley in Figure 10.1(b)). The pole-zero plot specifies a transform $X(z)$, except for a constant factor. In the region outside the circle with radius 0.8, $X(z)$ exists and is a valid frequency-domain representation of the signal. In general, the ROC of a z-transform is the region in the

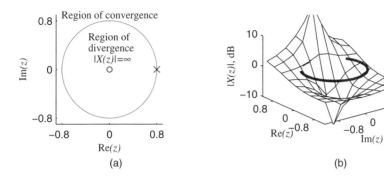

Figure 10.1 (a) The pole-zero plot of the z-transform $z/(z - 0.8)$ of $(0.8)^n u(n)$; (b) the magnitude of the z-transform

z-plane that is exterior to the smallest circle, centered at the origin, enclosing all its poles.

Example 10.4. Find the z-transform of the signal $e^{j\omega n} u(n)$. Deduce the z-transform of $\sin(\omega n) u(n)$.

Solution
Using the transform of $a^n u(n)$ with $a = e^{j\omega}$, we get

$$e^{j\omega n} u(n) \Longleftrightarrow \frac{z}{z - e^{j\omega}} \qquad |z| > 1$$

Since the magnitude of $a = e^{j\omega}$ is 1, the convergence condition is $|z| > 1$. Using the fact that $j2 \sin(\omega n) = (e^{j\omega n} - e^{-j\omega n})$, we get

$$j2X(z) = \frac{z}{z - e^{j\omega}} - \frac{z}{z - e^{-j\omega}} \qquad |z| > 1$$

$$\sin(\omega n) u(n) \Longleftrightarrow \frac{z \sin(\omega)}{(z - e^{j\omega})(z - e^{-j\omega})} = \frac{z \sin(\omega)}{z^2 - 2z \cos(\omega) + 1} \qquad |z| > 1 \qquad \square$$

Figure 10.2(a) shows the pole-zero plot and Figure 10.2(b) shows the magnitude of the z-transform $z \sin(\frac{\pi}{4})/(z^2 - 2z \cos(\frac{\pi}{4}) + 1)$ of the signal $\sin(\frac{\pi}{4}n)u(n)$. There is a zero at $z = 0$ and poles at $z = e^{j\frac{\pi}{4}}$ and $z = e^{-j\frac{\pi}{4}}$, a pair of complex-conjugate poles.

Example 10.5. Find the z-transform of the signal defined as

$$x(n) = \begin{cases} (0.2)^n & \text{for } 0 \le n \le 5 \\ (0.2)^n + (-0.6)^n & \text{for } 6 \le n \le 8 \\ (-0.6)^n & \text{for } 9 \le n < \infty \end{cases}$$

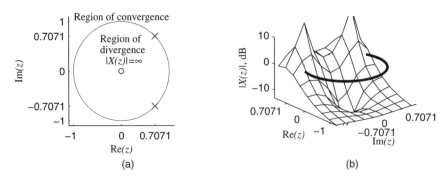

Figure 10.2 (a) The pole-zero plot of the z-transform $z\sin(\frac{\pi}{4})/[(z - e^{j\frac{\pi}{4}})(z - e^{-j\frac{\pi}{4}})]$ of $\sin(\frac{\pi}{4}n)u(n)$; (b) the magnitude of the z-transform

Solution
From the definition of the z-transform, we get

$$X(z) = \sum_{n=0}^{8}(0.2)^n z^{-n} + \sum_{n=0}^{\infty}(-0.6)^n z^{-n} - \sum_{n=0}^{5}(-0.6)^n z^{-n}$$

$$X(z) = \frac{z^9 - (0.2)^9}{z^8(z - 0.2)} + \frac{z}{z + 0.6} - \frac{z^6 - (-0.6)^6}{z^5(z + 0.6)} \qquad |z| > 0.6 \qquad \square$$

10.3 Properties of the z-Transform

Properties present the frequency-domain effect of time-domain characteristics and operations on signals and vice versa. In addition, they are used to find new transform pairs more easily.

10.3.1 Linearity

It is often advantageous to decompose a complex sequence into a linear combination of simpler sequences (as in Example 10.4) in the manipulation of sequences and their transforms. If $x(n) \Longleftrightarrow X(z)$ and $y(n) \Longleftrightarrow Y(z)$, then

$$ax(n) + by(n) \Longleftrightarrow aX(z) + bY(z)$$

where a and b are arbitrary constants. The z-transform of a linear combination of sequences is the same linear combination of the z-transforms of the individual sequences. This property is due to the linearity of the defining summation operation of the transform.

10.3.2 Left Shift of a Sequence

The shift property is used to express the transform of the shifted version, $x(n + m)u(n)$, of a sequence $x(n)$ in terms of its transform $X(z)$. If $x(n)u(n) \Longleftrightarrow X(z)$ and m is a positive integer, then

$$x(n + m)u(n) \Longleftrightarrow z^m X(z) - z^m \sum_{n=0}^{m-1} x(n)z^{-n}$$

Let the z-transform of the sequence $x(n + m)u(n)$ be $Y(z)$. Then,

$$z^{-m} Y(z) = x(m)z^{-m} + x(m + 1)z^{-m-1} + \cdots$$

By adding m terms, $\sum_{n=0}^{m-1} x(n)z^{-n}$, to both sides of the equation, we get

$$z^{-m} Y(z) + x(m - 1)z^{-m+1} + x(m - 2)z^{-m+2} + \cdots + x(0) = X(z)$$

$$Y(z) = z^m X(z) - z^m \sum_{n=0}^{m-1} x(n)z^{-n}$$

For example,

$$x(n + 1)u(n) \Longleftrightarrow zX(z) - zx(0) \text{ and } x(n + 2)u(n) \Longleftrightarrow z^2 X(z) - z^2 x(0) - zx(1)$$

Consider the sequence $x(n)$ with $x(-2) = 1, x(-1) = 2, x(0) = -1, x(1) = 1$, $x(2) = -2, x(3) = 2$ and $x(n) = 0$ otherwise, shown in Figure 10.3(a). The transform of $x(n)$ is $X(z) = -1 + z^{-1} - 2z^{-2} + 2z^{-3}$. The sequence $p(n)$, shown in Figure 10.3(b), is the left-shifted sequence $x(n + 1)$. The transform of $p(n)u(n) = x(n + 1)u(n)$ is

$$P(z) = 1 - 2z^{-1} + 2z^{-2} = zX(z) - zx(0) = z(-1 + z^{-1} - 2z^{-2} + 2z^{-3}) + z$$

(a)   (b)   (c)

Figure 10.3 (a) $x(n)$; (b) $p(n) = x(n + 1)$; (c) $q(n) = x(n - 1)$

10.3.3 Right Shift of a sequence

If $x(n)u(n) \Longleftrightarrow X(z)$ and m is a positive integer, then

$$x(n - m)u(n) \Longleftrightarrow z^{-m}X(z) + z^{-m} \sum_{n=1}^{m} x(-n)z^n$$

For example,

$$x(n - 1)u(n) \Longleftrightarrow z^{-1}X(z) + x(-1)$$
$$x(n - 2)u(n) \Longleftrightarrow z^{-2}X(z) + z^{-1}x(-1) + x(-2)$$

The sequence $q(n)$, shown in Figure 10.3(c), is the right shifted sequence $x(n - 1)$. The transform of $q(n)u(n) = x(n - 1)u(n)$ is

$$Q(z) = 2 - z^{-1} + z^{-2} - 2z^{-3} + 2z^{-4}$$
$$= 2 + z^{-1}(-1 + z^{-1} - 2z^{-2} + 2z^{-3}) = x(-1) + z^{-1}X(z)$$

In finding the response $y(n)$ of a system for $n \geq 0$, the initial conditions, such as $y(-1)$ and $y(-2)$, must be taken into account. The shift properties provide the way for the automatic inclusion of the initial conditions. The left shift property is more convenient for solving difference equations in advance operator form. Consider solving the difference equation $y(n) = x(n) + \frac{1}{2}y(n - 1)$ with the initial condition $y(-1) = 3$ and $x(n) = 0$. The solution is $y(n) = 3(\frac{1}{2})^{n+1}u(n)$, using time-domain method. Taking the z-transform of the difference equation, we get $Y(z) = X(z) + \frac{1}{2}(z^{-1}Y(z) + 3)$. Solving for $Y(z)$, $Y(z) = \frac{3}{2}z/(z - \frac{1}{2})$. The inverse transform of $Y(z)$ is $y(n) = \frac{3}{2}(\frac{1}{2})^n u(n)$, which is the same as that obtained earlier.

10.3.4 Convolution

If $x(n)u(n) \Longleftrightarrow X(z)$ and $h(n)u(n) \Longleftrightarrow H(z)$, then

$$y(n) = \sum_{m=0}^{\infty} h(m)x(n - m) \Longleftrightarrow Y(z) = H(z)X(z)$$

The DTFT of $x(n)r^{-n}$ is the z-transform $X(z)$ of $x(n)$. The convolution of $x(n)r^{-n}$ and $h(n)r^{-n}$ corresponds to $X(z)H(z)$ in the frequency-domain. The inverse DTFT of $X(z)H(z)$, therefore, is the convolution of $x(n)r^{-n}$ and $h(n)r^{-n}$ given by

$$\sum_{m=0}^{\infty} x(m)r^{-m}h(n - m)r^{(-n+m)} = r^{-n} \sum_{m=0}^{\infty} x(m)h(n - m) = r^{-n}(x(n) * h(n))$$

As finding the inverse z-transform is the same as finding the inverse DTFT in addition to multiplying the signal by r^n, as will be seen later, we get the convolution of $x(n)$ and $h(n)$ by finding the inverse z-transform of $X(z)H(z)$.

Consider the two sequences and their transforms $x(n) = (\frac{1}{2})^n u(n) \iff X(z) = z/(z - \frac{1}{2})$ and $h(n) = (\frac{1}{3})^n u(n) \iff H(z) = z/(z - \frac{1}{3})$. The convolution of the sequences, in the transform domain, is given by the product of their transforms,

$$X(z)H(z) = \frac{z}{z - \frac{1}{2}} \frac{z}{z - \frac{1}{3}} = \frac{3z}{z - \frac{1}{2}} - \frac{2z}{z - \frac{1}{3}}$$

The inverse transform of $X(z)H(z)$ is the convolution of the sequences in the time-domain and it is $(3(\frac{1}{2})^n - 2(\frac{1}{3})^n)u(n)$.

10.3.5 Multiplication by n

If $x(n)u(n) \iff X(z)$, then

$$nx(n)u(n) \iff -z\frac{d}{dz}X(z)$$

Differentiating the defining expression for $X(z)$ with respect to z and multiplying it by $-z$, we get

$$-z\frac{d}{dz}X(z) = -z\frac{d}{dz}\sum_{n=0}^{\infty} x(n)z^{-n} = \sum_{n=0}^{\infty} nx(n)z^{-n} = \sum_{n=0}^{\infty} (nx(n))z^{-n}$$

For example,

$$\delta(n) \iff 1 \quad \text{and} \quad n\delta(n) = 0 \iff 0$$

$$u(n) \iff \frac{z}{z-1} \quad \text{and} \quad nu(n) \iff \frac{z}{(z-1)^2}$$

10.3.6 Multiplication by a^n

If $x(n)u(n) \iff X(z)$, then

$$a^n x(n)u(n) \iff X\left(\frac{z}{a}\right)$$

From the z-transform definition, we get

$$X(z) = \sum_{n=0}^{\infty} a^n x(n)z^{-n} = \sum_{n=0}^{\infty} x(n)\left(\frac{z}{a}\right)^{-n} = X\left(\frac{z}{a}\right)$$

Multiplication of $x(n)$ by a^n corresponds to scaling the frequency variable z. For example,

$$u(n) \Longleftrightarrow \frac{z}{z-1} \quad \text{and} \quad (2)^n u(n) \Longleftrightarrow \frac{\frac{z}{2}}{(\frac{z}{2}-1)} = \frac{z}{z-2}$$

The pole at $z = 1$ in the transform of $u(n)$ is shifted to the point $z = 2$ in the transform of $(2)^n u(n)$.

With $a = -1$ and $x(n)u(n) \Longleftrightarrow X(z)$, $(-1)^n x(n)u(n) \Longleftrightarrow X(-z)$. For example, $u(n) \Longleftrightarrow \frac{z}{z-1}$ and $(-1)^n u(n) \Longleftrightarrow -z/(-z-1) = z/(z+1)$.

10.3.7 Summation

If $x(n)u(n) \Longleftrightarrow X(z)$, then $y(n) = \sum_{m=0}^{n} x(m) \Longleftrightarrow Y(z) = [z/(z-1)]X(z)$. The product $[z/(z-1)]X(z)$ corresponds to the convolution of $x(n)$ and $u(n)$ in the time-domain, which, of course, is equivalent to the sum of the first $n+1$ values of $x(n)$.

For example, $x(n) = (-1)^n u(n) \Longleftrightarrow \frac{z}{z+1}$. Then, $Y(z) = \frac{z}{z-1}\frac{z}{z+1} = \frac{1}{2}\left(\frac{z}{z-1} + \frac{z}{z+1}\right)$.

Taking the inverse z-transform, we get $y(n) = \frac{1}{2}(1 + (-1)^n)$.

10.3.8 Initial Value

Using this property, the initial value of $x(n)$, $x(0)$, can be determined directly from $X(z)$. If $x(n)u(n) \Longleftrightarrow X(z)$, then

$$x(0) = \lim_{z \to \infty} X(z) \quad \text{and} \quad x(1) = \lim_{z \to \infty} (z(X(z) - x(0)))$$

From the definition of the transform, we get

$$\lim_{z \to \infty} X(z) = \lim_{z \to \infty} (x(0) + x(1)z^{-1} + x(2)z^{-2} + x(3)z^{-3} + \cdots) = x(0)$$

As $z \to \infty$, each term, except $x(0)$, tends to zero. Let $X(z) = (z^2 - 2z + 5)/(z^2 + 3z - 2)$. Then,

$$x(0) = \lim_{z \to \infty} \frac{(z^2 - 2z + 5)}{(z^2 + 3z - 2)} = 1$$

Note that, when $z \to \infty$, only the terms of the highest power are significant.

10.3.9 Final Value

Using this property, the final value of $x(n)$, $x(\infty)$, can be determined directly from $X(z)$. If $x(n)u(n) \Longleftrightarrow X(z)$, then

$$\lim_{n \to \infty} x(n) = \lim_{z \to 1}((z-1)X(z))$$

provided the ROC of $(z-1)X(z)$ includes the unit-circle (otherwise, $x(n)$ has no limit as $n \to \infty$). Let $X(z) = (z^2 - 2z + 5)/(z^2 + 3z - 2)$. The property does not apply since the ROC of $(z-1)X(z)$ does not include the unit-circle. Let $X(z) = (z^2 - 2z + 5)/(z^2 - 1.5z + 0.5)$. Then,

$$\lim_{n \to \infty} x(n) = \lim_{z \to 1}(z-1)\frac{(z^2 - 2z + 5)}{(z^2 - 1.5z + 0.5)} = \lim_{z \to 1}\frac{(z^2 - 2z + 5)}{(z - 0.5)} = 8$$

The value $\lim_{n \to \infty} x(n)$, if it is nonzero, is solely due to the scaled unit-step component of $x(n)$. Multiplying $X(z)$ by $(z-1)$ and setting $z = 1$ is just finding the partial fraction coefficient of the unit-step component of $x(n)$.

10.3.10 Transform of Semiperiodic Functions

Consider the function $x(n)u(n)$ that is periodic of period N for $n \geq 0$, that is $x(n + N) = x(n)$, $n \geq 0$. Let $x_1(n) = x(n)u(n) - x(n - N)u(n - N) \Longleftrightarrow X_1(z)$. $x_1(n)$ is equal to $x(n)u(n)$ over its first period and is zero elsewhere. Then,

$$x(n)u(n) = x_1(n) + x_1(n - N) + x_1(n - 2N) + \cdots$$

Using the right shift property, the transform of $x(n)u(n)$ is

$$X(z) = X_1(z)(1 + z^{-N} + z^{-2N} + \cdots) = \frac{X_1(z)}{1 - z^{-N}} = X_1(z)\left(\frac{z^N}{z^N - 1}\right)$$

Let us find the transform of $x(n) = (-1)^n u(n)$ with period $N = 2$. $X_1(z) = 1 - z^{-1} = (z-1)/z$. From the property,

$$X(z) = \frac{z^2}{(z^2 - 1)}\frac{(z-1)}{z} = \frac{z}{(z+1)}$$

10.4 The Inverse z-Transform

Consider the transform pair $x(n)u(n) \Longleftrightarrow z/(z-2)$, $|z| > 2$. Multiplying the signal by $(\frac{1}{4})^n u(n)$ gives $x(n)(\frac{1}{4})^n u(n) \Longleftrightarrow z/(z - 0.5)$, $|z| > 0.5$, due to the multiplication by a^n property. Now, the ROC includes the unit circle in the z-plane. Let us substitute

$z = e^{j\omega}$ in $z/(z - 0.5)$ to get $1/(1 - 0.5e^{-j\omega})$. The inverse DTFT of this transform is the signal $0.5^n u(n) = x(n)(\frac{1}{4})^n u(n)$. Now, multiplying both sides by $4^n u(n)$ gives the original time-domain signal $x(n)u(n) = 2^n u(n)$. This way of finding the inverse z-transform gives us a clear understanding of how the z-transform is the generalized version of the DTFT.

The inverse z-transform relation enables us to find a sequence from its z-transform. The DTFT of $x(n)r^{-n}$ can be written as

$$X(re^{j\omega}) = \sum_{n=0}^{\infty} x(n)(re^{j\omega})^{-n}$$

The inverse DTFT of $X(re^{j\omega})$ is

$$x(n)r^{-n} = \frac{1}{2\pi} \int_{-\pi}^{\pi} X(re^{j\omega})e^{j\omega n} d\omega$$

Multiplying both sides by r^n, we get

$$x(n) = \frac{1}{2\pi} \int_{-\pi}^{\pi} X(re^{j\omega})(re^{j\omega})^n d\omega$$

Let $z = re^{j\omega}$. Then, $dz = jre^{j\omega}d\omega = jzd\omega$. Now, the inverse z-transform of $X(z)$, in terms of the variable z, is defined as

$$x(n) = \frac{1}{2\pi j} \oint_C X(z)z^{n-1} dz \tag{10.2}$$

with the integral evaluated, in the counterclockwise direction, along any simply connected closed contour C, encircling the origin, that lies in the ROC of $X(z)$. As ω varies from $-\pi$ to π, the variable z traverses the circle of radius r in the counterclockwise direction once.

We can use any appropriate contour of integration in evaluating the inverse z-transform because the transform values corresponding to the contour are taken in the inverse process. As can be seen from Figures 10.1(b) and Figure 10.2(b), the z-transform values vary with each of the infinite choices for the contour of integration.

10.4.1 Finding the Inverse z-Transform

While the most general way of finding the inverse z-transform is to evaluate the contour integral Equation (10.2), for most practical purposes, two other simpler methods are commonly used.

10.4.1.1 The Partial Fraction Method

In LTI system analysis, we are mostly encountered with the problem of inverting a z-transform that is a rational function (a ratio of two polynomials in z). In the partial fraction method, the rational function of the z-transform is decomposed into a linear combination of transforms such as those of $\delta(n)$, $a^n u(n)$, and $na^n u(n)$. Then, it is easy to find the inverse transform from a short table of transform pairs.

Consider finding the partial fraction expansion of $X(z) = z/(z - \frac{1}{5})(z - \frac{1}{4})$. As the partial fraction of the form $kz/(z - p)$ is more convenient, we first expand $X(z)/z$ and then multiply both sides by z.

$$\frac{X(z)}{z} = \frac{1}{(z - \frac{1}{5})(z - \frac{1}{4})} = \frac{A}{(z - \frac{1}{5})} + \frac{B}{(z - \frac{1}{4})}$$

Multiplying all the expressions by $(z - \frac{1}{5})$, we get

$$\left(z - \frac{1}{5}\right) \frac{X(z)}{z} = \frac{1}{(z - \frac{1}{4})} = A + \frac{B(z - \frac{1}{5})}{(z - \frac{1}{4})}$$

Letting $z = \frac{1}{5}$, we get $A = (z - \frac{1}{5})\frac{X(z)}{z}|_{z=\frac{1}{5}} = -20$. Similarly, $B = (z - \frac{1}{4})\frac{X(z)}{z}|_{z=\frac{1}{4}} = 20$.

Therefore,

$$X(z) = \frac{-20z}{(z - \frac{1}{5})} + \frac{20z}{(z - \frac{1}{4})}$$

The time-domain sequence $x(n)$ corresponding to $X(z)$ is given by

$$x(n) = \left(-20 \left(\frac{1}{5}\right)^n + 20 \left(\frac{1}{4}\right)^n\right) u(n)$$

The first four values of the sequence $x(n)$ are

$$x(0) = 0 \qquad x(1) = 1 \qquad x(2) = 0.45 \qquad x(3) = 0.1525$$

As the sum of the terms of a partial fraction will always produce a numerator polynomial whose order is less than that of the denominator, the order of the numerator polynomial of the rational function must be less than that of the denominator. This condition is satisfied by $X(z)/z$, as the degree of the numerator polynomial, for z-transforms of practical interest, is at the most equal to that of the denominator.

Example 10.6. Find the inverse z-transform of

$$X(z) = \frac{z^2}{(z - \frac{1}{2})(z + \frac{1}{3})}$$

Solution

$$\frac{X(z)}{z} = \left(\frac{z}{(z - \frac{1}{2})(z + \frac{1}{3})} \right) \quad \text{and} \quad X(z) = \frac{\frac{3}{5}z}{(z - \frac{1}{2})} + \frac{\frac{2}{5}z}{(z + \frac{1}{3})}$$

$$x(n) = \left(\frac{3}{5} \left(\frac{1}{2} \right)^n + \frac{2}{5} \left(-\frac{1}{3} \right)^n \right) u(n)$$

The first four values of the sequence $x(n)$ are

$$x(0) = 1, \quad x(1) = 0.1667, \quad x(2) = 0.1944, \quad x(3) = 0.0602 \qquad \square$$

The partial fraction method applies for complex poles also. Of course, the complex poles and their coefficients will always appear in conjugate pairs for $X(z)$ with real coefficients. Therefore, finding one of the coefficients of each pair of poles is sufficient.

Example 10.7. Find the inverse z-transform of

$$X(z) = \frac{z}{(z^2 - 2z + 4)}$$

Solution
Factorizing the denominator of $X(z)$ and finding the partial fraction, we get

$$\frac{X(z)}{z} = \left(\frac{1}{(z - (1 + j\sqrt{3}))(z - (1 - j\sqrt{3}))} \right)$$

$$X(z) = \left(\frac{\frac{z}{j2\sqrt{3}}}{z - (1 + j\sqrt{3})} + \frac{\frac{z}{-j2\sqrt{3}}}{z - (1 - j\sqrt{3})} \right)$$

$$x(n) = \left(\frac{1}{j2\sqrt{3}} \right) (1 + j\sqrt{3})^n + \left(\frac{1}{-j2\sqrt{3}} \right) (1 - j\sqrt{3})^n \qquad n = 0, 1, \ldots$$

The two terms of $x(n)$ form a complex conjugate pair. The conjugate of a complex number $z = x + jy$, denoted by z^*, is defined as $z^* = x - jy$, that is the imaginary part is negated. Now, $z + z^* = 2x$ (twice the real part of z or z^*). This result is very useful in simplifying expressions involving complex conjugate poles. Let $(a + jb)$

and $(a - jb)$ are a pair of complex-conjugate poles and $(c + jd)$ and $(c - jd)$ are their respective partial fraction coefficients. Then, the poles combine to produce the time-domain response $2A(r)^n \cos(\omega n + \theta)$, where $r = \sqrt{a^2 + b^2}$ and $\omega = \tan^{-1}(\frac{b}{a})$, and $A = \sqrt{c^2 + d^2}$ and $\theta = \tan^{-1}(\frac{d}{c})$. For the specific example, twice the real part of $(\frac{1}{j2\sqrt{3}})(1 + j\sqrt{3})^n$ or $(\frac{1}{-j2\sqrt{3}})(1 - j\sqrt{3})^n$ is

$$x(n) = \frac{1}{\sqrt{3}}(2)^n \cos\left(\frac{\pi}{3}n - \frac{\pi}{2}\right) u(n)$$

The first four values of the sequence $x(n)$ are

$$x(0) = 0 \qquad x(1) = 1 \qquad x(2) = 2 \qquad x(3) = 0 \qquad \square$$

Example 10.8. Find the inverse z-transform of

$$X(z) = \frac{(z^2 - 2z + 2)}{(z^2 - \frac{7}{12}z + \frac{1}{12})}$$

Solution

$$\frac{X(z)}{z} = \left(\frac{(z^2 - 2z + 2)}{z(z - \frac{1}{3})(z - \frac{1}{4})}\right) \quad \text{and} \quad X(z) = \left(24 + \frac{52z}{(z - \frac{1}{3})} - \frac{75z}{(z - \frac{1}{4})}\right)$$

$$x(n) = 24\delta(n) + \left(52\left(\frac{1}{3}\right)^n - 75\left(\frac{1}{4}\right)^n\right) u(n)$$

The first four values of the sequence $x(n)$ are

$$x(0) = 1 \qquad x(1) = -1.4167 \qquad x(2) = 1.0903 \qquad x(3) = 0.7541 \qquad \square$$

For a pole of order m, there must be m partial fraction terms corresponding to poles of order $m, m - 1, \ldots, 1$.

Example 10.9. Find the inverse z-transform of

$$X(z) = \frac{z^2}{(z - \frac{1}{3})^2(z - \frac{1}{2})}$$

Solution

$$\frac{X(z)}{z} = \left(\frac{z}{(z - \frac{1}{3})^2(z - \frac{1}{2})}\right) = \left(\frac{A}{(z - \frac{1}{3})^2} + \frac{B}{(z - \frac{1}{3})} + \frac{C}{(z - \frac{1}{2})}\right)$$

Now, A can be found to be -2 by substituting $z = \frac{1}{3}$ in the expression $z/(z - \frac{1}{2})$. C can be found to be 18 by substituting $z = \frac{1}{2}$ in the expression $z/(z - \frac{1}{3})^2$. One method to determine the value of B is to substitute a value for z, which is not equal to any of the poles. For example, by substituting $z = 0$ in the expression, the only unknown B is evaluated to be -18. Another method is to subtract the term $-2/(z - \frac{1}{3})^2$ from the expression $z/(z - \frac{1}{3})^2(z - \frac{1}{2})$ to get $3/(z - \frac{1}{2})(z - \frac{1}{3})$. Substituting $z = \frac{1}{3}$ in the expression $3/(z - \frac{1}{2})$, we get $B = -18$. Therefore,

$$X(z) = \left(-\frac{2z}{(z - \frac{1}{3})^2} - \frac{18z}{(z - \frac{1}{3})} + \frac{18z}{(z - \frac{1}{2})} \right)$$

$$x(n) = \left(-2n \left(\frac{1}{3} \right)^{n-1} - 18 \left(\frac{1}{3} \right)^n + 18 \left(\frac{1}{2} \right)^n \right) u(n)$$

The first four values of the sequence $x(n)$ are

$$x(0) = 0 \qquad x(1) = 1 \qquad x(2) = 1.1667 \qquad x(3) = 0.9167 \qquad \square$$

The next example is similar to Example 10.9 with the difference that a second-order pole occurs at $z = 0$.

Example 10.10. Find the inverse z-transform of

$$X(z) = \frac{z^2 + 1}{z^2(z - \frac{1}{3})}$$

Solution

$$\frac{X(z)}{z} = \left(\frac{z^2 + 1}{z^3(z - \frac{1}{3})} \right) = \left(\frac{A}{z^3} + \frac{B}{z^2} + \frac{C}{z} + \frac{D}{(z - \frac{1}{3})} \right)$$

$$X(z) = \left(\frac{-3}{z^2} + \frac{-9}{z} - 30 + \frac{30z}{(z - \frac{1}{3})} \right)$$

$$x(n) = -30\delta(n) - 9\delta(n - 1) - 3\delta(n - 2) + 30 \left(\frac{1}{3} \right)^n u(n)$$

The first four values of the sequence $x(n)$ are

$$x(0) = 0 \qquad x(1) = 1 \qquad x(2) = 0.3333 \qquad x(3) = 1.1111 \qquad \square$$

10.4.1.2 The Long Division Method

By dividing the numerator polynomial by the denominator polynomial, we can express a z-transform in a form that is similar to that of the defining series. Then, from inspection, the sequence values can be found. For example, the inverse z-transform of $X(z) = z/(z - 0.8)$ is obtained dividing z by $z - 0.8$. The quotient is

$$X(z) = 1 + 0.8z^{-1} + 0.64z^{-2} + 0.512z^{-3} + \cdots$$

Comparing with the definition of the z-transform, the time-domain values are $x(0) = 1$, $x(1) = 0.8$, $x(2) = 0.64$, $x(3) = 0.512$, and so on. These values can be verified from $x(n) = (0.8)^n u(n)$, which is the closed-form solution of the inverse z-transform (Example 10.3). This method is particularly useful when only the first few values of the time-domain sequence are required.

10.5 Applications of the z-Transform

10.5.1 Transfer Function and System Response

The input–output relationship of a LTI system is given by the convolution operation in the time-domain. Since convolution corresponds to multiplication in the frequency-domain, we get

$$y(n) = \sum_{m=0}^{\infty} x(m)h(n - m) \Longleftrightarrow Y(z) = X(z)H(z),$$

where $x(n)$, $h(n)$, and $y(n)$ are, respectively the system input, impulse response, and output, and $X(z)$, $H(z)$, and $Y(z)$ are their respective transforms. As input is transferred to output by multiplication with $H(z)$, $H(z)$ is called the transfer function of the system. The transfer function, which is the transform of the impulse response, characterizes a system in the frequency-domain just as the impulse response does in the time-domain. For stable systems, the frequency response $H(e^{j\omega})$ is obtained from $H(z)$ by replacing z by $e^{j\omega}$.

We can as well apply any input, with nonzero spectral amplitude for all values of z in the ROC, to the system, find the response, and the ratio of the z-transforms $Y(z)$ of the output and $X(z)$ of the input is $H(z) = \frac{Y(z)}{X(z)}$. Consider the system governed by the difference equation

$$y(n) = 2x(n) - 3y(n - 1)$$

The impulse response of the system is $h(n) = 2(-3)^n$. The transform of $h(n)$ is $H(z) = 2z/(z + 3)$. The output of this system, with initial condition zero, to the input $x(n) = 3u(n)$ is $y(n) = \frac{3}{2}(1 - (-3)^{(n+1)})u(n)$. The transform of $y(n)$ is

$$Y(z) = \frac{3}{2}\left(\frac{z}{z-1} + \frac{3z}{z+3}\right)$$

The transform of the input $x(n) = 3u(n)$ is $X(z) = 3z/(z-1)$. Now,

$$H(z) = \frac{Y(z)}{X(z)} = \frac{\frac{3}{2}\left(\frac{z}{z-1} + \frac{3z}{z+3}\right)}{\frac{3z}{z-1}} = \frac{2z}{z+3}$$

which is the same as the transform of the impulse response.

Since the transform of a delayed signal is its transform multiplied by a factor, we can just as well find the transfer function by taking the transform of the difference equation characterizing a system. Consider the difference equation of a causal LTI discrete system.

$$y(n) + a_{K-1}y(n-1) + a_{K-2}y(n-2) + \cdots + a_0 y(n-K)$$
$$= b_M x(n) + b_{M-1}x(n-1) + \cdots + b_0 x(n-M)$$

Taking the z-transform of both sides, we get, assuming initial conditions are all zero,

$$Y(z)(1 + a_{K-1}z^{-1} + a_{K-2}z^{-2} + \cdots + a_0 z^{-K})$$
$$= X(z)(b_M + b_{M-1}z^{-1} + \cdots + b_0 z^{-M})$$

The transfer function $H(z)$ is obtained as

$$H(z) = \frac{Y(z)}{X(z)} = \frac{b_M + b_{M-1}z^{-1} + \cdots + b_0 z^{-M}}{1 + (a_{K-1}z^{-1} + a_{K-2}z^{-2} + \cdots + a_0 z^{-K})}$$
$$= \frac{\sum_{l=0}^{M} b_{M-l}z^{-l}}{1 + \sum_{l=1}^{K} a_{K-l}z^{-l}}$$

The transfer function written in positive powers of z,

$$H(z) = \frac{z^{K-M}(b_M z^M + b_{M-1}z^{M-1} + \cdots + b_0)}{z^K + (a_{K-1}z^{K-1} + a_{K-2}z^{K-2} + \cdots + a_0)}$$

is more convenient for manipulation.

10.5.2 Characterization of a System by its Poles and Zeros

By using the pole-zero representation of the z-transform, the transfer function can be written as

$$H(z) = B \frac{z^{K-M}(z - z_1)(z - z_2) \cdots (z - z_M)}{(z - p_1)(z - p_2) \cdots (z - p_K)} = Bz^{K-M} \frac{\prod_{l=1}^{M}(z - z_l)}{\prod_{l=1}^{K}(z - p_l)}$$

where B is a constant. As the coefficients of the polynomials of $H(z)$ are real for practical systems, the zeros and poles are real-valued or they always occur as complex-conjugate pairs.

The pole-zero plot of the transfer function $H(z)$ of a system is a pictorial description of its characteristics, such as speed of response, frequency selectivity, and stability. Poles with magnitudes much smaller than one results in a fast-responding system, with its transient response decaying rapidly. On the other hand, poles with magnitudes closer to one result in a sluggish system. Complex-conjugate poles located inside the unit-circle result in an oscillatory transient response that decays with time. The frequency of oscillation is higher for poles located in the second and third quadrants of the unit-circle. Complex-conjugate poles located on the unit-circle result in a steady oscillatory transient response. Poles located on the positive real axis inside the unit circle result in exponentially decaying transient response. Alternating positive and negative samples is the transient response due to poles located on the negative real axis. The frequency components of an input signal with frequencies close to a zero will be suppressed while those close to a pole will be readily transmitted. Poles located symmetrically about the positive real axis inside the unit-circle and close to the unit-circle in the passband result in a lowpass system that more readily transmits low-frequency signals than high-frequency signals. Zeros located symmetrically about the negative real axis in the stopband further enhance the lowpass character of the frequency response. On the other hand, poles located symmetrically about the negative real axis inside the unit-circle and close to the unit-circle in the passband result in a highpass system that more readily transmits high-frequency signals than low-frequency signals. For example, a system with its pole-zero plot such as that shown in Figure 10.1 is a lowpass filter. The stability of a system can also be determined from its pole-zero plot, as presented later.

Example 10.11. Find the zero-input, zero-state, transient, steady-state, and complete responses of the system governed by the difference equation

$$y(n) = 2x(n) - x(n - 1) + 3x(n - 2) + \frac{9}{20}y(n - 1) - \frac{1}{20}y(n - 2)$$

with the initial conditions $y(-1) = 3$ and $y(-2) = 2$ and, the input $x(n) = u(n)$, the unit-step function.

Solution

The z-transforms of the terms of the difference equation are

$$x(n) \Longleftrightarrow \frac{z}{z-1} \qquad x(n-1) \Longleftrightarrow \frac{1}{z-1} \qquad x(n-2) \Longleftrightarrow \frac{1}{z(z-1)}$$

$$y(n) \Longleftrightarrow Y(z) \qquad y(n-1) \Longleftrightarrow y(-1) + z^{-1}Y(z) = z^{-1}Y(z) + 3$$

$$y(n-2) \Longleftrightarrow y(-2) + z^{-1}y(-1) + z^{-2}Y(z) = z^{-2}Y(z) + 3z^{-1} + 2$$

Substituting the corresponding transform for each term in the difference equation and factoring, we get

$$\frac{Y(z)}{z} = \frac{2z^2 - z + 3}{(z-1)(z-\frac{1}{5})(z-\frac{1}{4})} + \frac{(\frac{5}{4}z - \frac{3}{20})}{(z-\frac{1}{5})(z-\frac{1}{4})}$$

The first term on the right-hand side is $H(z)X(z)/z$ and corresponds to the zero-state response. The second term is due to the initial conditions and corresponds to the zero-input response.

Expanding into partial fractions, we get

$$\frac{Y(z)}{z} = \frac{\frac{20}{3}}{(z-1)} + \frac{72}{(z-\frac{1}{5})} - \frac{\frac{230}{3}}{(z-\frac{1}{4})} - \frac{2}{(z-\frac{1}{5})} + \frac{\frac{13}{4}}{(z-\frac{1}{4})}$$

Taking the inverse z-transform, we get the complete response.

$$y(n) = \overbrace{\frac{20}{3} + 72\left(\frac{1}{5}\right)^n - \frac{230}{3}\left(\frac{1}{4}\right)^n}^{\text{zero-state}} \overbrace{-2\left(\frac{1}{5}\right)^n + \frac{13}{4}\left(\frac{1}{4}\right)^n}^{\text{zero-input}}, \quad n = 0, 1, \ldots$$

$$y(n) = \frac{20}{3} + 70\left(\frac{1}{5}\right)^n - \frac{881}{12}\left(\frac{1}{4}\right)^n, \quad n = 0, 1, \ldots$$

The first four values of $y(n)$ are

$$y(0) = 3.2500 \qquad y(1) = 2.3125 \qquad y(2) = 4.8781 \qquad y(3) = 6.0795$$

The responses are shown in Figure 10.4. The zero-input response (a) is $-2(\frac{1}{5})^n + \frac{13}{4}(\frac{1}{4})^n$, the response due to initial conditions alone. The zero-state response (b) is $\frac{20}{3} + 72(\frac{1}{5})^n - \frac{230}{3}(\frac{1}{4})^n$, the response due to input alone. The transient response (e) is $72(\frac{1}{5})^n - \frac{230}{3}(\frac{1}{4})^n - 2(\frac{1}{5})^n + \frac{13}{4}(\frac{1}{4})^n = 70(\frac{1}{5})^n - \frac{881}{12}(\frac{1}{4})^n$, the response that decays with time. The steady-state response (f) is $\frac{20}{3}u(n)$, the response after the transient response has died out completely. Either the sum of the zero-input and zero-state

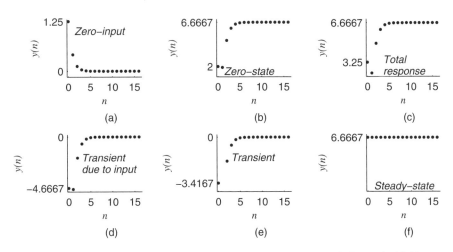

Figure 10.4   Various components of the response of the system in Example 10.11

components (a) and (b) or the sum of the transient and steady-state components (e) and (f) of the response is the complete response (c) of the system. Either the difference of the transient and zero-input components (e) and (a) or the difference of the zero-state and steady-state components (b) and (f) of the response is the transient response (d) of the system due to input alone. The initial and final values of $y(n)$ are 3.25 and $20/3$, respectively. These values can be verified by applying the initial and final value properties to $Y(z)$. We can also verify that the initial conditions at $n = -1$ and at $n = -2$ are satisfied by the zero-input component of the response.   □

The same set of coefficients is involved in both the difference equation and transfer function models of a system. Therefore, either of the models can be used to determine the complete response of a system. In formulating the transfer function model, we have assumed that the initial conditions are zero. However, it should be noted that, with appropriately chosen input that yields the same output as the initial conditions, we can use the transfer function concept, even for problems with nonzero initial conditions. Consider the transform of the output obtained in the example in presenting the right-shift property, $Y(z) = \frac{3}{2}z/(z - \frac{1}{2})$. This equation can be considered as $Y(z) = H(z)X'(z)$ with $H(z) = z/(z - \frac{1}{2})$ and $X'(z) = \frac{3}{2}$. $X'(z)$ corresponds to the time-domain input $\frac{3}{2}\delta(n)$, which produces the same response that results from the initial condition.

10.5.3 System Stability

The zero-input response of a system depends solely on the locations of its poles. A system is considered stable if its zero-input response, due to finite initial conditions, converges, marginally stable if its zero-input response tends to a constant value or oscillates with a constant amplitude, and unstable if its zero-input response diverges. Commonly used marginally stable systems are oscillators, which produce a bounded zero-input response. The response corresponding to each pole p of a system is of the form $r^n e^{jn\theta}$, where the magnitude and phase of the pole are r and θ, respectively. If $r < 1$, then r^n tends to zero as n tends to ∞. If $r > 1$, then r^n tends to ∞ as n tends to ∞. If $r = 1$, then $r^n = 1$ for all n. However, the response tends to infinity, for poles of order more than one lying on the unit-circle, as the expression for the response includes a factor that is a function of n. Poles of any order lying inside the unit-circle do not cause instability. Therefore, we conclude that, from the locations of the poles of a system,

- All the poles, of any order, of a stable system must lie inside the unit-circle. That is, the ROC of $H(z)$ must include the unit-circle.
- Any pole lying outside the unit-circle or any pole of order more than one lying on the unit-circle makes a system unstable.
- A system is marginally stable if it has no poles outside the unit-circle and has poles of order one on the unit-circle.

Figure 10.5 shows pole locations of some transfer functions and the corresponding impulse responses. If all the poles of a system lie inside the unit-circle, the bounded-input bounded-output stability condition (Chapter 4) is satisfied. However, the converse is not necessarily true, since the impulse response is an external description of a system and may not include all its poles. The bounded-input bounded-output stability condition is not satisfied by a marginally stable system.

10.5.4 Realization of Systems

To implement a system, a realization diagram has to be derived. Several realizations of a system are possible, each realization differing in such characteristics as the amount of arithmetic required, sensitivity to coefficient quantization, etc. The z-transform of the output of a Nth order system is given as

$$Y(z) = X(z)H(z) = X(z)\frac{N(z)}{D(z)} = \frac{X(z)(b_N + b_{N-1}z^{-1} + \cdots + b_0 z^{-N})}{1 + a_{N-1}z^{-1} + \cdots + a_0 z^{-N}}$$

Let $R(z) = \frac{X(z)}{D(z)}$. Then, $Y(z) = R(z)N(z)$. Now, the system structure can be realized as a cascade of two systems. The first system, $R(z) = X(z)/D(z)$, has only poles with input $x(n)$ and output $r(n)$. The second system, $Y(z) = R(z)N(z)$, has only zeros with input $r(n)$ and output $y(n)$, where

$$r(n) = x(n) - \sum_{k=1}^{N} a_{N-k}r(n-k) \quad \text{and} \quad y(n) = \sum_{k=0}^{N} b_{N-k}r(n-k)$$

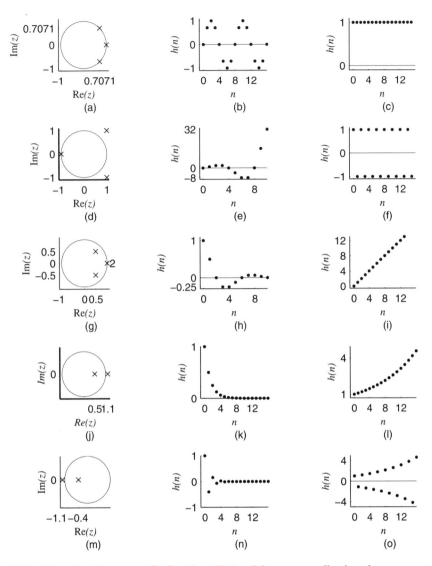

Figure 10.5 The poles of some transfer functions $H(z)$ and the corresponding impulse responses $h(n)$: (a) $H(z) = (z/\sqrt{2})/(z^2 - \sqrt{2}z + 1) = 0.5jz/[z - ((1/\sqrt{2}) - (j/\sqrt{2}))] - 0.5jz/[z - ((1/\sqrt{2}) + (j/\sqrt{2}))]$ and $H(z) = z/(z - 1)$; (b) $h(n) = \sin(\frac{\pi}{4}n)u(n)$; (c) $h(n) = u(n)$; (d) $H(z) = z/(z^2 - 2z + 2) = 0.5jz/[z - (1 - j1)] - 0.5jz/[z - (1 + j1)]$; and $H(z) = z/(z + 1)$; (e) $h(n) = (\sqrt{2})^n \sin(\frac{\pi}{4}n)u(n)$; (f) $h(n) = (-1)^n u(n)$; (g) $H(z) = z(z - 0.5)/(z^2 - z + 0.5) = 0.5z/[z - (0.5 - 0.5j)] + 0.5z/[z - (0.5 + 0.5j)]$; $H(z) = z/(z - 1)^2$; (h) $h(n) = (\frac{1}{\sqrt{2}})^n \cos(\frac{\pi}{4}n)u(n)$ and (i) $h(n) = nu(n)$.; (j) $H(z) = z/(z - 0.5)$ and $H(z) = z/(z - 1.1)$; (k) $h(n) = (0.5)^n u(n)$; (l) $h(n) = (1.1)^n u(n)$; (m) $H(z) = z/(z + 0.4)$, and $H(z) = z/z + 1.1$; (n) $h(n) = (-0.4)^n u(n)$ and (o) $h(n) = (-1.1)^n u(n)$.

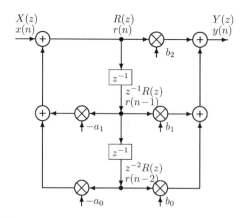

Figure 10.6   Realization of a second-order system

Both the systems can share a set of delay units as the term $r(n - k)$ is common. The realization of a second-order system is shown in Figure 10.6. This realization is known as the canonical form I realization, implying the use of the minimum number of delay elements. A transposed form of a system structure is obtained by: (i) reversing the directions of all the signal flow paths; (ii) replacing the junction points by adders and vice versa; and (iii) interchanging the input and output points. The transposed form of the system in Figure 10.6 is shown in Figure 10.7. This realization is known as the canonical form II realization. This form is derived as follows.

$$H(z) = \frac{Y(z)}{X(z)} = \frac{b_2 z^2 + b_1 z + b_0}{z^2 + a_1 z + a_0} = \frac{b_2 + b_1 z^{-1} + b_0 z^{-2}}{1 + a_1 z^{-1} + a_0 z^{-2}}$$

$$Y(z)(1 + a_1 z^{-1} + a_0 z^{-2}) = X(z)(b_2 + b_1 z^{-1} + b_0 z^{-2})$$

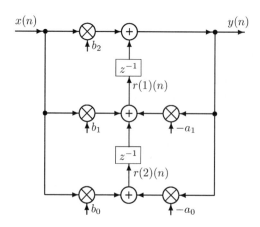

Figure 10.7   Transposed form of the system structure shown in Figure 10.6

$$Y(z) = b_2 X(z) + z^{-1}(b_1 X(z) - a_1 Y(z)) + z^{-2}(b_0 X(z) - a_0 Y(z))$$
$$= b_2 X(z) + z^{-1}\{(b_1 X(z) - a_1 Y(z)) + z^{-1}(b_0 X(z) - a_0 Y(z))\}$$
$$= b_2 X(z) + z^{-1}\{(b_1 X(z) - a_1 Y(z)) + z^{-1} r(2)(z)\}$$
$$= b_2 X(z) + z^{-1} r(1)(z)$$

Therefore, the following difference equations characterize this system structure.

$$y(n) = b_2 x(n) + r(1)(n-1)$$
$$r(1)(n) = b_1 x(n) - a_1 y(n) + r(2)(n-1)$$
$$r(2)(n) = b_0 x(n) - a_0 y(n)$$

These realizations have the advantage of using the coefficients of the transfer function directly.

While this type of realization is applicable to system of any order, it becomes more sensitive to coefficient quantization due to the tendency of the poles and zeros to occur in clusters. Therefore, usually, a higher-order system is decomposed into first- and second-order sections connected in cascade or parallel. In the cascade form, the transfer function is decomposed into a product of first- and second-order transfer functions.

$$H(z) = H_1(z) H_2(z) \cdots H_m(z)$$

In the parallel form, the transfer function is decomposed into a sum of first- and second-order transfer functions.

$$H(z) = g + H_1(z) + H_2(z) +, \cdots, + H_m(z),$$

where g is a constant. Each section is independent and clustering of poles and zeros is avoided as the maximum number of poles and zeros in each section is limited to two. Each second-order section is realized as shown in Figure 10.6 or 10.7.

10.5.5 Feedback Systems

In feedback systems, a fraction of the output signal is fed back and subtracted from the input signal to form the effective input signal. By using negative feedback, we can change the speed of response, reduce sensitivity, improve stability, and increase the range of operation of a system at the cost of reducing the open-loop gain. Consider the feedback system shown in Figure 10.8. The feedback signal $R(z)$ can be expressed as

$$R(z) = F(z) Y(z)$$

Figure 10.8   Two systems connected in a feedback configuration

where $F(z)$ is the feedback transfer function of the system and $Y(z)$ is the output. Now, the error signal $E(z)$ is

$$E(z) = X(z) - R(z) = X(z) - F(z)Y(z)$$

The output $Y(z)$ is expressed as

$$Y(z) = G(z)E(z) = G(z)(X(z) - F(z)Y(z)),$$

where $G(z)$ is the forward transfer function of the system. Therefore, the transfer function of the feedback system is given as

$$\frac{Y(z)}{X(z)} = \frac{G(z)}{1 + G(z)F(z)}$$

If $G(z)$ is very large, the transfer function of the feedback system approximates to the inverse of the feedback transfer function of the system.

$$\frac{Y(z)}{X(z)} \simeq \frac{1}{F(z)}$$

Consider the system with the transfer function

$$G(z) = \frac{z}{z - \frac{3}{2}}$$

$G(z)$ has a pole at $z = \frac{3}{2}$ and, therefore, the system is unstable. We can make a stable feedback system, using this system in the forward path and another suitable system in the feedback path. Let the transfer function of the system in the feedback path be

$$F(z) = \frac{1}{z - \frac{1}{5}}$$

Then, the transfer function of the feedback system is

$$\frac{G(z)}{1 + G(z)F(z)} = \frac{\frac{z}{z - \frac{3}{2}}}{1 + \frac{z}{z - \frac{3}{2}}\frac{1}{z - \frac{1}{5}}} = \frac{z(z - \frac{1}{5})}{z^2 - \frac{7}{10}z + \frac{3}{10}}$$

Now, both the poles of this system lie inside the unit-circle and, therefore, the system is stable.

10.6 Summary

- In this chapter, the theory of the one-sided z-transform, its properties and some applications have been described. As practical systems are causal, the one-sided z-transform is mostly used in practice.
- The z-transform is a generalized version of Fourier analysis. The basis waveforms consist of sinusoids with varying amplitudes or exponentials with complex exponents. The larger set of basis waveforms makes this transform suitable for the analysis of a larger class of signals and systems.
- The z-transform corresponding to a one-dimensional sequence is two-dimensional (a surface), since it is a function of two variables (the real and imaginary parts of the complex frequency). In the frequency-domain, a sequence is uniquely specified by its z-transform along with its ROC. The spectral values along any simply connected closed contour, encircling the origin, in the ROC can be used to reconstruct the corresponding time-domain sequence.
- The inverse z-transform is defined by a contour integral. However, for most practical purposes, the partial fraction method along with a short list of z-transform pairs is adequate to find the inverse z-transform.
- The z-transform is essential for the design, and transient and stability analysis of discrete LTI systems. The z-transform of the impulse response of a system, the transfer function, is a frequency-domain model of the system.

Further Reading

1. Sundararajan, D., *Digital Signal Processing, Theory and Practice*, World Scientific, Singapore, 2003.
2. Lathi, B. P., *Linear Systems and Signals*, Oxford University Press, New York, 2004.

Exercises

10.1 The nonzero values of a sequence $x(n)$ are specified as $\{x(-2) = 1, x(0) = 2, x(3) = -4\}$. Find the unilateral z-transform of
    10.1.1 $x(n - 3)$.
    10.1.2 $x(n - 1)$.
    *10.1.3 $x(n)$.

10.1.4 $x(n+1)$.

10.1.5 $x(n+2)$.

10.1.6 $x(n+4)$.

10.2 Find the nonzero values of the inverse z-transform of

10.2.1 $X(z) = 2 - 3z^{-2} + z^{-4}$.

10.2.2 $X(z) = z^{-2} - 2z^{-5}$.

10.2.3 $X(z) = -2 + 3z^{-1} - z^{-10}$.

*10.2.4 $X(z) = 1 + z^{-1} - z^{-2}$.

10.2.5 $X(z) = z^{-2} + 2z^{-3}$.

10.3 Using the z-transform of $u(n)$ and $nu(n)$, and the shift property, find the z-transform of $x(n)$.

10.3.1 $x(n) = u(n-3) - u(n-5)$.

10.3.2 $x(n) = nu(n-3)$.

10.3.3 $x(n) = n$, $0 \le n \le 4$ and $x(n) = 0$ otherwise.

*10.3.4 $x(n) = (n-2)u(n)$.

10.4 The nonzero values of two sequences $x(n)$ and $h(n)$ are given. Using the z-transform, find the convolution of the sequences $y(n) = x(n) * h(n)$.

*10.4.1 $\{x(0) = 2, x(2) = 3, x(4) = -2\}$ and $\{h(1) = 2, h(3) = -4\}$.

10.4.2 $\{x(1) = 3, x(4) = -4\}$ and $\{h(0) = -2, h(3) = 3\}$.

10.4.3 $\{x(2) = 3, x(4) = -2\}$ and $\{h(1) = 4, h(2) = 1\}$.

10.4.4 $\{x(0) = -4, x(3) = -1\}$ and $\{h(0) = 1, h(2) = -2\}$.

10.4.5 $\{x(2) = 3, x(4) = -1\}$ and $\{h(1) = 2, h(3) = 2\}$.

10.5 Using the multiplication by n property, find the z-transform of $x(n)$.

10.5.1 $x(n) = nu(n)$.

*10.5.2 $x(n) = n2^n u(n)$.

10.5.3 $x(n) = nu(n-2)$.

10.6 Using the multiplication by a^n property, find the z-transform of $x(n)$.

10.6.1 $x(n) = 3^n u(n)$.

*10.6.2 $x(n) = n4^n u(n)$.

10.6.3 $x(n) = 2^n \cos(n)u(n)$.

10.7 Using the summation property, find the sum $y(n) = \sum_{m=0}^{n} x(m)$.

10.7.1 $x(n) = \cos(\frac{2\pi}{4}n)u(n)$.

*10.7.2 $x(n) = \sin(\frac{2\pi}{4}n)u(n)$.

10.7.3 $x(n) = e^{(j\frac{2\pi}{4}n)}u(n)$.

10.7.4 $x(n) = u(n)$.

10.7.5 $x(n) = (n)u(n)$.

10.8 Find the initial and final values of the sequence $x(n)$ corresponding to the transform $X(z)$, using the initial and final value properties.

10.8.1 $X(z) = \frac{z(3z+2)}{(z-\frac{1}{2})(z+\frac{1}{4})}$.

10.8.2 $X(z) = \frac{3z}{(z-2)(z+3)}$.

*10.8.3 $X(z) = \frac{2z(z+3)}{(z-\frac{1}{2})(z-1)}$.

10.8.4 $X(z) = \frac{z}{(z-1)^2}$.

10.8.5 $X(z) = \frac{2z^2}{(z-1)(z+2)}$.

10.9   Given the sample values of the first period, find the z-transform of the semiperi-
        odic function $x(n)u(n)$.

      10.9.1 $\{1, 0, -1, 0\}$.

      10.9.2 $\{0, 1, 0, -1\}$.

      10.9.3 $\{1, j, -1, -j\}$.

    *10.9.4 $\{1, 1, -1, -1\}$.

      10.9.5 $\{0, 1, 2, 1\}$.

10.10   Find the inverse z-transform of $X(z)$ using the inverse DTFT.

     10.10.1 $X(z) = \frac{z}{z-5}$.

     10.10.2 $X(z) = \frac{z}{z-0.8}$.

     10.10.3 $X(z) = \frac{z}{(z-1)^2}$.

10.11   Find the inverse z-transform of

$$X(z) = \frac{z(2z+3)}{\left(z^2 - \frac{2}{15}z - \frac{1}{15}\right)}$$

    List the first four values of $x(n)$.

10.12   Find the inverse z-transform of

$$X(z) = \frac{(3z-1)}{\left(z^2 - \frac{11}{12}z + \frac{1}{6}\right)}$$

    List the first four values of $x(n)$.

*10.13 Find the inverse z-transform of

$$X(z) = \frac{z(z+2)}{(z^2 + 2z + 2)}$$

    List the first four values of $x(n)$.

10.14   Find the inverse z-transform of

$$X(z) = \frac{2z^2 + 1}{(z^2 - z - 6)}$$

    List the first four values of $x(n)$.

10.15   Find the inverse z-transform of

$$X(z) = \frac{z}{\left(z^3 + \frac{3}{2}z^2 - \frac{1}{2}\right)}$$

    List the first four values of $x(n)$.

10.16   Find the inverse z-transform of

$$X(z) = \frac{z^2 - 1}{z^2(z + \frac{1}{3})}$$

List the first four values of $x(n)$.

10.17   Find the first four values of the inverse z-transform of $X(z)$ by the long division method.

10.17.1 $X(z) = \frac{2z^2+2z-3}{z^2-z+1}$.

10.17.2 $X(z) = \frac{z}{z^2+2z-2}$.

10.17.3 $X(z) = \frac{3z^2-z+2}{2z^2+z-3}$.

10.18   Using the z-transform, derive the closed-form expression of the impulse response $h(n)$ of the system governed by the difference equation

$$y(n) = x(n) + 2x(n-1) + x(n-2) + 3y(n-1) - 2y(n-2)$$

with input $x(n)$ and output $y(n)$.

List the first four values of $h(n)$.

10.19   Given the difference equation of a system and the input signal $x(n)$, find the steady-state response of the system.

10.19.1 $y(n) = x(n) + 0.8y(n-1)$ and $x(n) = 2e^{j(\frac{2\pi}{4}n+\frac{\pi}{6})}u(n)$.

*10.19.2 $y(n) = x(n) + 0.7y(n-1)$ and $x(n) = 3\cos(\frac{2\pi}{4}n - \frac{\pi}{3})u(n)$.

10.19.3 $y(n) = x(n) + 0.5y(n-1)$ and $x(n) = 4\sin(\frac{2\pi}{4}n + \frac{\pi}{4})u(n)$.

10.20   Using the z-transform, derive the closed-form expression of the complete response of the system governed by the difference equation

$$y(n) = 2x(n) - x(n-1) + x(n-2) + \frac{7}{6}y(n-1) - \frac{1}{3}y(n-2)$$

with the initial conditions $y(-1) = 2$ and $y(-2) = -3$ and, the input $x(n) = u(n)$, the unit-step function.

List the first four values of $y(n)$.

Deduce the expressions for the zero-input, zero-state, transient, and steady-state responses of the system.

*10.21 Using the z-transform, derive the closed-form expression of the complete response of the system governed by the difference equation

$$y(n) = x(n) + 2x(n-1) - x(n-2) + \frac{5}{4}y(n-1) - \frac{3}{8}y(n-2)$$

with the initial conditions $y(-1) = 2$ and $y(-2) = 1$ and, the input $x(n) = (-1)^n u(n)$.

List the first four values of $y(n)$.
Deduce the expressions for the zero-input, zero-state, transient, and steady-state responses of the system.

10.22 Using the z-transform, derive the closed-form expression of the complete response of the system governed by the difference equation

$$y(n) = 3x(n) - 3x(n - 1) + x(n - 2) + \frac{7}{12}y(n - 1) - \frac{1}{12}y(n - 2)$$

with the initial conditions $y(-1) = 1$ and $y(-2) = 2$ and, the input $x(n) = nu(n)$, the unit-ramp function.
List the first four values of $y(n)$.
Deduce the expressions for the zero-input, zero-state, transient, and steady-state responses of the system.

10.23 Using the z-transform, derive the closed-form expression of the complete response of the system governed by the difference equation

$$y(n) = x(n) - 3x(n - 1) + 2x(n - 2) + y(n - 1) - \frac{2}{9}y(n - 2)$$

with the initial conditions $y(-1) = 3$ and $y(-2) = 2$ and, the input $x(n) = (\frac{1}{4})^n u(n)$.
List the first four values of $y(n)$.
Deduce the expressions for the zero-input, zero-state, transient, and steady-state responses of the system.

10.24 Using the z-transform, derive the closed-form expression of the complete response of the system governed by the difference equation

$$y(n) = x(n) + x(n - 1) - x(n - 2) + \frac{3}{4}y(n - 1) - \frac{1}{8}y(n - 2)$$

with the initial conditions $y(-1) = 1$ and $y(-2) = 2$ and, the input $x(n) = 2\cos(\frac{2\pi}{4}n - \frac{\pi}{6})u(n)$.
List the first four values of $y(n)$.
Deduce the expressions for the zero-input, zero-state, transient, and steady-state responses of the system.

10.25 Using the z-transform, derive the closed-form expression of the impulse response of the cascade system consisting of systems governed by the given difference equations with input $x(n)$ and output $y(n)$.
List the first four values of the impulse response of the cascade system.
*10.25.1 $y(n) = 2x(n) - x(n - 1) + \frac{1}{4}y(n - 1)$ and
$y(n) = 3x(n) + x(n - 1) - \frac{1}{3}y(n - 1)$.

10.25.2 $y(n) = x(n) + x(n-1) - \frac{2}{3}y(n-1)$ and
$y(n) = 2x(n) - x(n-1) - \frac{1}{5}y(n-1)$.

10.25.3 $y(n) = x(n) + 2x(n-1) + \frac{1}{3}y(n-1)$ and
$y(n) = 3x(n) + 2x(n-1) + \frac{1}{2}y(n-1)$.

10.26 Using the z-transform, derive the closed-form expression of the impulse response of the combined system, connected in parallel, consisting of systems governed by the given difference equations with input $x(n)$ and output $y(n)$. List the first four values of the impulse response of the parallel system.

10.26.1 $y(n) = 2x(n) - x(n-1) + \frac{1}{4}y(n-1)$ and
$y(n) = 3x(n) + x(n-1) - \frac{1}{3}y(n-1)$.

10.26.2 $y(n) = x(n) + x(n-1) - \frac{2}{3}y(n-1)$ and
$y(n) = 2x(n) - x(n-1) - \frac{1}{5}y(n-1)$.

*10.26.3 $y(n) = x(n) + 2x(n-1) + \frac{1}{3}y(n-1)$ and
$y(n) = 3x(n) + 2x(n-1) + \frac{1}{2}y(n-1)$.

11

The Laplace Transform

The generalization of the Fourier transform for continuous signals, by including sinu-soids with exponentially varying amplitudes in the set of basis signals, is called the Laplace transform. This generalization makes the transform analysis applicable to a larger class of signals and systems. In Section 11.1, we develop the Laplace transform starting from the definition of the Fourier transform. In Section 11.2, the properties of the Laplace transform are described. In Section 11.3, the inverse Laplace transform is derived. Typical applications of the Laplace transform are presented in Section 11.4.

11.1 The Laplace Transform

We assume, in this chapter, that all the signals are causal, that is $x(t) = 0$, $t < 0$, unless otherwise specified. This leads to the one-sided or unilateral version of the Laplace transform, which is mostly used for practical system analysis. If a signal $x(t)u(t)$ is not Fourier transformable, then its exponentially weighted version, $x(t)e^{-\sigma t}$, may be Fourier transformable for the positive real quantity $\sigma > 0$. If $x(t)u(t)$ is Fourier transformable, $x(t)e^{-\sigma t}$ may still be transformable for some values of $\sigma < 0$. The Fourier transform of this signal is

$$\int_0^\infty (x(t)e^{-\sigma t})e^{-j\omega t}\,\mathrm{d}t$$

By combining the exponential factors, we get,

$$X(\sigma + j\omega) = \int_0^\infty x(t)e^{-(\sigma+j\omega)t}\,\mathrm{d}t \qquad (11.1)$$

This equation can be interpreted as the generalized Fourier transform of the signal $x(t)$ using exponentials with complex exponents or sinusoids with varying amplitudes

A Practical Approach to Signals and Systems   D. Sundararajan
© 2008 John Wiley & Sons (Asia) Pte Ltd

as the basis signals. Therefore, a signal may be decomposed in terms of constant-amplitude sinusoids or exponentially decaying sinusoids or exponentially growing sinusoids, or an infinite combination of these types of sinusoids. By substituting $s = \sigma + j\omega$, we get the defining equation of the Laplace transform of $x(t)$ as

$$X(s) = \int_{0^-}^{\infty} x(t)e^{-st}dt$$

Note that the lower limit is assumed, in this book, to be 0^-, where $t = 0^-$ is the instant immediately before $t = 0$. This implies that a jump discontinuity or an impulse component of the function $x(t)$ at $t = 0$ is included in the integral. In addition, this lower limit enables the use of the initial conditions at $t = 0^-$ directly. In practical applications, we are more likely to know the initial conditions before the input signal is applied, rather than after.

While $X(j\omega)$ is the FT of $x(t)$, $X(s) = X(\sigma + j\omega)$ is the FT of $x(t)e^{-\sigma t}$ for all values of σ for which $\int_{0^-}^{\infty} |x(t)e^{-\sigma t}|dt < \infty$. If the value zero is included in these values of σ, then $X(j\omega)$ can be obtained from $X(s)$ by the substitution $s = j\omega$. The Laplace transform of $x(t)$, $X(s)$, exists for $\text{Re}(s) > \sigma_0$ if $|x(t)| \leq Me^{\sigma_0 t}$ for some constants M and σ_0. For example, the signal e^{t^2} has no Laplace transform. In essence, the Laplace transform of a signal, whether it is converging or not, is the FT of all its versions, obtained by multiplying it by a real exponential of the form $e^{-\sigma t}$, so that the modified signal is guaranteed to converge.

The advantages of the Laplace transform include: the pictorial description of the behavior of the system obtained by the use of the complex frequency; the ability to analyze unstable systems or systems with exponentially growing inputs; automatic inclusion of the initial conditions of the system in finding the output; and easier manipulation of the expressions involving the variable s rather than $j\omega$.

Example 11.1. Find the Laplace transform of the unit-impulse signal, $\delta(t)$.

Solution
Using the Laplace transform definition, we get

$$X(s) = \int_{0^-}^{\infty} \delta(t)e^{-st}dt = 1 \qquad \text{for all } s \qquad \text{and} \qquad \delta(t) \Longleftrightarrow 1, \text{ for all } s$$

This transform pair can also be obtained by applying a limit process to any function that degenerates into an impulse and its transform.   □

The exponential signal, $e^{-at}u(t)$, is fundamental to the study of linear continuous systems, as it is more convenient to express the natural response of systems in that form.

Example 11.2. Find the Laplace transform of the real exponential signal, $x(t) = e^{-at}u(t)$. Deduce the Laplace transform of the unit step signal, $x(t) = u(t)$.

Solution
Using the Laplace transform definition, we get

$$X(s) = \int_{0^-}^{\infty} e^{-at}u(t)e^{-st}dt = \int_{0^-}^{\infty} e^{-at}e^{-st}dt$$

$$= \int_{0^-}^{\infty} e^{-(s+a)t}dt = -\frac{e^{-(s+a)t}}{s+a}\bigg|_{0^-}^{\infty} = \frac{1}{s+a} - \frac{e^{-(s+a)t}}{s+a}\bigg|_{t=\infty}$$

For the integral to converge, $\lim_{t\to\infty} e^{-(s+a)t}$ must be equal to zero. This implies that the real part of $(s+a)$ is greater than zero and, hence, the convergence condition is $\text{Re}(s) > -a$. This condition describes a region in the s-plane (a complex plane used for displaying the Laplace transform) that lies to the right of the vertical line characterized by the equation $\text{Re}(s) = -a$. Note that the Fourier transform of $e^{-at}u(t)$ does not exist for negative values of a, whereas the Laplace transform exists for all values of a as long as $\text{Re}(s) > -a$. Therefore, we get the Laplace transform pair

$$e^{-at}u(t) \Longleftrightarrow \frac{1}{s+a} \qquad \text{Re}(s) > -a$$

This transform pair remains the same for complex-valued a with the convergence condition, $\text{Re}(s) > \text{Re}(-a)$.
With $a = 0$, we get the transform pair

$$u(t) \Longleftrightarrow \frac{1}{s} \qquad \text{Re}(s) > 0 \qquad \qquad \square$$

The region, consisting of the set of all values of s in the s-plane for which the defining integral of the Laplace transform converges, is called the region of convergence (ROC). For the signal in Example 11.2, the region to the right of the vertical line at $\text{Re}(s) = \text{Re}(-a)$ is the ROC.

The frequency content of a signal is usually displayed by the locations of zeros and poles, and the magnitude of its Laplace transform. Figure 11.1(a) shows the pole-zero plot and Figure 11.1(b) shows the magnitude of the Laplace transform $X(s) = 1/(s+2)$ of the signal $e^{-2t}u(t)$. When $s = -2$, $|X(s)| = \infty$. This point marked by the cross in Figure 11.1(a) is called a pole of $X(s)$ (the peak in Figure 11.1b). Except for a constant factor, the Laplace transform of a signal can be reconstructed from its pole-zero plot. For all values of s in the ROC (the region to the right of the dotted vertical line at $\text{Re}(s) = -2$ shown in Figure 11.1a), $X(s)$ exists and is a valid representation of the signal. In general, the ROC of a Laplace transform is the region in the s plane that is to the right of the vertical line passing through the rightmost pole location. If

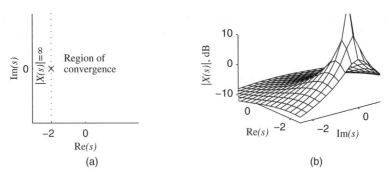

Figure 11.1   (a) The pole-zero plot of the Laplace transform, $1/(s + 2)$, of the signal, $e^{-2t}u(t)$; (b) the magnitude of the Laplace transform

the ROC includes the imaginary axis, $\text{Re}(s) = 0$, in the s plane (as in Figure 11.1a), then the FT can be obtained from the Laplace transform by replacing s with $j\omega$.

Example 11.3.   Find the Laplace transform of the signal $e^{j\omega_0 t}u(t)$. Deduce the Laplace transform of $\cos(\omega_0 t)u(t)$.

Solution
Using the transform of $e^{-at}u(t)$ with $a = -j\omega_0$, we get

$$e^{j\omega_0 t}u(t) \Longleftrightarrow \frac{1}{s - j\omega_0} \qquad \text{Re}(s - j\omega_0) = \text{Re}(s) > 0$$

Using the fact that $2\cos(\omega_0 t) = (e^{j\omega_0 t} + e^{-j\omega_0 t})$, we get

$$2X(s) = \frac{1}{s - j\omega_0} + \frac{1}{s + j\omega_0} \qquad \text{Re}(s) > 0$$

$$\cos(\omega_0 t)u(t) \Longleftrightarrow \frac{s}{s^2 + \omega_0^2} \qquad \text{Re}(s) > 0 \qquad \square$$

Figure 11.2(a) shows the pole-zero plot and Figure 11.2(b) shows the magnitude of the Laplace transform, $s/[s^2 + (\frac{\pi}{4})^2]$ of the signal $\cos(\frac{\pi}{4}t)u(t)$. When $s = \pm j\frac{\pi}{4}$, $|X(s)| = \infty$. These points marked by the symbol $\times$ in Figure 11.2(a) are the poles of $X(s)$ (the peaks in Figure 11.2(b)). When $s = 0$, $X(s) = 0$. This point marked by the symbol o in Figure 11.2(a) is the zero of $X(s)$ (the valley in Figure 11.2(b)).

11.1.1 Relationship Between the Laplace Transform and the z-Transform

A relationship between the FT of a sampled signal and the DTFT of the corresponding discrete signal was derived in Chapter 9. Now, we derive a similar relationship between

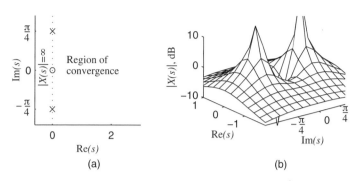

Figure 11.2 (a) The pole-zero plot of the Laplace transform $s/[s^2 + \left(\frac{\pi}{4}\right)^2]$ of the signal, $\cos(\frac{\pi}{4}t)u(t)$; (b) the magnitude of the Laplace transform

the Laplace transform and the z-transform. The sampled version of a signal $x(t)u(t)$ is $x_s(t) = \sum_{n=0}^{\infty} x(n)\delta(t - n)$, with a sampling interval of one second. As the Laplace transform of $\delta(t - n)$ is e^{-sn} and due to the linearity property of the Laplace transform, the Laplace transform of the sampled signal $x_s(t)$ is given by

$$X_s(s) = \sum_{n=0}^{\infty} x(n)e^{-sn}$$

With $z = e^s$, this equation becomes

$$X_s(s) = \sum_{n=0}^{\infty} x(n)z^{-n}$$

The right-hand side of this equation is the z-transform of $x(n)$.

For example, let $x(t) = e^{-2t}u(t)$. Then, the corresponding discrete signal is $x(n) = e^{-2n}u(n)$ with its z-transform $z/(z - e^{-2})$ Now, the Laplace transform of the sampled version of $x(t)$, $x_s(t) = \sum_{n=0}^{\infty} e^{-2n}\delta(t - n)$, is $e^s/(e^s - e^{-2})$, which is obtained from $z/(z - e^{-2})$ by the substitution $z = e^s$.

11.2 Properties of the Laplace Transform

Properties present the frequency-domain effect of time-domain characteristics and operations on signals and vice versa. In addition, they are used to find new transform pairs more easily.

11.2.1 Linearity

The Laplace transform of a linear combination of signals is the same linear combination of their individual Laplace transforms. If $x(t) \Longleftrightarrow X(s)$ and $y(t) \Longleftrightarrow Y(s)$, then

$ax(t) + by(t) \Longleftrightarrow aX(s) + bY(s)$, where a and b are arbitrary constants. This property is due to the linearity of the defining integral of the Laplace transform. We use this property often to decompose a time-domain function in finding its Laplace transform (as in Example 11.3) and to decompose a transform in order to find its inverse.

11.2.2 Time Shifting

If $x(t)u(t) \Longleftrightarrow X(s)$, then

$$x(t - t_0)u(t - t_0) \qquad t_0 \geq 0 \Longleftrightarrow e^{-st_0}X(s)$$

Now, $e^{-st_0} = e^{-(\sigma + j\omega)t_0} = e^{-\sigma t_0}e^{-j\omega t_0}$. The term $e^{-j\omega t_0}$ is the linear shift of the phase of sinusoids, as in the case of the Fourier analysis. Due to the fact that the basis functions are sinusoids with varying amplitudes, we need another factor $e^{-\sigma t_0}$ to set the amplitude of the sinusoids appropriately so that the reconstructed waveform is the exact time-shifted version of $x(t)$.

Consider the waveform $x(t)u(t) = e^{-0.1t}u(t)$ and its shifted version $e^{-0.1(t-8)}$ $u(t - 8)$. The Laplace transforms of the two functions are, respectively, $1/(s + 0.1)$ and $e^{-8s}/(s + 0.1)$.

This property holds only for causal signals and for right shift only. Remember that the transform of the shifted signal is expressed in terms of that of the original signal, which is assumed to be zero for $t < 0$. To find the transform of signals such as $x(t - t_0)u(t)$ and $x(t)u(t - t_0)$, express the signal so that the arguments of the signal and the unit-step signal are the same and then apply the property. Of course, the transform can also be computed using the defining integral.

11.2.3 Frequency Shifting

If $x(t)u(t) \Longleftrightarrow X(s)$, then

$$e^{s_0 t}x(t)u(t) \Longleftrightarrow X(s - s_0)$$

Multiplying the signal $x(t)$ by the exponential $e^{s_0 t}$ amounts to changing the complex frequency of its spectral components by s_0. Therefore, the spectrum $X(s)$ is shifted in the s-plane by the amount s_0.

Consider finding the transform of the signal $e^{2t}u(t)$. This signal can be considered as the unit-step, $u(t)$, multiplied by the exponential with $s_0 = 2$. Therefore, according to this property, the transform of $e^{2t}u(t)$ is the transform of $u(t)$, which is $\frac{1}{s}$, with the substitution $s = s - 2$, that is $1/(s - 2)$.

11.2.4 Time-differentiation

The time-differentiation property is used to express the transform of the derivative, $dx(t)/dt$, of a signal $x(t)$ in terms of its transform $X(s)$. If $x(t) \Longleftrightarrow X(s)$, then

$$\frac{dx(t)}{dt} \Longleftrightarrow sX(s) - x(0^-)$$

As the signal, in the frequency-domain, is expressed in terms of exponentials e^{st} and the derivative of the exponential is se^{st}, the differentiation of a signal in time-domain corresponds to multiplication of its transform by the frequency variable s, in addition to a constant term due to the initial value of the signal at $t = 0^-$. The point is that two signals $x(t)$ and $x(t)u(t)$ have the same unilateral Laplace transform. However, the Laplace transforms of their derivatives will be different if $x(t)$ and $x(t)u(t)$ have different values of discontinuity at $t = 0$. The derivative of $x(t)$ with a different value of $x(0^-)$ differs from that of $x(t)u(t)$, only at $t = 0$. $sX(s)$ is the derivative of $x(t)u(t)$, that is the derivative of $x(t)$ with $x(0^-) = 0$.

A signal $x(t)$, with step discontinuities, for example, at $t = 0$ of height $(x(0^+) - x(0^-))$ and at $t = t_1 > 0$ of height $(x(t_1^+) - x(t_1^-))$ can be expressed as

$$x(t) = x_c(t) + (x(0^+) - x(0^-))u(t) + (x(t_1^+) - x(t_1^-))u(t - t_1),$$

where $x_c(t)$ is $x(t)$ with the discontinuities removed and $x(t_1^+)$ and $x(t_1^-)$ are, respectively, the right- and left-hand limits of $x(t)$ at $t = t_1$. The derivative of $x(t)$ is given by generalized function theory as

$$\frac{dx(t)}{dt} = \frac{dx_c(t)}{dt} + (x(0^+) - x(0^-))\delta(t) + (x(t_1^+) - x(t_1^-))\delta(t - t_1),$$

where $dx_c(t)/dt$ is the ordinary derivative of $x_c(t)$ at all t except at $t = 0$ and $t = t_1$. The Laplace transform of $dx(t)/dt$ is given by

$$sX_c(s) + (x(0^+) - x(0^-)) + (x(t_1^+) - x(t_1^-))e^{-st_1} = sX(s) - x(0^-)$$

Consider the signal, shown in Figure 3.8(a),

$$x(t) = u(-t - 1) + e^{-t}(u(t + 1) - u(t - 1)) + 2t(u(t - 1) - u(t - 2))$$

$$+ \cos\left(\frac{\pi}{2}t\right)u(t - 2)$$

$$= x_c(t) + 1.7183u(t + 1) + 1.6321u(t - 1) - 5u(t - 2),$$

and its derivative

$$\frac{dx(t)}{dt} = 0 - e^{-t}(u(t+1) - u(t-1)) + 2(u(t-1) - u(t-2))$$

$$- \left(\frac{\pi}{2}\right) \sin\left(\frac{\pi}{2}t\right) u(t-2)$$

$$+ 1.7183\delta(t+1) + 1.6321\delta(t-1) - 5\delta(t-2),$$

shown in Figure 3.8(b). The transform of $x(t)$ is

$$X(s) = \frac{1}{s+1} - \frac{e^{-1}e^{-s}}{s+1} + \frac{2e^{-s}}{s^2} + \frac{2e^{-s}}{s} - \frac{2e^{-2s}}{s^2} - \frac{4e^{-2s}}{s} - \frac{se^{-2s}}{s^2 + \left(\frac{\pi}{2}\right)^2}$$

Remember that the value of $x(t)$ for $t < 0$ is ignored in computing the unilateral Laplace transform. The term $\cos(\frac{\pi}{2}t)u(t-2)$ can be rewritten as $-\cos(\frac{\pi}{2}(t-2))$ $u(t-2)$ and then the time-shifting theorem can be applied to find its transform. The initial value is $x(0^-) = 1$. Now,

$$sX(s) - x(0^-) = \frac{s}{s+1} - \frac{se^{-1}e^{-s}}{s+1} + \frac{2e^{-s}}{s} + 2e^{-s} - \frac{2e^{-2s}}{s} - 4e^{-2s} - \frac{s^2 e^{-2s}}{s^2 + \left(\frac{\pi}{2}\right)^2} - 1$$

The transform of $dx(t)/dt$ is

$$-\frac{1}{s+1} + \frac{e^{-1}e^{-s}}{s+1} + \frac{2e^{-s}}{s} - \frac{2e^{-2s}}{s} + \frac{\left(\frac{\pi}{2}\right)^2 e^{-2s}}{s^2 + \left(\frac{\pi}{2}\right)^2} + (2 - e^{-1})e^{-s} - 5e^{-2s}$$

$$= sX(s) - x(0^-)$$

This property can be extended, by repeated application, to find the transform of higher-order derivatives. For example,

$$\frac{d}{dt}\left(\frac{dx(t)}{dt}\right) = \frac{d^2x(t)}{dt^2} \iff$$

$$s(sX(s) - x(0^-)) - \frac{dx(t)}{dt}\Big|_{t=0^-} = s^2 X(s) - sx(0^-) - \frac{dx(t)}{dt}\Big|_{t=0^-}$$

One common application of this property is in the modeling of system components such as an inductor. The relationship between the current $i(t)$ through an inductor of value L henries and the voltage $v(t)$ across it is $v(t) = L(di(t)/dt)$. Assuming the initial value of current in the inductor is $i(0^-)$, using this property, we get the Laplace transform of the voltage across the inductor as $V(s) = L(sI(s) - i(0^-))$.

The application of time-differentiation and linearity properties reduces a differential equation to an algebraic equation, which can be easily solved. Consider solving the differential equation $dy(t)/dt + \frac{1}{2}y(t) = 0$, with the initial condition $y(0^-) = 3$. The solution using time-domain method (Chapter 5) is $y(t) = 3e^{-\frac{1}{2}t}u(t)$. Taking the transform of the differential equation, we get $sY(s) - 3 + \frac{1}{2}Y(s) = 0$. Solving for $Y(s)$, $Y(s) = 3/(s + \frac{1}{2})$. Finding the inverse transform, we get the same solution.

11.2.5 Integration

If $x(t) \Longleftrightarrow X(s)$, then

$$\int_{0^-}^{t} x(\tau) \, d\tau \Longleftrightarrow \frac{1}{s} X(s)$$

As the signal, in the frequency-domain, is expressed in terms of exponentials e^{st} and the integral of the exponential is e^{st}/s, the integration of a signal in time-domain corresponds to a division of its transform by the frequency variable s. From another point of view, the product $(1/s)X(s)$ corresponds to the convolution of $x(t)$ and $u(t)$ in the time-domain, which, of course, is equivalent to the integral of $x(t)$ from 0 to t. For example, the transform of the unit-step signal, which is the integral of the unit impulse function with $X(s) = 1$, is $1/s$. Similarly, $tu(t) \Longleftrightarrow 1/s^2$.

Consider the function $\sin(t)u(t)$ with its transform $1/(s^2 + 1)$. Using this property,

$$\int_{0^-}^{t} \sin(\tau) d\tau \Longleftrightarrow \frac{1}{s(s^2 + 1)}$$

Finding the inverse transform, we get $(1 - \cos(t))u(t)$, which can be verified to be the time-domain integral of the sine function.

As the definite integral $\int_{-\infty}^{0^-} x(\tau) \, d\tau$ is a constant,

$$\int_{-\infty}^{t} x(\tau) \, d\tau = \int_{-\infty}^{0^-} x(\tau) \, d\tau + \int_{0^-}^{t} x(\tau) \, d\tau \Longleftrightarrow \frac{1}{s} \int_{-\infty}^{0^-} x(\tau) \, d\tau + \frac{1}{s} X(s)$$

One common application of this property is in the modeling of system components such as a capacitor. The relationship between the current $i(t)$ through a capacitor of value C farads and the voltage $v(t)$ across it is $v(t) = \frac{1}{C} \int_{0^-}^{t} i(\tau) d\tau + v(0^-)$, where $v(0^-)$ is the initial voltage across the capacitor. Using this property, we get the Laplace transform of the voltage across the capacitor as

$$V(s) = \frac{I(s)}{sC} + \frac{v(0^-)}{s}$$

11.2.6 Time Scaling

If $x(t)u(t) \Longleftrightarrow X(s)$, then

$$x(at)u(at) \Longleftrightarrow \frac{1}{a} X\left(\frac{s}{a}\right) \qquad a > 0$$

The Laplace transform of $x(at)u(at)$, from the definition, is

$$\int_{0^-}^{\infty} x(at)u(at)e^{-st}\,dt$$

Substituting $at = \tau$, we get $t = \tau/a$ and $dt = d\tau/a$. Note that $u(at) = u(t)$, $a > 0$. With these changes, the transform becomes

$$\frac{1}{a}\int_{0^-}^{\infty} x(\tau)e^{-\frac{s}{a}\tau}\,d\tau = \frac{1}{a} X\left(\frac{s}{a}\right)$$

Compression (expansion) of a signal in the time-domain, by changing t to at, results in the expansion (compression) of its spectrum with the change s to s/a, in addition to scaling by $1/a$ (to take into account the change in energy).
Consider the transform pair

$$e^{-2t}\sin(t)u(t) \Longleftrightarrow \frac{1}{s^2 + 4s + 5} = \frac{1}{(s+2-j)(s+2+j)}$$

The two poles are located at $-2 + j1$ and $-2 - j1$. With $a = 2$, we get

$$e^{-4t}\sin(2t)u(2t) \Longleftrightarrow \frac{1}{2}\frac{1}{\left(\frac{s}{2}\right)^2 + 4\left(\frac{s}{2}\right) + 5} = \frac{2}{(s+4-j2)(s+4+j2)}$$

The two poles are located at $-4 + j2$ and $-4 - j2$.

11.2.7 Convolution in Time

If $x(t)u(t) \Longleftrightarrow X(s)$ and $h(t)u(t) \Longleftrightarrow H(s)$, then

$$y(t) = x(t)u(t) * h(t)u(t) = \int_{0}^{\infty} x(\tau)h(t-\tau)\,d\tau \Longleftrightarrow X(s)H(s)$$

The FT of $x(t)e^{-\sigma t}$ is the Laplace transform $X(s)$ of $x(t)$. The convolution of $x(t)e^{-\sigma t}$ and $h(t)e^{-\sigma t}$ corresponds to $X(s)H(s)$ in the frequency-domain. The inverse FT of

$X(s)H(s)$, therefore, is the convolution of $x(t)e^{-\sigma t}$ and $h(t)e^{-\sigma t}$ given by

$$\int_0^\infty x(\tau)e^{-\sigma\tau}h(t-\tau)e^{-\sigma(t-\tau)}d\tau = e^{-\sigma t}\int_0^\infty x(\tau)h(t-\tau)d\tau = e^{-\sigma t}(x(t) * h(t))$$

As finding the inverse Laplace transform is the same as finding the inverse FT in addition to multiplying the signal by $e^{\sigma t}$, as will be seen later, we get the convolution of $x(t)$ and $h(t)$ by finding the inverse Laplace transform of $X(s)H(s)$.

Consider the convolution of $e^{2t}u(t)$ and $e^{-2t}u(t)$. The inverse of the product of their transforms,

$$\frac{1}{(s-2)(s+2)} = \frac{1}{4}\left(\frac{1}{(s-2)} - \frac{1}{(s+2)}\right),$$

is the convolution output $\frac{1}{4}(e^{2t} - e^{-2t})u(t)$.

11.2.8 Multiplication by t

If $x(t)u(t) \Longleftrightarrow X(s)$, then

$$tx(t)u(t) \Longleftrightarrow -\frac{dX(s)}{ds}$$

Differentiating the defining expression for $-X(s)$ with respect to s, we get

$$-\frac{dX(s)}{ds} = -\frac{d}{ds}\left(\int_{0^-}^\infty x(t)u(t)e^{-st}dt\right) = \int_{0^-}^\infty tx(t)e^{-st}dt$$

In general,

$$t^n x(t)u(t) \Longleftrightarrow (-1)^n \frac{d^n X(s)}{ds^n} \qquad n = 0, 1, 2, \ldots$$

For example, $t\delta(t) = 0 \Longleftrightarrow -d(1)/ds = 0$. Another example is $tu(t) \Longleftrightarrow -d(1/s)/ds = 1/s^2$.

11.2.9 Initial Value

If only the initial and final values of a function $x(t)$ are required, these values can be found directly, from $X(s)$, using the following properties rather than finding the function $x(t)$ by inverting $X(s)$.

If $x(t) \Longleftrightarrow X(s)$ and the degree of the numerator polynomial of $X(s)$ is less than that of the denominator polynomial, then

$$x(0^+) = \lim_{s \to \infty} sX(s)$$

As $s \to \infty$, the value of any term with a higher-order denominator tends to zero and

$$\lim_{s \to \infty} sX(s) = \lim_{s \to \infty} \left(\frac{sA_1}{s - s_1} + \frac{sA_2}{s - s_2} + \cdots + \frac{sA_N}{s - s_N} \right) = A_1 + A_2 + \cdots + A_N$$

The inverse transform of $X(s)$, as $t \to 0$, is

$$x(t) = A_1 e^{s_1 t} + A_2 e^{s_2 t} + \cdots + A_N e^{s_N t}$$

The right-hand limit of $x(t)$, as $t \to 0$, is

$$x(0^+) = A_1 + A_2 + \cdots + A_N = \lim_{s \to \infty} sX(s)$$

Similarly,

$$\frac{dx(t)}{dt}\Big|_{t=0^+} = \lim_{s \to \infty} (s^2 X(s) - sx(0^-))$$

11.2.10 Final Value

If $x(t) \Longleftrightarrow X(s)$ and the ROC of $sX(s)$ includes the $j\omega$ axis, then

$$\lim_{t \to \infty} x(t) = \lim_{s \to 0} sX(s)$$

As $t \to \infty$, the value $x(\infty)$, if it is nonzero, is solely due to the scaled unit-step component of $x(t)$. Multiplying $X(s)$ by s and setting $s = 0$ is just finding the partial fraction coefficient of the unit-step component of $x(t)$.

The initial and final values from the transform $1/[s(s + 2)]$ are

$$x(0^+) = \lim_{s \to \infty} \frac{1}{s + 2} = 0 \quad \text{and} \quad \lim_{t \to \infty} x(t) = \lim_{s \to 0} \frac{1}{s + 2} = \frac{1}{2}$$

11.2.11 Transform of Semiperiodic Functions

Consider the function $x(t)u(t)$ that is periodic of period T for $t \geq 0$, that is $x(t + T) = x(t)$, $t \geq 0$. Let $x_1(t) = x(t)u(t) - x(t - T)u(t - T) \Longleftrightarrow X_1(s)$. $x_1(t)$ is equal to $x(t)u(t)$ over its first period and is zero elsewhere. Then,

$$x(t)u(t) = x_1(t) + x_1(t - T) + \cdots + x_1(t - nT) + \cdots$$

Using the time-shifting property, the transform of $x(t)u(t)$ is

$$X(s) = X_1(s)(1 + e^{-sT} + \cdots + e^{-nsT} + \cdots) = \frac{X_1(s)}{1 - e^{-sT}}$$

Let us find the transform of a semiperiodic square wave, the first period of which is defined as

$$x_1(t) = \begin{cases} 1 & \text{for } 0 < t < 2 \\ 0 & \text{for } 2 < t < 4 \end{cases}$$

As $x_1(t) = (u(t) - u(t - 2))$, $X_1(s) = (1 - e^{-2s})/s$. From the property,

$$X(s) = \frac{1}{(1 - e^{-4s})} \frac{(1 - e^{-2s})}{s} = \frac{1}{s(1 + e^{-2s})}$$

11.3 The Inverse Laplace Transform

Consider the transform pair $x(t)u(t) \Longleftrightarrow 1/(s - 4)$, $\text{Re}(s) > 4$. Multiplying $x(t)u(t)$ by $e^{-5t}u(t)$ gives $x(t)e^{-5t}u(t) \Longleftrightarrow 1/[(s + 5) - 4] = 1/(s + 1)$ $\text{Re}(s) > -1$, due to the frequency-shift property. Now, the ROC includes the $j\omega$ axis in the s-plane. Let us substitute $s = j\omega$ in $1/(s + 1)$ to get $1/(j\omega + 1)$. The inverse FT of this transform is the signal $e^{-t}u(t) = x(t)e^{-5t}u(t)$. Now, multiplying both sides by $e^{5t}u(t)$ gives the original time-domain signal $x(t)u(t) = e^{4t}u(t)$. This way of finding the inverse Laplace transform gives us a clear understanding of how the Laplace transform is the generalized version of the FT.

The inverse FT of $X(\sigma + j\omega)$, defined in (Equation 11.1), is given as

$$x(t)e^{-\sigma t} = \frac{1}{2\pi} \int_{-\infty}^{\infty} X(\sigma + j\omega)e^{j\omega t} d\omega$$

Multiplying both sides by $e^{\sigma t}$, we get

$$x(t) = \frac{1}{2\pi} \int_{-\infty}^{\infty} X(\sigma + j\omega)e^{(\sigma + j\omega)t} d\omega$$

The complex frequency $(\sigma + j\omega)$ can be replaced by a complex variable $s = (\sigma + j\omega)$ with the limits of the integral changed to $\sigma - j\infty$ and $\sigma + j\infty$. As $ds = jd\omega$, we get the inverse Laplace transform of $X(s)$ as

$$x(t) = \frac{1}{2\pi j} \int_{\sigma-j\infty}^{\sigma+j\infty} X(s)e^{st}ds$$

where σ is any real value that lies in the ROC of $X(s)$. Note that the integral converges to the value zero for $t < 0$ and to the mid-point value at any discontinuity of $x(t)$. This equation is not often used for finding the inverse transform, as it requires integration in the complex plane. The partial fraction method, which is essentially the same as that described in Chapter 10, is commonly used. The difference is that the partial fraction terms are of the form $k/(s - p)$ in contrast to $kz/(z - p)$, as shown in Example 11.4.

11.4 Applications of the Laplace Transform

11.4.1 Transfer Function and System Response

Consider the second-order differential equation of a causal LTI continuous system relating the input $x(t)$ and the output $y(t)$,

$$\frac{d^2y(t)}{dt^2} + a_1\frac{dy(t)}{dt} + a_0y(t) = b_2\frac{d^2x(t)}{dt^2} + b_1\frac{dx(t)}{dt} + b_0x(t)$$

Taking the Laplace transform of both sides, we get, assuming initial conditions are all zero,

$$(s^2 + a_1s + a_0)Y(s) = (b_2s^2 + b_1s + b_0)X(s)$$

The transfer function $H(s)$, which is the ratio of the transforms of the output and the input signals with the initial conditions zero, is obtained as

$$H(s) = \frac{Y(s)}{X(s)} = \frac{b_2s^2 + b_1s + b_0}{s^2 + a_1s + a_0} = \frac{\sum_{l=0}^{2} b_l s^l}{s^2 + \sum_{l=0}^{1} a_l s^l}$$

In general,

$$H(s) = \frac{Y(s)}{X(s)} = \frac{b_M s^M + b_{M-1}s^{M-1} + \cdots + b_1s + b_0}{s^N + a_{N-1}s^{N-1} + \cdots + a_1s + a_0}$$

If the input to the system is the unit-impulse signal, then its transform is one and $H(s) = Y(s)$. That is, the transform of the impulse response is the transfer function of the system. For stable systems, the frequency response $H(j\omega)$ is obtained from $H(s)$ by replacing s by $j\omega$.

11.4.2 Characterization of a System by its Poles and Zeros

The numerator and denominator polynomials of the transfer function can be factored to get

$$H(s) = K \frac{(s - z_1)(s - z_2) \cdots (s - z_M)}{(s - p_1)(s - p_2) \cdots (s - p_N)} = K \frac{\prod_{l=1}^{M}(s - z_l)}{\prod_{l=1}^{N}(s - p_l)},$$

where K is a constant. As the coefficients of the polynomials of $H(s)$ are real for practical systems, the zeros and poles are real-valued or they always occur as complex-conjugate pairs.

The pole-zero plot of the transfer function $H(s)$ of a system is a pictorial description of its characteristics, such as speed of response, frequency selectivity, and stability. Poles located farther from the imaginary axis in the left half of the s-plane result in a fast-responding system with its transient response decaying rapidly. On the other hand, poles located close to the imaginary axis in the left half of the s-plane result in a sluggish system. Complex-conjugate poles located in the left-half of the s-plane result in an oscillatory transient response that decays with time. Complex-conjugate poles located on the imaginary axis result in a steady oscillatory transient response. Poles located on the positive real axis in the left-half of the s-plane result in an exponentially decaying transient response. The frequency components of an input signal with frequencies close to a zero will be suppressed while those close to a pole will be readily transmitted. Poles located symmetrically about the negative real axis in the left half of the s-plane and close to the imaginary axis in the passband results in a lowpass system that more readily transmits low-frequency signals than high-frequency signals. Zeros placed in the stopband further enhances the lowpass character of the frequency response. For example, pole-zero plots of some lowpass filters are shown in Figures 11.1 and 11.12. The stability of a system can also be determined from its pole-zero plot, as presented later.

Example 11.4. Find the zero-input, zero-state, transient, steady-state, and complete responses of the system governed by the differential equation

$$\frac{d^2 y(t)}{dt^2} + 4\frac{dy(t)}{dt} + 4y(t) = \frac{d^2 x(t)}{dt^2} + \frac{dx(t)}{dt} + 2x(t)$$

with the initial conditions

$$y(0^-) = 2 \qquad \text{and} \qquad \frac{dy(t)}{dt}\Big|_{t=0^-} = 3$$

and the input $x(t) = u(t)$, the unit-step function.

Solution

The Laplace transforms of the terms of the differential equation are

$$x(t) \Longleftrightarrow \frac{1}{s} \qquad \frac{dx(t)}{dt} \Longleftrightarrow 1 \qquad \frac{d^2x(t)}{dt^2} \Longleftrightarrow s$$

$$y(t) \Longleftrightarrow Y(s) \qquad \frac{dy(t)}{dt} \Longleftrightarrow sY(s) - 2 \qquad \frac{d^2y(t)}{dt^2} \Longleftrightarrow s^2Y(s) - 2s - 3$$

Substituting the corresponding transform for each term in the differential equation and solving for $Y(s)$, we get

$$Y(s) = \frac{s^2 + s + 2}{s(s^2 + 4s + 4)} + \frac{2s + 11}{s^2 + 4s + 4}$$

The first term on the right-hand side is $H(s)X(s)$ and corresponds to the zero-state response. The second term is due to the initial conditions and corresponds to the zero-input response. Expanding into partial fractions, we get

$$Y(s) = \frac{0.5}{s} + \frac{0.5}{(s+2)} - \frac{2}{(s+2)^2} + \frac{2}{(s+2)} + \frac{7}{(s+2)^2}$$

Taking the inverse Laplace transform, we get the complete response.

$$y(t) = (\overbrace{0.5 + 0.5e^{-2t} - 2te^{-2t}}^{\text{zero-state}} + \overbrace{2e^{-2t} + 7te^{-2t}}^{\text{zero-input}})u(t)$$
$$= (0.5 + 2.5e^{-2t} + 5te^{-2t})u(t)$$

The steady-state response is $0.5u(t)$ and the transient response is $(2.5e^{-2t} + 5te^{-2t})u(t)$. The initial and final values of $y(t)$ are 3 and 0.5, respectively. These values can be verified by applying the initial and final value properties to $Y(s)$. We can also verify that the initial conditions at $t = 0^-$ are satisfied by the zero-input component of the response.   □

11.4.3 System Stability

The zero-input response of a system depends solely on the locations of its poles. A system is considered stable if its zero-input response due to finite initial conditions, converges, marginally stable if its zero-input response tends to a constant value or oscillates with a constant amplitude, and unstable if its zero-input response diverges. Commonly used marginally stable systems are oscillators, which produce a bounded zero-input response. The response corresponding to each pole p of a system is of the form e^{at}, where a is the location of the pole in the s-plane. If the real part of a is less than zero, then e^{at} tends to zero as t tends to ∞. If the real part of a is greater than zero, then e^{at} tends to ∞ as t tends to ∞. If the real part of a is equal to zero, then e^{at} remains bounded as t tends to ∞. However, the response tends to infinity, for poles of order more than one lying on the imaginary axis of the s-plane, as the expression for the response includes a factor that is a function of t. Poles of any order lying to the left of the imaginary axis of the s-plane do not cause instability. Figure 11.3 shows pole locations of some transfer functions and the corresponding impulse responses.

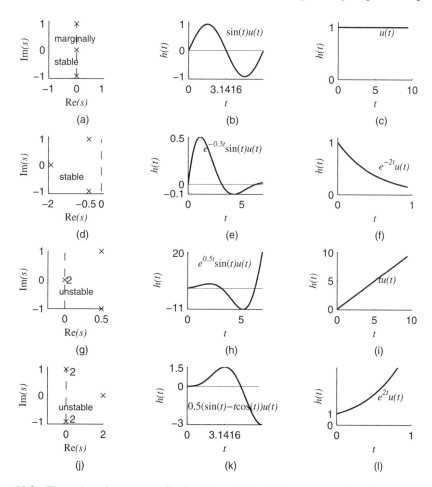

Figure 11.3 The poles of some transfer functions $H(s)$ and the corresponding impulse responses $h(t)$. The imaginary axis is shown by a dashed line. (a) $H(s) = 1/(s^2 + 1) = 1/[(s + j)(s - j)]$ and $H(s) = 1/s$; (b) $h(t) = \sin(t)u(t)$ and (c) $h(t) = u(t)$; (d) $H(s) = 1/[(s + 0.5)^2 + 1] = 1/[(s + 0.5 + j)(s + 0.5 - j)]$ and $H(s) = 1/(s + 2)$; (e) $h(t) = e^{-0.5t} \sin(t)u(t)$ and (f) $h(t) = e^{-2t}u(t)$; (g) $H(s) = 1/[(s - 0.5)^2 + 1] = 1/[(s - 0.5 + j)(s - 0.5 - j)]$ and $H(s) = 1/s^2$; (h) $h(t) = e^{0.5t} \sin(t)u(t)$; (i) $h(t) = tu(t)$; (j) $H(s) = 1/[(s^2 + 1)^2] = 1/[(s + j)^2(s - j)^2]$ and $H(s) = 1/(s - 2)$; (k) $h(t) = 0.5(\sin(t) - t\cos(t))u(t)$ and (l) $h(t) = e^{2t}u(t)$

Therefore, we conclude that, in terms of the locations of the poles of a system,:

- All the poles, of any order, of a stable system must lie to the left of the imaginary axis of the s-plane. That is, the ROC of $H(s)$ must include the imaginary axis.
- Any pole lying to the right of the imaginary axis or any pole of order more than one lying on the imaginary axis makes a system unstable.
- A system is marginally stable if it has no poles to the right of the imaginary axis and has poles of order one on the imaginary axis.

If all the poles of a system lie to the left of the imaginary axis of the s-plane, the bounded-input bounded-output stability condition (Chapter 5) is satisfied. However, the converse is not necessarily true, since the impulse response is an external description of a system and may not include all its poles. The bounded-input bounded-output stability condition is not satisfied by a marginally stable system.

11.4.4 Realization of Systems

Most of the transfer functions of practical continuous and discrete systems are rational functions of two polynomials, the difference being that the independent variable is s in $H(s)$ and it is z in $H(z)$. For example,

$$H(s) = \frac{b_2 s^2 + b_1 s + b_0}{s^2 + a_1 s + a_0} \quad \text{and} \quad H(z) = \frac{b_2 z^2 + b_1 z + b_0}{z^2 + a_1 z + a_0}$$

We realized discrete systems using multipliers, adders, and delay units, in Chapter 10. By comparison of the corresponding difference and differential equations, we find that the only difference being that integrators are required in realizing continuous systems instead of delay units. Therefore, the realization of continuous-time systems is the same as that for discrete systems, described in Chapter 10, except that delay units are replaced by integrators. Figure 11.4 shows the realization of a second-order continuous system. Integrators with feedback are used to simulate differential equations.

11.4.5 Frequency-domain Representation of Circuits

By replacing each element in a circuit, along with their initial conditions, by the corresponding frequency-domain representation, we can analyze the circuit in a way similar to a resistor network. This procedure is quite effective for circuits with nonzero initial conditions compared with writing the differential equation and then finding the Laplace transform.

In time-domain representation, a capacitor with initial voltage $v(0^-)$ is modeled as a uncharged capacitor in series with a voltage source $v(0^-)u(t)$. The voltage–current

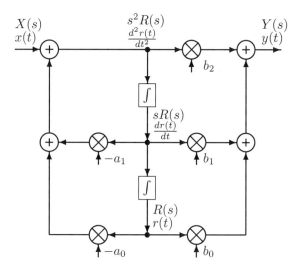

Figure 11.4   The realization of a second-order continuous system

relationship of a capacitor is

$$v(t) = \frac{1}{C}\int_{-\infty}^{t} i(\tau)\mathrm{d}\tau = \frac{1}{C}\int_{-\infty}^{0^-} i(\tau)\mathrm{d}\tau + \frac{1}{C}\int_{0^-}^{t} i(\tau)\mathrm{d}\tau = v(0^-) + \frac{1}{C}\int_{0^-}^{t} i(\tau)\mathrm{d}\tau$$

Taking the Laplace transform, the voltage across the capacitor is given as

$$V(s) = \frac{I(s)}{sC} + \frac{v(0^-)}{s}$$

The capacitor is modeled as an impedance $\frac{1}{sC}$ in series with an ideal voltage source $v(0^-)/s$. By taking the factor $1/sC$ out, an alternate representation is obtained as

$$V(s) = \frac{1}{sC}\left(I(s) + Cv(0^-)\right)$$

The voltage across the capacitor is due to the current $(I(s) + Cv(0^-))$ flowing through it. This representation, in the time-domain, implies an uncharged capacitor in parallel with an impulsive current source $Cv(0^-)\delta(t)$.

In time-domain representation, an inductor with initial current $i(0^-)$ is modeled as an inductor, with no initial current, in series with an impulsive voltage source $Li(0^-)\delta(t)$. The voltage–current relationship of an inductor is

$$v(t) = L\frac{\mathrm{d}i(t)}{\mathrm{d}t}$$

Figure 11.5   A RCL circuit

Taking the Laplace transform, the voltage across the inductor is given as

$$V(s) = L(sI(s) - i(0^-))$$

The inductor is modeled as an impedance sL in series with an ideal voltage source $-Li(0^-)$. By taking the factor sL out, an alternate representation is obtained as

$$V(s) = sL\left(I(s) - \frac{i(0^-)}{s}\right)$$

The voltage across the inductor is due to the current $\{I(s) - [i(0^-)/s]\}$ flowing through it.

Example 11.5.   Find the current in the circuit, shown in Figure 11.5, with the initial current through the inductor $i(0^-) = 3$A and the initial voltage across capacitor $v(0^-) = 2$ volts and, the input $x(t) = 3u(t)$ V.

Solution
The Laplace transform representation of the circuit in Figure 11.5 is shown in Figure 11.6. The sum of the voltages in the circuit is

$$\frac{3}{s} + 6 - \frac{2}{s} = \frac{6s + 1}{s}$$

Figure 11.6   The Laplace transform representation of the circuit in Figure 11.5

The circuit impedance is

$$\frac{5}{3} + 2s + \frac{1}{3s} = \frac{6s^2 + 5s + 1}{3s}$$

Dividing the voltage by the impedance, we get the current in the circuit as

$$I(s) = \frac{6s + 1}{s} \frac{3s}{6s^2 + 5s + 1} = \frac{(3s + \frac{1}{2})}{s^2 + \frac{5}{6}s + \frac{1}{6}}$$

Expanding into partial fractions, we get

$$I(s) = \frac{6}{s + \frac{1}{2}} - \frac{3}{s + \frac{1}{3}}$$

Finding the inverse Laplace transform, we get the current in the circuit as

$$i(t) = (6e^{-\frac{1}{2}t} - 3e^{-\frac{1}{3}t})u(t) \qquad \qquad \square$$

11.4.6 Feedback Systems

Consider the two systems connected in a feedback configuration, shown in Figure 11.7. The feedback signal $R(s)$ can be expressed as $R(s) = F(s)Y(s)$, where $F(s)$ is the feedback transfer function of the system and $Y(s)$ is the output. Now, the error signal $E(s)$ is

$$E(s) = X(s) - R(s) = X(s) - F(s)Y(s)$$

The output $Y(s)$ is expressed as

$$Y(s) = G(s)E(s) = G(s)(X(s) - F(s)Y(s))$$

Figure 11.7   Two systems connected in a feedback configuration

where $G(s)$ is the forward transfer function of the system. Therefore, the transfer function of the feedback system is given as

$$H(s) = \frac{Y(s)}{X(s)} = \frac{G(s)}{1 + G(s)F(s)}$$

If $G(s)$ is very large, the transfer function of the feedback system approximates to the inverse of the feedback transfer function of the system.

$$H(s) = \frac{Y(s)}{X(s)} \approx \frac{1}{F(s)}$$

11.4.6.1 Operational Amplifier Circuits

The frequency-domain representation of a scalar multiplier unit using an operational amplifier is shown in Figure 11.8. Operational amplifier circuits, shown in Figure 11.8 by a triangular symbol, are very large gain (of the order of 10^6) amplifiers with almost infinite input impedance and zero output impedance. There are two input terminals, indicated by the symbols $+$ and $-$ (called, respectively, the noninverting and inverting input terminals), and one output terminal. The output voltage is specified as $v_0 = A(v_+ - v_-)$. As the gain A is very large, the voltage at the inverting terminal, in Figure 11.8, is very small and can be considered as virtual ground. Further, the large input impedance makes the input terminal current negligible. Therefore, the currents in the forward and feedback paths must be almost equal and

$$\frac{X(s)}{R1} \approx -\frac{Y(s)}{R2}$$

The transfer function of the circuit is, therefore,

$$H(s) = \frac{Y(s)}{X(s)} \approx -\frac{R2}{R1}$$

Figure 11.8   Realization of a scalar multiplier unit using an operational amplifier

In general, the elements in the circuit can be impedances and the transfer function is expressed as

$$H(s) = \frac{Y(s)}{X(s)} \approx -\frac{Z2(s)}{Z1(s)}$$

The transfer function of the integrator circuit, shown in Figure 11.9, is

$$H(s) = -\frac{Z2(s)}{Z1(s)} = -\frac{1}{sRC}$$

This is an ideal integrator with gain $-1/RC$. Let $x(t) = u(t)$, the unit-step signal. Then, $X(s) = 1/s$ and

$$Y(s) = -\frac{1}{RCs}\frac{1}{s} = -\frac{1}{RCs^2}$$

The inverse transform of $Y(s)$ is $y(t) = -\frac{1}{RC}tu(t)$, as the integral of unit-step is the unit-ramp. Compare this response with that of a passive RC network, $y(t) = (1 - e^{-\frac{t}{RC}})u(t) \approx \frac{1}{RC}tu(t)$. Due to the large gain of the amplifier and the feedback, we get an ideal response. In addition, the amplifier, due to its large input impedance, does not load the source of the input signal much and can feed several circuits at the output.

The output $Y(s)$ of the summer, shown in Figure 11.10, is given as

$$Y(s) = -\left(\frac{R_f}{R1}X1(s) + \frac{R_f}{R2}X2(s)\right)$$

Remembering that the basic elements of a continuous system are scalar multipliers, integrators, and summers, we can build any system, however complex it may be, using the three operational amplifier circuits described.

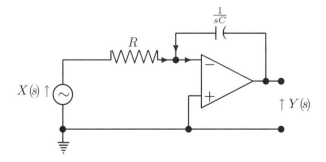

Figure 11.9 Realization of an integrator unit using an operational amplifier

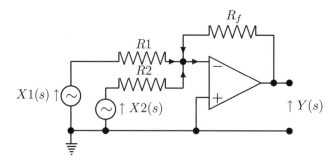

Figure 11.10   Realization of a summer unit using an operational amplifier

11.4.7 Analog Filters

We present, in this subsection, an example of the design of lowpass filters. The rectangle, shown in Figure 11.13 by dashed line, is the magnitude of the frequency response of an ideal analog lowpass filter. As the ideal filter is practically unrealizable, actual filters approximate the ideal filters to a desirable accuracy. While there are several types of filters with different characteristics, we describe the commonly used Butterworth filter.

11.4.7.1 Butterworth Filters

While active filters and digital filters are more commonly used, the word filter instantaneously reminds us the resister–capacitor lowpass filter circuit shown in Figure 11.11. The impedance, $1/sC$, of the capacitor is small at higher frequencies compared with that at lower frequencies. Therefore, the voltage across it is composed of high-frequency components with smaller amplitudes than low-frequency components compared with those of the input voltage. The reverse is the case for the voltage across the resistor. For example, there is no steady-state current with dc input (frequency = 0) and, therefore, all the input voltage appears across the capacitor.

In the Laplace transform model of the RC circuit, the input voltage is $X(s)$. The circuit impedance is $R + 1/(sC)$. Therefore, the current in the circuit is

Figure 11.11   The representation of a resister-capacitor filter circuit in the frequency-domain

$X(s)/[R + (1/sC)]$. The output voltage $Y(s)$ across the capacitor is

$$Y(s) = \left(\frac{X(s)}{R + \frac{1}{sC}} \right) \left(\frac{1}{sC} \right)$$

Therefore, the transfer function is

$$H(s) = \frac{Y(s)}{X(s)} = \frac{1}{1 + sRC}$$

Letting $s = j\omega$, we get the frequency response of the filter as

$$H(j\omega) = \frac{1}{1 + j\omega RC}$$

Let the cutoff frequency of the filter be $\omega_c = \frac{1}{RC} = 1$ radian/second. Then,

$$H(j\omega) = \frac{1}{1 + j\frac{\omega}{\omega_c}} = \frac{1}{1 + j\omega} \quad \text{and} \quad |H(j\omega)| = \frac{1}{\sqrt{1 + \omega^2}}$$

The filter circuit is a first-order system and a first-order lowpass Butterworth filter. For a Butterworth filter of order N, the magnitude of the frequency response, with $\omega_c = 1$, is

$$|H(j\omega)| = \frac{1}{\sqrt{1 + \omega^{(2N)}}}$$

The filter with $\omega_c = 1$ is called the normalized filter. From the transfer function of this filter, we can find the transfer function of other types of filters, such as highpass, with arbitrary cutoff frequencies using appropriate frequency transformations.

To find the transfer function of the normalized Butterworth filter, we substitute $\omega = s/j$ in the expression for the squared magnitude of the frequency response and get

$$|H(j\omega)|^2 = H(j\omega)H(-j\omega) = H(s)H(-s) = \frac{1}{1 + \omega^{(2N)}} = \frac{1}{1 + \left(\frac{s}{j}\right)^{(2N)}}$$

The poles of $H(s)H(-s)$ are obtained by solving the equation

$$s^{2N} = -(j)^{2N} = e^{j\pi(2n-1)}(e^{j\frac{\pi}{2}})^{(2N)} = e^{j\pi(2n-1+N)}$$

where n is an integer. Note that $e^{j\pi(2n-1)} = -1$ for an integral n and $e^{j\frac{\pi}{2}} = j$. As the transfer function $H(s)$ is to represent a stable system, all its poles must lie in the left half of the s-plane. Therefore, the poles p_n of $H(s)$, which are the N roots (of the $2N$th roots of $-(j)^{2N}$) with negative real parts, are specified as

$$p_n = e^{\frac{j\pi}{2N}(2n+N-1)} \qquad n = 1, 2, \ldots, N$$

The transfer function is given by

$$H(s) = \frac{1}{\prod_{n=1}^{N}(s - p_n)}$$

For $N = 1$, the pole is $p_1 = e^{\frac{j\pi}{2(1)}(2(1)+1-1)} = e^{j\pi} = -1$. The transfer function is specified as

$$H(s) = \frac{1}{(s + 1)}$$

The pole locations of the filter for $N = 2$ and $N = 3$ are shown in Figure 11.12. The symmetrically located poles are equally spaced around the left half of the unit circle. There is a pole on the real axis for N odd.

Consider the magnitude of the frequency response of normalized Butterworth lowpass filters shown in Figure 11.13. As the frequency response is an even function of ω, the figure shows the response for the positive half of the frequency range only. In both the passband and the stopband, the gain is monotonically decreasing. The asymptotic falloff rate, beyond the 3-dB frequency, is $-6N$ dB per octave (as the frequency is doubled) or $-20N$ dB per decade (as the frequency becomes ten times) approximately, where N is the order of the filter. Normalized filters of any order have the -3 dB ($-10 \log_{10}(2)$ to be more precise) or $1/\sqrt{2}$ response point at the same frequency, $\omega_{3dB} = 1$ radian per second. An higher-order filter approximates the ideal response, shown by the dashed line, more closely compared with a lower-order filter.

Figure 11.12   Pole locations of: (a) second-; (b) third-order normalized lowpass Butterworth analog filters

Figure 11.13 (a) The magnitude of the frequency response of the first- and second-order normalized lowpass Butterworth analog filters; (b) the passbands are shown on an expanded scale

11.5 Summary

- In this chapter, the theory of the one-sided Laplace transform, its properties, and some of its applications have been described. As practical systems are causal, the one-sided Laplace transform is mostly used in practice.
- The Laplace transform is a generalization of the Fourier transform. The basis waveforms include sinusoids with varying amplitudes or exponentials with complex exponents. The larger set of basis waveforms makes this transform suitable for the analysis of a larger class of signals and systems.
- The Laplace transform corresponding to a signal is a surface, since it is a function of two variables (the real and imaginary parts of the complex frequency). The Laplace transform of a signal, along with its ROC, uniquely represents the signal in the frequency-domain. The spectral values along any straight line in the ROC can be used to reconstruct the corresponding time-domain signal.
- The inverse Laplace transform is defined by an integral in the complex plane. However, the partial fraction method, along with a short list of Laplace transform pairs, is adequate for most practical purposes to find the inverse Laplace transform.
- The Laplace transform is essential for the design, and transient and stability analysis of continuous LTI systems. The Laplace transform of the impulse response of a system, the transfer function, is a frequency-domain model of the system.

Further Reading

1. Lathi, B. P., *Linear Systems and Signals*, Oxford University Press, New York, 2004.

Exercises

11.1   Find the Laplace transform of the unit-impulse signal, $\delta(t)$, by applying a limiting process to the rectangular pulse, defined as

$$x(t) = \begin{cases} \frac{1}{2a} & \text{for } -a < t < a \\ 0 & \text{otherwise} \end{cases} \quad a > 0,$$

and its transform, as a tends to zero.

11.2   Find the Laplace transform of the function $x(t)$ using the time-shifting property and the transforms of $u(t)$, $tu(t)$, and $t^2u(t)$.

    11.2.1 $x(t) = u(t - 5)$.

    11.2.2 $x(t) = 2$, $0 \le t \le 4$ and $x(t) = 0$ otherwise.

    11.2.3 $x(t) = 4$, $1 \le t \le 3$ and $x(t) = 0$ otherwise.

    *11.2.4 $x(t) = tu(t - 2)$.

    11.2.5 $x(t) = (t - 3)u(t)$.

    11.2.6 $x(t) = 2t^2u(t - 2)$.

11.3   Find the Laplace transform of the function $x(t)$ using the frequency-shifting property.

    11.3.1 $x(t) = e^{-2t}\cos(3t)u(t)$.

    11.3.2 $x(t) = e^{-3t}\sin(2t)u(t)$.

11.4   Find the derivative $dx(t)/dt$ of $x(t)$. Verify that the Laplace transform of $dx(t)/dt$ is $sX(s) - x(0^-)$.

    11.4.1 $x(t) = \cos(2t)$.

    11.4.2 $x(t) = \cos(3t)u(t)$.

    11.4.3 $x(t) = u(t) - u(t - 1)$.

    11.4.4 $x(t) = 3t(u(t - 2) - u(t - 4))$.

    11.4.5

$$x(t) = \begin{cases} (t - 1) & \text{for } t < 1 \\ 2 & \text{for } 1 < t < 3 \\ \cos(\frac{\pi}{3}t) & \text{for } t > 3 \end{cases}$$

    11.4.6

$$x(t) = \begin{cases} 2e^t & \text{for } t < 0 \\ 3\sin(t) & \text{for } 0 < t < \frac{\pi}{2} \\ 4u(t - \frac{\pi}{2}) & \text{for } t > \frac{\pi}{2} \end{cases}$$

11.5   Given the Laplace transform $X(s)$ of $x(t)$, find the transform of $x(at)$ using the scaling property. Find the location of the poles and zeros of the two transforms. Find $x(t)$ and $x(at)$.

    11.5.1 $X(s) = \frac{s+4}{s^2+5s+6}$ and $a = \frac{1}{2}$.

    *11.5.2 $X(s) = \frac{s-1}{s^2+3s+2}$ and $a = 2$.

    11.5.3 $X(s) = \frac{s-2}{s^2+1}$ and $a = 3$.

11.6   Using the Laplace transform, find the convolution, $y(t) = x(t) * h(t)$, of the functions $x(t)$ and $h(t)$.

    11.6.1 $x(t) = e^{-2t}u(t)$ and $h(t) = u(t)$.

    11.6.2 $x(t) = u(t)$ and $h(t) = u(t)$.

    11.6.3 $x(t) = e^{3t}u(t)$ and $h(t) = e^{-4t}u(t)$.

11.6.4 $x(t) = e^{-2t}u(t)$ and $x(t) = e^{-2t}u(t)$.

*11.6.5 $x(t) = te^{-t}u(t)$ and $h(t) = e^{-t}u(t)$.

11.6.6 $x(t) = 2u(t-2)$ and $h(t) = 3u(t-3)$.

11.6.7 $x(t) = 2e^{-(t-2)}u(t-2)$ and $h(t) = 5u(t)$.

11.7   Find the Laplace transform of the function $x(t)$ using the multiplication by t property.

11.7.1 $x(t) = 4t\cos(2t)u(t)$.

11.7.2 $x(t) = 5t\sin(3t)u(t)$.

11.8   Find the initial and final values of the function $x(t)$ corresponding to the transform $X(s)$, using the initial and final value properties.

11.8.1 $X(s) = \frac{s+2}{(s+3)}$.

11.8.2 $X(s) = \frac{2}{s+3}$.

11.8.3 $X(s) = \frac{2}{s^2+1}$.

*11.8.4 $X(s) = \frac{3s^2+3s+2}{s(s^2+3s+2)}$.

11.8.5 $X(s) = \frac{s+2}{s(s-2)}$.

11.8.6 $X(s) = \frac{s+1}{(s-1)}$.

11.9   Find the Laplace transform of the semiperiodic signal $x(t)u(t)$, the first period of which is defined as follows.

11.9.1

$$x_1(t) = \begin{cases} 1 & \text{for } 0 < t < 2 \\ -1 & \text{for } 2 < t < 4 \end{cases}$$

11.9.2

$$x_1(t) = t \quad \text{for } 0 < t < 5$$

* 11.9.3

$$x_1(t) = \begin{cases} t & \text{for } 0 < t < 2 \\ 4-t & \text{for } 2 < t < 4 \end{cases}$$

11.9.4

$$x_1(t) = \sin(\omega t) \quad \text{for } 0 < t < \tfrac{\pi}{\omega}$$

11.10   Find the inverse Laplace transform of $X(s)$ using the inverse FT.

11.10.1 $x(t)u(t) \Longleftrightarrow X(s) = 1/s^2$.

11.10.2 $x(t)u(t) \Longleftrightarrow X(s) = 1/(s-2)$.

11.10.3 $x(t)u(t) \Longleftrightarrow X(s) = 1/(s+2)$.

11.11   Find the inverse Laplace transform of

$$X(s) = \frac{s}{(s^2 + 3s + 2)}$$

11.12   Find the inverse Laplace transform of

$$X(s) = \frac{3s^2 + 2s + 3}{(s^2 + 5s + 6)}$$

11.13   Find the inverse Laplace transform of

$$X(s) = \frac{2s + 4}{(s^2 + 1)}$$

11.14   Find the inverse Laplace transform of

$$X(s) = \frac{s + 3}{(s^3 + 4s^2 + 5s + 2)}$$

*11.15 Find the inverse Laplace transform of

$$X(s) = \frac{s + 2}{(s^3 + s^2)}$$

11.16   Find the inverse Laplace transform of

$$X(s) = \frac{s + 2e^{-3s}}{(s + 2)(s + 3)}$$

11.17   Find the inverse Laplace transform of

$$X(s) = \frac{se^{-s}}{(s + 1)(s + 3)}$$

11.18   Using the Laplace transform, derive the closed-form expression for the impulse response $h(t)$ of the system, with input $x(t)$ and output $y(t)$, governed by the given differential equation.

11.18.1 $\frac{d^2 y(t)}{dt^2} + 6\frac{dy(t)}{dt} + 8y(t) = \frac{dx(t)}{dt} + x(t)$.

11.18.2 $\frac{d^2 y(t)}{dt^2} + 3\frac{dy(t)}{dt} + 2y(t) = x(t)$.

*11.18.3 $\frac{d^2 y(t)}{dt^2} - 4\frac{dy(t)}{dt} + 3y(t) = x(t)$.

11.19   Using the Laplace transform, find the zero-input, zero-state, transient, steady-state, and complete responses of the system governed by the differential equation

$$\frac{d^2y(t)}{dt^2} + 6\frac{dy(t)}{dt} + 8y(t) = 2\frac{dx(t)}{dt} + 3x(t)$$

with the initial conditions $y(0^-) = 2$ and

$$\frac{dy(t)}{dt}\Big|_{t=0^-} = 3$$

and the input $x(t) = u(t)$, the unit-step function. Find the initial and final values of the complete and zero-state responses.

*11.20 Using the Laplace transform, find the zero-input, zero-state, transient, steady-state, and complete responses of the system governed by the differential equation

$$\frac{d^2y(t)}{dt^2} + 2\frac{dy(t)}{dt} + y(t) = x(t)$$

with the initial conditions $y(0^-) = 3$ and

$$\frac{dy(t)}{dt}\Big|_{t=0^-} = -2$$

and the input $x(t) = e^{-2t}u(t)$. Find the initial and final values of the complete and zero-state responses.

11.21 Using the Laplace transform, find the zero-input, zero-state, transient, steady-state, and complete responses of the system governed by the differential equation

$$\frac{d^2y(t)}{dt^2} + 5\frac{dy(t)}{dt} + 6y(t) = x(t)$$

with the initial conditions

$$y(0^-) = -1 \quad \text{and} \quad \frac{dy(t)}{dt}\Big|_{t=0^-} = -2$$

and the input $x(t) = tu(t)$. Find the initial and final values of the complete and zero-state responses.

11.22 Using the Laplace transform, find the zero-input, zero-state, transient, steady-state, and complete responses of the system governed by the differential equation

$$\frac{d^2 y(t)}{dt^2} + 7\frac{dy(t)}{dt} + 12y(t) = x(t)$$

with the initial conditions $y(0^-) = 2$ and

$$\frac{dy(t)}{dt}\Big|_{t=0^-} = -3$$

and the input $x(t) = 2\cos(\frac{2\pi}{4}t - \frac{\pi}{6})u(t)$. Find the initial and final values of the complete and zero-state responses.

11.23 Given the differential equation of a system and the input signal $x(t)$, find the steady-state response of the system.

*11.23.1 $\frac{dy(t)}{dt} + 0.5y(t) = x(t)$ and $x(t) = 3\cos(0.5t - \frac{\pi}{3})u(t)$.

11.23.2 $\frac{dy(t)}{dt} + y(t) = 2x(t)$ and $x(t) = 2\sin(t + \frac{\pi}{4})u(t)$.

11.23.3 $\frac{dy(t)}{dt} + y(t) = x(t)$ and $x(t) = 3e^{j(\sqrt{3}t - \frac{\pi}{6})}u(t)$.

11.24 Using the Laplace transform, derive the closed-form expression of the impulse response of the cascade system consisting of systems, with input $x(t)$ and output $y(t)$, governed by the given differential equations.

11.24.1 $\frac{dy(t)}{dt} + 2y(t) = \frac{dx(t)}{dt} + x(t)$ and $\frac{dy(t)}{dt} + 3y(t) = 2\frac{dx(t)}{dt} + 3x(t)$.

*11.24.2 $\frac{dy(t)}{dt} - y(t) = x(t)$ and $\frac{dy(t)}{dt} = x(t)$.

11.24.3 $\frac{dy(t)}{dt} + 3y(t) = 2\frac{dx(t)}{dt} - x(t)$ and $\frac{dy(t)}{dt} + 2y(t) = 3\frac{dx(t)}{dt} + 2x(t)$.

11.25 Using the Laplace transform, derive the closed-form expression of the impulse response of the combined system, connected in parallel, consisting of systems, with input $x(t)$ and output $y(t)$, governed by the given differential equations.

11.25.1 $\frac{dy(t)}{dt} + 2y(t) = \frac{dx(t)}{dt} + x(t)$ and $\frac{dy(t)}{dt} + 3y(t) = 2\frac{dx(t)}{dt} + 3x(t)$.

11.25.2 $\frac{dy(t)}{dt} - y(t) = x(t)$ and $\frac{dy(t)}{dt} = x(t)$.

*11.25.3 $\frac{dy(t)}{dt} + 3y(t) = 2\frac{dx(t)}{dt} - x(t)$ and $\frac{dy(t)}{dt} + 2y(t) = 3\frac{dx(t)}{dt} + 2x(t)$.

11.26 Using the Laplace transform representation of the circuit elements, find the current in the series resistor-inductor circuit, with $R = 2\,\Omega$, $L = 3$ H, and the initial current through the inductor $i(0^-) = 4$ A, excited by the input voltage $x(t) = 10u(t)$ V.

11.27 Using the Laplace transform representation of the circuit elements, find the current in the series resistor-inductor circuit, with $R = 3\,\Omega$, $L = 4$ H, and the initial current through the inductor $i(0^-) = 1$ A, excited by the input voltage $x(t) = 10\delta(t)$ V.

11.28   Using the Laplace transform representation of the circuit elements, find the voltage across the capacitor in the series resistor-capacitor circuit, with $R = 2$ Ω, $C = 1$ F, and the initial voltage across capacitor $v(0^-) = 1$ V, excited by the input voltage $x(t) = e^{-t}u(t)$ V.

*11.29   Using the Laplace transform representation of the circuit elements, find the voltage across the capacitor in the series resistor-capacitor circuit, with $R = 4$ Ω, $C = 2$ F, and the initial voltage across capacitor $v(0^-) = 3$ V, excited by the input voltage $x(t) = \delta(t)$ V.

11.30   Find the response of a differentiator to unit-step input signal: (i) if the circuit is realized using resistor R and capacitor C; (ii) if the circuit is realized using resistor R and capacitor C, and an operational amplifier.

11.31   Find the transfer function $H(s)$ of fourth- and fifth-order Butterworth normalized lowpass filters.

12

State-space Analysis of Discrete Systems

So far, we studied three types of modeling of systems, the difference equation model, the convolution–summation model, and the transfer function model. Using these models, we concentrated on finding the output of a system corresponding to an input. However, in any system, there are internal variables. For example, the values of currents and voltages at various parts of a circuit are internal variables. The values of these variables are of interest in the analysis and design of systems. These values could indicate whether the components of a system work in their linear range and within their power ratings. Therefore, we need a model that also explicitly includes the internal description of systems. This type of model, which is a generalization of the other models of systems, is called the state-space model. In addition, it is easier to extend this model to the analysis of nonlinear and time-varying systems.

In Section 12.1, we study the state-space model of some common realizations of systems. The time-domain and frequency-domain solutions of the state equations are presented, respectively, in Sections 12.2 and 12.3. The linear transformation of state vector to obtain different realizations of systems is described in Section 12.4.

12.1 The State-space Model

Consider the state-space model, shown in Figure 12.1, of a second-order discrete system characterized by the difference equation

$$y(n) + a_1 y(n-1) + a_0 y(n-2) = b_2 x(n) + b_1 x(n-1) + b_0 x(n-2)$$

In addition to the input $x(n)$ and the output $y(n)$, we have shown two internal variables (called state variables), $q_1(n)$ and $q_2(n)$, of the system. State variables are a

A Practical Approach to Signals and Systems   D. Sundararajan
© 2008 John Wiley & Sons (Asia) Pte Ltd

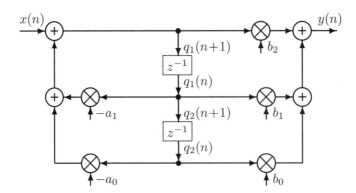

Figure 12.1   A state-space model of the canonical form I of a second-order discrete system

minimal set of variables (N for a Nth-order system) of a system so that a knowledge of the values of these variables (the state of the system) at $n = k$ and those of the input for $n \geq k$ will enable the determination of the values of the state variables for all $n > k$ and the output for all $n \geq k$. An infinite number of different sets, each of N state variables, are possible for a particular Nth-order system.

From Figure 12.1, we can write down the following state equations defining the state variables $q_1(n)$ and $q_2(n)$.

$$q_1(n + 1) = -a_1q_1(n) - a_0q_2(n) + x(n)$$
$$q_2(n + 1) = q_1(n)$$

The $(n + 1)$th sample value of each state variable is expressed in terms of the nth sample value of all the state variables and the input. This form of the first-order difference equation is called the standard form. A second-order difference equation characterizing the system, shown in Figure 12.1, has been decomposed into a set of two simultaneous first-order difference equations. Selecting state variables as the output of the delay elements is a natural choice, since a delay element is characterized by a first-order difference equation. With that choice, we can write down a state equation at the input of each delay element. However, the state variables need not correspond to quantities those are physically observable in a system. In the state-space model of a system, in general, an Nth-order difference equation characterizing a system is decomposed into a set of N simultaneous first-order difference equations of a standard form. With a set of N simultaneous difference equations, we can solve for N unknowns. These are the N internal variables, called the state variables, of the system. The output is expressed as a linear combination of the state variables and the input. The concepts of impulse response, convolution, and transform analysis are all equally applicable to the state-space model. The difference is that, as the system is modeled using matrix and vector quantities, the system analysis involves matrix and vector quantities. One of the advantages of the state-space model is the easier modeling of systems with multiple inputs and outputs. For simplicity, we describe systems with single input and single output only. The output $y(n)$ of the system, shown in Figure 12.1, is given by

$$y(n) = -b_2 a_1 q_1(n) - b_2 a_0 q_2(n)$$
$$+ b_1 q_1(n) + b_0 q_2(n) + b_2 x(n)$$

The output equation is an algebraic (not a difference) equation. We can write the state and output equations, using vectors and matrices, as

$$\begin{bmatrix} q_1(n+1) \\ q_2(n+1) \end{bmatrix} = \begin{bmatrix} -a_1 & -a_0 \\ 1 & 0 \end{bmatrix} \begin{bmatrix} q_1(n) \\ q_2(n) \end{bmatrix} + \begin{bmatrix} 1 \\ 0 \end{bmatrix} x(n)$$

$$y(n) = \begin{bmatrix} b_1 - b_2 a_1 & b_0 - b_2 a_0 \end{bmatrix} \begin{bmatrix} q_1(n) \\ q_2(n) \end{bmatrix} + b_2 x(n)$$

Let us define the state vector $q(n)$ as

$$q(n) = \begin{bmatrix} q_1(n) \\ q_2(n) \end{bmatrix}$$

Then, with

$$A = \begin{bmatrix} -a_1 & -a_0 \\ 1 & 0 \end{bmatrix} \qquad B = \begin{bmatrix} 1 \\ 0 \end{bmatrix} \qquad C = \begin{bmatrix} b_1 - b_2 a_1 & b_0 - b_2 a_0 \end{bmatrix} \qquad D = b_2,$$

the general state-space model description is given as

$$q(n+1) = Aq(n) + Bx(n)$$
$$y(n) = Cq(n) + Dx(n)$$

A block diagram representation of the state-space model of an Nth-order system, with single input and single output, is shown in Figure 12.2. Parallel lines terminating with an arrowhead indicate that the signal is a vector quantity.

Example 12.1. Derive the state-space model of the system governed by the difference equation

$$y(n) - 2y(n-1) + 3y(n-2) = 2x(n) - 3x(n-1) + 4x(n-2)$$

Assign the state variables as shown in Figure 12.1.

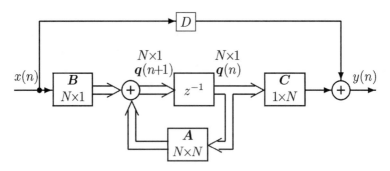

Figure 12.2   Block diagram representation of the state-space model of an Nth-order system, with single input and single output

Solution

With

$$A = \begin{bmatrix} -a_1 & -a_0 \\ 1 & 0 \end{bmatrix} = \begin{bmatrix} 2 & -3 \\ 1 & 0 \end{bmatrix} \qquad B = \begin{bmatrix} 1 \\ 0 \end{bmatrix}$$

$$C = \begin{bmatrix} b_1 - b_2 a_1 & b_0 - b_2 a_0 \end{bmatrix} = \begin{bmatrix} 1 & -2 \end{bmatrix} \qquad D = b_2 = 2$$

the state-space model of the system is

$$q(n+1) = \begin{bmatrix} q_1(n+1) \\ q_2(n+1) \end{bmatrix} = \begin{bmatrix} 2 & -3 \\ 1 & 0 \end{bmatrix} \begin{bmatrix} q_1(n) \\ q_2(n) \end{bmatrix} + \begin{bmatrix} 1 \\ 0 \end{bmatrix} x(n)$$

$$y(n) = \begin{bmatrix} 1 & -2 \end{bmatrix} \begin{bmatrix} q_1(n) \\ q_2(n) \end{bmatrix} + 2x(n) \qquad \qquad \square$$

While there are several realizations of a system, some realizations are more commonly used. The realization, shown in Figure 12.1, is called canonical form I. There is a dual realization that can be derived by using the transpose operation of a matrix. This realization, shown in Figure 12.3, is called canonical form II and is characterized by the matrices defined, in terms of those of canonical form I, as

$$\overline{A} = A^{\mathrm{T}} \qquad \overline{B} = C^{\mathrm{T}} \qquad \overline{C} = B^{\mathrm{T}} \qquad \overline{D} = D$$

The state-space model of the canonical form II of the system in Example 12.1 is

$$q(n+1) = \begin{bmatrix} q_1(n+1) \\ q_2(n+1) \end{bmatrix} = \begin{bmatrix} 2 & 1 \\ -3 & 0 \end{bmatrix} \begin{bmatrix} q_1(n) \\ q_2(n) \end{bmatrix} + \begin{bmatrix} 1 \\ -2 \end{bmatrix} x(n)$$

$$y(n) = \begin{bmatrix} 1 & 0 \end{bmatrix} \begin{bmatrix} q_1(n) \\ q_2(n) \end{bmatrix} + 2x(n)$$

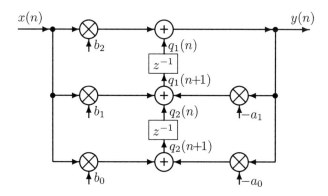

Figure 12.3 A state-space model of the canonical form II of a second-order discrete system

12.1.1 Parallel Realization

Consider a system characterized by the transfer function

$$H(z) = \frac{z^3 + z^2 - z + 1}{(z + 1)(z^2 + 2z + 3)}$$

The transfer function can be expanded into partial fractions as

$$H(z) = 1 + \frac{1}{(z + 1)} + \frac{-3z - 5}{(z^2 + 2z + 3)}$$

The state-space model, shown in Figure 12.4 using canonical form I, is

$$\mathbf{q}(n + 1) = \begin{bmatrix} q_1(n + 1) \\ q_2(n + 1) \\ q_3(n + 1) \end{bmatrix} = \begin{bmatrix} -1 & 0 & 0 \\ 0 & -2 & -3 \\ 0 & 1 & 0 \end{bmatrix} \begin{bmatrix} q_1(n) \\ q_2(n) \\ q_3(n) \end{bmatrix} + \begin{bmatrix} 1 \\ 1 \\ 0 \end{bmatrix} x(n)$$

$$y(n) = \begin{bmatrix} 1 & -3 & -5 \end{bmatrix} \begin{bmatrix} q_1(n) \\ q_2(n) \\ q_3(n) \end{bmatrix} + x(n)$$

Consider the transfer function with a repeated pole

$$H(z) = \frac{2z^3 - z^2 + 3z - 1}{(z + 1)(z + 2)^2}$$

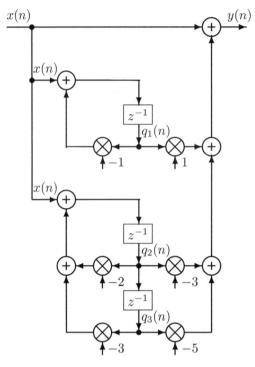

Figure 12.4 A state-space model of the parallel realization of a third-order discrete system, using canonical form I

The transfer function can be expanded into partial fractions as

$$H(z) = 2 + \frac{-7}{(z+1)} + \frac{-4}{(z+2)} + \frac{27}{(z+2)^2}$$

$$= 2 + \frac{-7}{(z+1)} + \frac{1}{(z+2)}\left(-4 + \frac{27}{(z+2)}\right)$$

The state-space model, shown in Figure 12.5 using canonical form I, is

$$\boldsymbol{q}(n+1) = \begin{bmatrix} q_1(n+1) \\ q_2(n+1) \\ q_3(n+1) \end{bmatrix} = \begin{bmatrix} -1 & 0 & 0 \\ 0 & -2 & 0 \\ 0 & 1 & -2 \end{bmatrix} \begin{bmatrix} q_1(n) \\ q_2(n) \\ q_3(n) \end{bmatrix} + \begin{bmatrix} 1 \\ 1 \\ 0 \end{bmatrix} x(n)$$

$$y(n) = \begin{bmatrix} -7 & -4 & 27 \end{bmatrix} \begin{bmatrix} q_1(n) \\ q_2(n) \\ q_3(n) \end{bmatrix} + 2x(n)$$

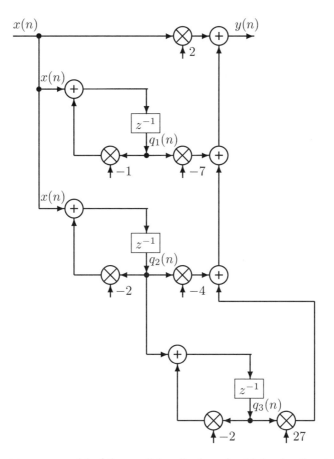

Figure 12.5 A state-space model of the parallel realization of a third-order discrete system with a repeated pole, using canonical form I

12.1.2 Cascade Realization

Consider the transfer function

$$H(z) = \frac{(4z + 1)(z^2 + 3z + 2)}{(z + 1)(z^2 + 2z + 3)}$$

In cascade form, several realizations are possible, depending on the grouping of the numerators and denominators and the order of the sections in the cascade. Let us assume that the first and second sections, respectively, have the transfer functions

$$H_1(z) = \frac{(4z + 1)}{(z + 1)} \quad \text{and} \quad H_2(z) = \frac{(z^2 + 3z + 2)}{(z^2 + 2z + 3)}$$

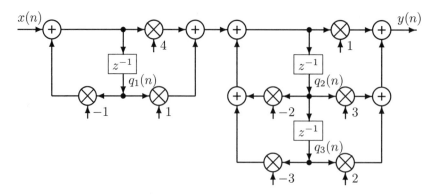

Figure 12.6 A state-space model of the cascade realization of a third-order discrete system, using canonical form I

The state-space model, shown in Figure 12.6 using canonical form I, is

$$q(n+1) = \begin{bmatrix} q_1(n+1) \\ q_2(n+1) \\ q3(n+1) \end{bmatrix} = \begin{bmatrix} -1 & 0 & 0 \\ -3 & -2 & -3 \\ 0 & 1 & 0 \end{bmatrix} \begin{bmatrix} q_1(n) \\ q_2(n) \\ q3(n) \end{bmatrix} + \begin{bmatrix} 1 \\ 4 \\ 0 \end{bmatrix} x(n)$$

$$y(n) = \begin{bmatrix} -3 & 1 & -1 \end{bmatrix} \begin{bmatrix} q_1(n) \\ q_2(n) \\ q3(n) \end{bmatrix} + 4x(n)$$

12.2 Time-domain Solution of the State Equation

12.2.1 Iterative Solution

The system response can be found iteratively as in the case of the difference equation model.

Example 12.2. Find the outputs $y(0)$, $y(1)$, and $y(2)$ of the system, described by the state-space model given in Example 12.1, iteratively with the initial conditions $y(-1) = 2$ and $y(-2) = 3$ and the input $u(n)$, the unit-step function.

Solution
The input is $x(-2) = 0$, $x(-1) = 0$, $x(0) = 1$, $x(1) = 1$, and $x(2) = 1$. We need the initial state vector values $q_1(0)$ and $q_2(0)$ to start the iteration. Therefore, we have to derive these values from the initial output conditions $y(-1) = 2$ and $y(-2) = 3$. Using the state equations, we get

$$q_1(-1) = 2q_1(-2) - 3q_2(-2)$$
$$q_2(-1) = q_1(-2)$$

Using the output equations, we get

$$y(-2) = q_1(-2) - 2q_2(-2) = 3$$
$$y(-1) = q_1(-1) - 2q_2(-1) = 2$$

Solving these equations, we find $q_1(-1) = \frac{16}{3}$ and $q_2(-1) = \frac{5}{3}$. Now,

$$q_1(0) = 2\left(\frac{16}{3}\right) - 3\left(\frac{5}{3}\right) = \frac{17}{3}$$

$$q_2(0) = q_1(-1) = \frac{16}{3}$$

At $n = 0$, we get

$$y(0) = \begin{bmatrix} 1 & -2 \end{bmatrix} \begin{bmatrix} \frac{17}{3} \\ \frac{16}{3} \end{bmatrix} + 2 = -3$$

$$q(1) = \begin{bmatrix} q_1(1) \\ q_2(1) \end{bmatrix} = \begin{bmatrix} 2 & -3 \\ 1 & 0 \end{bmatrix} \begin{bmatrix} \frac{17}{3} \\ \frac{16}{3} \end{bmatrix} + \begin{bmatrix} 1 \\ 0 \end{bmatrix} 1 = \begin{bmatrix} -\frac{11}{3} \\ \frac{17}{3} \end{bmatrix}$$

At $n = 1$, we get

$$y(1) = \begin{bmatrix} 1 & -2 \end{bmatrix} \begin{bmatrix} -\frac{11}{3} \\ \frac{17}{3} \end{bmatrix} + 2 = -13$$

$$q(2) = \begin{bmatrix} q_1(2) \\ q_2(2) \end{bmatrix} = \begin{bmatrix} 2 & -3 \\ 1 & 0 \end{bmatrix} \begin{bmatrix} -\frac{11}{3} \\ \frac{17}{3} \end{bmatrix} + \begin{bmatrix} 1 \\ 0 \end{bmatrix} 1 = \begin{bmatrix} -\frac{70}{3} \\ -\frac{11}{3} \end{bmatrix}$$

At $n = 2$, we get

$$y(2) = \begin{bmatrix} 1 & -2 \end{bmatrix} \begin{bmatrix} -\frac{70}{3} \\ -\frac{11}{3} \end{bmatrix} + 2 = -14$$

$\square$

12.2.2 Closed-form Solution

In the state-space model also, convolution–summation gives the zero-state response of a system in the time-domain. Substituting $n = 0$ in the state equation, we get

$$q(1) = Aq(0) + Bx(0)$$

Similarly, for $n = 1$ and $n = 2$, we get

$$
\begin{aligned}
q(2) &= Aq(1) + Bx(1) \\
&= A(Aq(0) + Bx(0)) + Bx(1) \\
&= A^2 q(0) + ABx(0) + Bx(1) \\
q(3) &= Aq(2) + Bx(2) \\
&= A(A^2 q(0) + ABx(0) + Bx(1)) + Bx(2) \\
&= A^3 q(0) + A^2 Bx(0) + ABx(1) + Bx(2)
\end{aligned}
$$

Proceeding in this way, we get the general expression for the state vector as

$$
q(n) = A^n q(0) + A^{n-1} Bx(0) + A^{n-2} Bx(1) + \cdots + Bx(n-1)
$$

$$
= \overbrace{A^n q(0)}^{q_{zi}(n)} + \overbrace{\sum_{m=0}^{n-1} A^{n-1-m} Bx(m)}^{q_{zs}(n)} \qquad n = 1, 2, 3, \ldots
$$

The first and the second expressions on the right-hand side are, respectively, the zero-input and zero-state components of the state vector $q(n)$. The second expression is the convolution-summation $A^{n-1} u(n-1) * Bx(n)$. Convolution of two matrices is similar to the multiplication operation of two matrices with the multiplication of the elements replaced by the convolution of the elements. Once we know the state vector, we get the output of the system using the output equation as

$$
y(n) = Cq(n) + Dx(n)
$$

$$
= \overbrace{CA^n q(0)}^{y_{zi}(n)} + \overbrace{\sum_{m=0}^{n-1} CA^{n-1-m} Bx(m) + Dx(n)}^{y_{zs}(n)} \qquad n = 1, 2, 3, \ldots
$$

The term $CA^n q(0)$ is the zero-input component and the other two terms constitute the zero-state component of the system response $y(n)$. The zero-input response of the system depends solely on the matrix A^n. This matrix is called the state transition or fundamental matrix of the system. This matrix, for an Nth-order system, is evaluated, using the Cayley–Hamilton theorem, as

$$
A^n = c_0 I + c_1 A + c_2 A^2 + \cdots + c_{N-1} A^{(N-1)}
$$

where

$$
\begin{bmatrix} c_0 \\ c_1 \\ \cdots \\ c_{N-1} \end{bmatrix} = \begin{bmatrix} 1 & \lambda_1 & \lambda_1^2 & \cdots & \lambda_1^{N-1} \\ 1 & \lambda_2 & \lambda_2^2 & \cdots & \lambda_2^{N-1} \\ & & \cdots & & \\ 1 & \lambda_N & \lambda_N^2 & \cdots & \lambda_N^{N-1} \end{bmatrix}^{-1} \begin{bmatrix} \lambda_1^n \\ \lambda_2^n \\ \cdots \\ \lambda_N^n \end{bmatrix}
$$

and $\lambda_1, \lambda_2, \ldots, \lambda_N$ are the N distinct characteristic values of A. The characteristic equation of the matrix A is $\det(zI - A) = 0$, where the abbreviation det stands for determinant and I is the identity matrix of the same size of that of A. The expanded form of $\det(zI - A)$ is a polynomial in z called the characteristic polynomial of A. The roots, which are the solutions of the characteristic equation, of this polynomial are the characteristic values of A.

For a value λ_r repeated m times, the first row corresponding to that value will remain the same as for a distinct value and the $m - 1$ successive rows will be successive derivatives of the first row with respect to λ_r. For example, with the first value of a forth-order system repeating two times, we get

$$
\begin{bmatrix} c_0 \\ c_1 \\ c_2 \\ c_3 \end{bmatrix} = \begin{bmatrix} 1 & \lambda_1 & \lambda_1^2 & \lambda_1^3 \\ 0 & 1 & 2\lambda_1 & 3\lambda_1^2 \\ 1 & \lambda_2 & \lambda_2^2 & \lambda_2^3 \\ 1 & \lambda_3 & \lambda_3^2 & \lambda_3^3 \end{bmatrix}^{-1} \begin{bmatrix} \lambda_1^n \\ n\lambda_1^{n-1} \\ \lambda_2^n \\ \lambda_3^n \end{bmatrix}
$$

Example 12.3. Derive the characteristic polynomial and determine the characteristic roots of the system with the state-space model as given in Example 12.1.

Solution

$$
A = \begin{bmatrix} 2 & -3 \\ 1 & 0 \end{bmatrix} \qquad (zI - A) = z \begin{bmatrix} 1 & 0 \\ 0 & 1 \end{bmatrix} - \begin{bmatrix} 2 & -3 \\ 1 & 0 \end{bmatrix} = \begin{bmatrix} z - 2 & 3 \\ -1 & z \end{bmatrix}
$$

The characteristic polynomial of the system, given by the determinant of this matrix, is

$$
z^2 - 2z + 3
$$

The characteristic roots, which are the roots of this polynomial, are

$$
\lambda_1 = 1 + j\sqrt{2} \quad \text{and} \quad \lambda_2 = 1 - j\sqrt{2}
$$

$\square$

Example 12.4. Find a closed-form expression for the output $y(n)$ of the system, described by the state-space model given in Example 12.1, using the time-domain method, with the initial conditions $y(-1) = 2$ and $y(-2) = 3$ and the input $u(n)$, the unit-step function.

Solution
The initial state vector was determined, from the given initial output conditions, in Example 12.2 as

$$q_1(0) = \frac{17}{3} \qquad q_2(0) = \frac{16}{3}$$

The characteristic values, as determined in Example 12.3, are

$$\lambda_1 = 1 + j\sqrt{2} \quad \text{and} \quad \lambda_2 = 1 - j\sqrt{2}$$

The transition matrix is given by

$$A^n = c_0 I + c_1 A$$

$$= c_0 \begin{bmatrix} 1 & 0 \\ 0 & 1 \end{bmatrix} + c_1 \begin{bmatrix} 2 & -3 \\ 1 & 0 \end{bmatrix} = \begin{bmatrix} c_0 + 2c_1 & -3c_1 \\ c_1 & c_0 \end{bmatrix}$$

where

$$\begin{bmatrix} c_0 \\ c_1 \end{bmatrix} = \begin{bmatrix} 1 & \lambda_1 \\ 1 & \lambda_2 \end{bmatrix}^{-1} \begin{bmatrix} \lambda_1^n \\ \lambda_2^n \end{bmatrix}$$

$$= \frac{j}{2\sqrt{2}} \begin{bmatrix} 1 - j\sqrt{2} & -1 - j\sqrt{2} \\ -1 & 1 \end{bmatrix} \begin{bmatrix} (1 + j\sqrt{2})^n \\ (1 - j\sqrt{2})^n \end{bmatrix}$$

$$= \frac{j}{2\sqrt{2}} \begin{bmatrix} (1 - j\sqrt{2})(1 + j\sqrt{2})^n + (-1 - j\sqrt{2})(1 - j\sqrt{2})^n \\ -(1 + j\sqrt{2})^n + (1 - j\sqrt{2})^n \end{bmatrix}$$

$$A^n = \frac{j}{2\sqrt{2}} \begin{bmatrix} -(1+j\sqrt{2})^{(n+1)} + (1 - j\sqrt{2})^{(n+1)} & 3(1+j\sqrt{2})^n - 3(1-j\sqrt{2})^n \\ -(1+j\sqrt{2})^n + (1-j\sqrt{2})^n & 3(1+j\sqrt{2})^{(n-1)} - 3(1-j\sqrt{2})^{(n-1)} \end{bmatrix}$$

As a check on A^n, verify that $A^n = I$ with $n = 0$ and $A^n = A$ with $n = 1$.
The zero-input component of the state vector is

$$q_{zi}(n) = A^n q(0) = \frac{j}{6\sqrt{2}} \begin{bmatrix} (31 - j17\sqrt{2})(1 + j\sqrt{2})^n - (31 + j17\sqrt{2})(1 - j\sqrt{2})^n \\ (-1 - j16\sqrt{2})(1 + j\sqrt{2})^n + (1 - j16\sqrt{2})(1 - j\sqrt{2})^n \end{bmatrix}$$

Using the fact that the sum a complex number and its conjugate is twice the real part of either of the numbers, we get

$$q_{zi}(n) = \begin{bmatrix} \frac{17}{3}(\sqrt{3})^n \cos(\tan^{-1}(\sqrt{2})n) - \frac{31}{3\sqrt{2}}(\sqrt{3})^n \sin(\tan^{-1}(\sqrt{2})n)) \\ \frac{16}{3}(\sqrt{3})^n \cos(\tan^{-1}(\sqrt{2})n) + \frac{1}{3\sqrt{2}}(\sqrt{3})^n \sin(\tan^{-1}(\sqrt{2})n)) \end{bmatrix}$$

The zero-input response $y_{zi}(n)$ is given by

$$CA^n q(0) = \begin{bmatrix} 1 & -2 \end{bmatrix} \begin{bmatrix} \frac{17}{3}(\sqrt{3})^n \cos(\tan^{-1}(\sqrt{2})n) - \frac{31}{3\sqrt{2}}(\sqrt{3})^n \sin(\tan^{-1}(\sqrt{2})n)) \\ \frac{16}{3}(\sqrt{3})^n \cos(\tan^{-1}(\sqrt{2})n) + \frac{1}{3\sqrt{2}}(\sqrt{3})^n \sin(\tan^{-1}(\sqrt{2})n)) \end{bmatrix}$$

$$= (-5(\sqrt{3})^n \cos(\tan^{-1}(\sqrt{2})n) - \frac{11}{\sqrt{2}}(\sqrt{3})^n \sin(\tan^{-1}(\sqrt{2})n))u(n)$$

The first four values of the zero-input response $y_{zi}(n)$ are

$$y_{zi}(0) = -5 \qquad y_{zi}(1) = -16 \qquad y_{zi}(2) = -17 \qquad y_{zi}(3) = 14$$

The zero-state component of the state vector is

$$q_{zs}(n) = \sum_{m=0}^{n-1} A^{n-1-m} Bx(m)$$

The convolution-summation, $A^{n-1}u(n-1) * Bx(n)$, can be evaluated, using the shift theorem of convolution (Chapter 4), by evaluating $A^n u(n) * Bx(n)$ first and then replacing n by $n - 1$.

$$Bx(n) = \begin{bmatrix} 1 \\ 0 \end{bmatrix} u(n) = \begin{bmatrix} u(n) \\ 0 \end{bmatrix}$$

$$A^n * Bx(n) = \frac{j}{2\sqrt{2}} \begin{bmatrix} (-(1+j\sqrt{2})^{(n+1)} + (1 - j\sqrt{2})^{(n+1)}) * u(n) \\ (-(1+j\sqrt{2})^n + (1 - j\sqrt{2})^n) * u(n) \end{bmatrix}$$

Since the first operand of the convolutions is the sum of two complex conjugate expressions and the convolution of $p(n)$ and $u(n)$ is equivalent to the sum of the first

$n + 1$ values of $p(n)$, we get

$$A^n * Bx(n)$$

$$= \begin{bmatrix} 2\,\mathrm{Re}\left\{\left(-\frac{1}{2} - \frac{j}{2\sqrt{2}}\right)\sum_{m=0}^{n}(1 + j\sqrt{2})^m\right\} \\ 2\,\mathrm{Re}\left\{\left(-\frac{j}{2\sqrt{2}}\right)\sum_{m=0}^{n}(1 + j\sqrt{2})^m\right\} \end{bmatrix}$$

$$= \begin{bmatrix} 2\,\mathrm{Re}\left\{\left(\frac{1}{2} - \frac{j}{2\sqrt{2}}\right)\left(\frac{1-(1+j\sqrt{2})^{n+1}}{1-(1+j\sqrt{2})}\right)\right\} \\ 2\,\mathrm{Re}\left\{\left(-\frac{j}{2\sqrt{2}}\right)\left(\frac{1-(1+j\sqrt{2})^{n+1}}{1-(1+j\sqrt{2})}\right)\right\} \end{bmatrix}$$

$$= \begin{bmatrix} \frac{1}{2} - \frac{1}{2}(\sqrt{3})^{(n+1)}\cos(\tan^{-1}(\sqrt{2})(n+1)) + \frac{1}{\sqrt{2}}(\sqrt{3})^{(n+1)}\sin(\tan^{-1}(\sqrt{2})(n+1)) \\ \frac{1}{2} - \frac{1}{2}(\sqrt{3})^{(n+1)}\cos(\tan^{-1}(\sqrt{2})(n+1)) \end{bmatrix}$$

Replacing $n = n - 1$, we get

$$q_{zs}(n) = A^{n-1} * Bx(n)$$

$$= \begin{bmatrix} \frac{1}{2} - \frac{1}{2}(\sqrt{3})^n\cos(\tan^{-1}(\sqrt{2})n) + \frac{1}{\sqrt{2}}(\sqrt{3})^n\sin(\tan^{-1}(\sqrt{2})n) \\ \frac{1}{2} - \frac{1}{2}(\sqrt{3})^n\cos(\tan^{-1}(\sqrt{2})) \end{bmatrix}$$

The zero-state response is given by multiplying the state vector with the C vector and adding the input signal as

$$y_{zs}(n) =$$

$$\begin{bmatrix} 1 & -2 \end{bmatrix}\begin{bmatrix} \frac{1}{2} - \frac{1}{2}(\sqrt{3})^n\cos(\tan^{-1}(\sqrt{2})n) + \frac{1}{\sqrt{2}}(\sqrt{3})^n\sin(\tan^{-1}(\sqrt{2})n) \\ \frac{1}{2} - \frac{1}{2}(\sqrt{3})^n\cos(\tan^{-1}(\sqrt{2})) \end{bmatrix} u(n-1)$$

$$+ 2u(n)$$

$$= \left(-\frac{1}{2} + \frac{1}{2}(\sqrt{3})^n\cos(\tan^{-1}(\sqrt{2})n) + \frac{1}{\sqrt{2}}(\sqrt{3})^n\sin(\tan^{-1}(\sqrt{2})n)\right)u(n-1) + 2u(n)$$

$$= \left(1.5 + \frac{1}{2}(\sqrt{3})^n\cos(\tan^{-1}(\sqrt{2})n) + \frac{1}{\sqrt{2}}(\sqrt{3})^n\sin(\tan^{-1}(\sqrt{2})n)\right)u(n)$$

The first four values of the zero-state response $y_{zs}(n)$ are

$$y_{zs}(0) = 2 \qquad y_{zs}(1) = 3 \qquad y_{zs}(2) = 3 \qquad y_{zs}(3) = 0$$

Adding the zero-input and the zero-state components, we get the total response of the system as

$$y(n) = 1.5 - 4.5(\sqrt{3})^n \cos(\tan^{-1}(\sqrt{2})n)$$

$$-\frac{10}{\sqrt{2}}(\sqrt{3})^n \sin(\tan^{-1}(\sqrt{2})n) \qquad n = 0, 1, 2, \ldots$$

The first four values of the total response $y(n)$ are

$$y(0) = -3, \quad y(1) = -13, \quad y(2) = -14, \quad y(3) = 14 \qquad \square$$

12.2.3 The Impulse Response

The impulse response, $h(n)$, is the output of an initially relaxed system with the input $x(n) = \delta(n)$ and is given by

$$h(n) = \sum_{m=0}^{n-1} C A^{n-1-m} B x(m) + D x(n) = C A^{n-1} B u(n-1) + D\delta(n)$$

Example 12.5. Find the closed-form expression for the impulse response of the system, described by the state-space model given in Example 12.1, using the time-domain method.

Solution
The impulse response is given by

$$h(n) = C A^{n-1} B u(n-1) + D\delta(n)$$

$$= \begin{bmatrix} 1 & -2 \end{bmatrix} \frac{j}{2\sqrt{2}} \begin{bmatrix} (-(1+j\sqrt{2})^n + (1-j\sqrt{2})^n) \\ (-(1+j\sqrt{2})^{(n-1)} + (1-j\sqrt{2})^{(n-1)}) \end{bmatrix} + 2\delta(n)$$

$$= \frac{1}{-j2\sqrt{2}} \left[(1-j\sqrt{2})(1+j\sqrt{2})^{n-1} + (-1-j\sqrt{2})(1-j\sqrt{2})^{n-1} \right] + 2\delta(n)$$

$$= 2\delta(n) + ((\sqrt{3})^{n-1} \cos((\tan^{-1}(\sqrt{2}))(n-1))$$

$$-\frac{1}{\sqrt{2}}(\sqrt{3})^{n-1} \sin((\tan^{-1}(\sqrt{2}))(n-1)))u(n-1) \qquad n = 0, 1, 2, \ldots$$

The first four values of the impulse response $h(n)$ are

$$h(0) = 2, \quad h(1) = 1, \quad h(2) = 0, \quad h(3) = -3 \qquad \square$$

12.3 Frequency-domain Solution of the State Equation

The z-transform of a vector function, such as $q(n)$, is defined to be the vector function $Q(z)$, where the elements are the transforms of the corresponding elements of $q(n)$. Taking the z-transform of the state equation, we get

$$z\, Q(z) - zq(0) = A\, Q(z) + BX(z)$$

We have used the left shift property of the z-transform and $q(0)$ is the initial state vector. Solving for $Q(z)$, we get

$$Q(z) = \overbrace{(zI - A)^{-1}zq(0)}^{Q_{zi}(z)} + \overbrace{(zI - A)^{-1}BX(z)}^{Q_{zs}(z)}$$

The inverse z-transforms of the first and the second expressions on the right-hand side yield, respectively, the zero-input and zero-state components of the state vector $q(n)$. Taking the z-transform of the output equation, we get

$$Y(z) = C\, Q(z) + DX(z)$$

Now, substituting for $Q(z)$, we get

$$Y(z) = \overbrace{Cz(zI - A)^{-1}q(0)}^{Y_{zi}(z)} + \overbrace{(C(zI - A)^{-1}B + D)X(z)}^{Y_{zs}(z)}$$

The inverse z-transforms of the first and the second expressions on the right-hand side yield, respectively, the zero-input and zero-state components of the system response $y(n)$. The inverse z-transform of $(z(zI - A)^{-1})$, by correspondence with the equation for state vector in time-domain, is A^n, the transition or fundamental matrix of the system. With the system initial conditions zero, the transfer function is given by

$$H(z) = \frac{Y(z)}{X(z)} = (C(zI - A)^{-1}B + D)$$

Example 12.6. Find a closed-form expression for the output $y(n)$ of the system, described by the state-space model given in Example 12.1, using the frequency-domain method, with the initial conditions $y(-1) = 2$ and $y(-2) = 3$ and the input $u(n)$, the unit-step function.

Solution
The initial state vector

$$q(0) = \begin{bmatrix} \frac{17}{3} \\ \frac{16}{3} \end{bmatrix}$$

is derived in Example 12.2 from the given initial output conditions.

$$(z\mathbf{I} - \mathbf{A}) = \begin{bmatrix} z-2 & 3 \\ -1 & z \end{bmatrix} \quad \text{and} \quad (z\mathbf{I} - \mathbf{A})^{-1} = \begin{bmatrix} \frac{z}{z^2-2z+3} & -\frac{3}{z^2-2z+3} \\ \frac{1}{z^2-2z+3} & \frac{z-2}{z^2-2z+3} \end{bmatrix}$$

As a check on $(z\mathbf{I} - \mathbf{A})^{-1}$, we use the initial value theorem of the z-transform to verify that

$$\lim_{z\to\infty} z(z\mathbf{I} - \mathbf{A})^{-1} = \mathbf{I} = \mathbf{A}^0$$

The transform of the zero-input component of the state vector is

$$Q_{zi}(z) = z(z\mathbf{I} - \mathbf{A})^{-1}q(0)$$

$$= z \begin{bmatrix} \frac{z}{z^2-2z+3} & -\frac{3}{z^2-2z+3} \\ \frac{1}{z^2-2z+3} & \frac{z-2}{z^2-2z+3} \end{bmatrix} \begin{bmatrix} \frac{17}{3} \\ \frac{16}{3} \end{bmatrix} = \frac{z}{3} \begin{bmatrix} \frac{17z-48}{z^2-2z+3} \\ \frac{16z-15}{z^2-2z+3} \end{bmatrix}$$

$$= \begin{bmatrix} \frac{(\frac{17}{6}+j\frac{31}{6\sqrt{2}})z}{z-1-j\sqrt{2}} + \frac{(\frac{17}{6}-j\frac{31}{6\sqrt{2}})z}{z-1+j\sqrt{2}} \\ \frac{(\frac{8}{3}-j\frac{1}{6\sqrt{2}})z}{z-1-j\sqrt{2}} + \frac{(\frac{8}{3}+j\frac{1}{6\sqrt{2}})z}{z-1+j\sqrt{2}} \end{bmatrix}$$

Finding the inverse z-transform and simplifying, we get the zero-input component of the state vector as

$$q_{zi}(n) = \begin{bmatrix} \frac{17}{3}(\sqrt{3})^n \cos(\tan^{-1}(\sqrt{2})n) - \frac{31}{3\sqrt{2}}(\sqrt{3})^n \sin(\tan^{-1}(\sqrt{2})n)) \\ \frac{16}{3}(\sqrt{3})^n \cos(\tan^{-1}(\sqrt{2})n) + \frac{1}{3\sqrt{2}}(\sqrt{3})^n \sin(\tan^{-1}(\sqrt{2})n)) \end{bmatrix} u(n)$$

The transform of the zero-state component of the state vector is

$$Q_{zs}(z) = z(z\mathbf{I} - \mathbf{A})^{-1}BX(z)$$

$$= \begin{bmatrix} \frac{z}{z^2-2z+3} & -\frac{3}{z^2-2z+3} \\ \frac{1}{z^2-2z+3} & \frac{z-2}{z^2-2z+3} \end{bmatrix} \begin{bmatrix} \frac{z}{z-1} \\ 0 \end{bmatrix} = z \begin{bmatrix} \frac{z}{(z-1)(z^2-2z+3)} \\ \frac{1}{(z-1)(z^2-2z+3)} \end{bmatrix}$$

$$= \begin{bmatrix} \frac{(\frac{1}{2})z}{z-1} - \frac{\frac{1}{4}(1+j\sqrt{2})z}{z-1-j\sqrt{2}} - \frac{\frac{1}{4}(1-j\sqrt{2})z}{z-1+j\sqrt{2}} \\ \frac{(\frac{1}{2})z}{z-1} - \frac{\frac{1}{4}z}{z-1-j\sqrt{2}} - \frac{\frac{1}{4}z}{z-1+j\sqrt{2}} \end{bmatrix}$$

Finding the inverse z-transform and simplifying, we get the zero-state component of the state vector as

$$q_{zs}(n) = \begin{bmatrix} \frac{1}{2} - \frac{1}{2}(\sqrt{3})^n \cos(\tan^{-1}(\sqrt{2})n) + \frac{1}{\sqrt{2}}(\sqrt{3})^n \sin(\tan^{-1}(\sqrt{2})n) \\ \frac{1}{2} - \frac{1}{2}(\sqrt{3})^n \cos(\tan^{-1}(\sqrt{2})) \end{bmatrix} u(n)$$

Using the output equation, the output can be computed as given in Example 12.4.

□

Example 12.7. Find a closed-form expression the impulse response of the system, described by the state-space model given in Example 12.1, using the frequency-domain method.

Solution
The transfer function of a system is given by

$$H(z) = (C(zI - A)^{-1}B + D)$$

$$H(z) = \begin{bmatrix} 1 & -2 \end{bmatrix} \begin{bmatrix} \frac{z}{z^2-2z+3} & -\frac{3}{z^2-2z+3} \\ \frac{1}{z^2-2z+3} & \frac{z-2}{z^2-2z+3} \end{bmatrix} \begin{bmatrix} 1 \\ 0 \end{bmatrix} + 2 = \frac{z-2}{(z^2-2z+3)} + 2$$

Expanding into partial fractions, we get

$$H(z) = 2 + \frac{0.5 + j\frac{1}{2\sqrt{2}}}{z - 1 - j\sqrt{2}} + \frac{0.5 - j\frac{1}{2\sqrt{2}}}{z - 1 + j\sqrt{2}}$$

Finding the inverse z-transform and simplifying, we get

$$h(n) = 2\delta(n) + ((\sqrt{3})^{n-1} \cos((\tan^{-1}(\sqrt{2}))(n-1))$$

$$- \frac{1}{\sqrt{2}}(\sqrt{3})^{n-1} \sin((\tan^{-1}(\sqrt{2}))(n-1)))u(n-1), \quad n = 0, 1, 2, \ldots$$

□

12.4 Linear Transformation of State Vectors

For a specific input–output relationship of a system, the system can have different internal structures. By a linear transformation of a state vector, we can obtain another vector, implying a different internal structure of the system. Let us find the state-space model of a system with state vector q using another state vector $\bar{q}$ such that $\bar{q} = Pq$ and $q = P^{-1}\bar{q}$, where P is the $N \times N$ transformation matrix and P^{-1} exists. With the new state vector, the state equation can be written as

$$P^{-1}\bar{q}(n+1) = AP^{-1}\bar{q}(n) + Bx(n)$$

Premultiplying by P, we get

$$\bar{q}(n+1) = PAP^{-1}\bar{q}(n) + PBx(n)$$

With $\bar{A} = PAP^{-1}$ and $\bar{B} = PB$, the state equation can be written as

$$\bar{q}(n+1) = \bar{A}\bar{q}(n) + \bar{B}x(n)$$

With $\bar{C} = CP^{-1}$, the output equation can be written as

$$y(n) = \bar{C}\bar{q}(n) + Dx(n)$$

Some properties of A and $\bar{A}$ matrices can be used to check the computation of $\bar{A}$. The determinants of A and $\bar{A}$ are equal. The determinants of $(zI - A)$ and $(zI - \bar{A})$ are the same. The traces (sum of the diagonal elements) of A and $\bar{A}$ are equal.

Example 12.8. Derive the state-space model of the system in Example 12.1 with the new state vector that is related to old state vector as

$$\bar{q}_1(n) = 2q_1(n) + q_2(n)$$
$$\bar{q}_2(n) = q_1(n) - q_2(n)$$

Verify that the transfer function remains the same, using either state-space model.

Solution

$$P = \begin{bmatrix} 2 & 1 \\ 1 & -1 \end{bmatrix} \qquad P^{-1} = \begin{bmatrix} \frac{1}{3} & \frac{1}{3} \\ \frac{1}{3} & -\frac{2}{3} \end{bmatrix}$$

$$\bar{A} = PAP^{-1} = \begin{bmatrix} 2 & 1 \\ 1 & -1 \end{bmatrix}\begin{bmatrix} 2 & -3 \\ 1 & 0 \end{bmatrix}\begin{bmatrix} \frac{1}{3} & \frac{1}{3} \\ \frac{1}{3} & -\frac{2}{3} \end{bmatrix} = \begin{bmatrix} -\frac{1}{3} & \frac{17}{3} \\ -\frac{2}{3} & \frac{7}{3} \end{bmatrix}$$

$$\bar{B} = PB = \begin{bmatrix} 2 & 1 \\ 1 & -1 \end{bmatrix}\begin{bmatrix} 1 \\ 0 \end{bmatrix} = \begin{bmatrix} 2 \\ 1 \end{bmatrix}$$

$$\bar{C} = CP^{-1} = \begin{bmatrix} 1 & -2 \end{bmatrix}\begin{bmatrix} \frac{1}{3} & \frac{1}{3} \\ \frac{1}{3} & -\frac{2}{3} \end{bmatrix} = \begin{bmatrix} -\frac{1}{3} & \frac{5}{3} \end{bmatrix}$$

Figure 12.7 The state-space model of a second-order discrete system with the new state vector

The state-space model of a second-order discrete system with the new state vector is shown in Figure 12.7. The transfer function, computed using the new state-space model, is

$$H(z) = \begin{bmatrix} -\frac{1}{3} & \frac{5}{3} \end{bmatrix} \begin{bmatrix} z + \frac{1}{3} & -\frac{17}{3} \\ \frac{2}{3} & z - \frac{7}{3} \end{bmatrix}^{-1} \begin{bmatrix} 2 \\ 1 \end{bmatrix} + 2 = \frac{z-2}{(z^2 - 2z + 3)} + 2$$

which is the same as that obtained in Example 12.7.   □

12.5 Summary

- In this chapter, state-space analysis of discrete systems has been presented.
- The state-space model of a system is a generalization of input–output models, such as the transfer function.
- State-space model, in addition to the input and the output, includes N internal variables, called state variables, of an Nth-order system. All the outputs of the system are expressed in terms of the state variables and the input.
- A system, characterized by an Nth-order difference equation, is characterized, in the state-space model, by a set of N simultaneous first-order difference equations and a set of output equations.
- Solution of the N first-order difference equations yields the values of the N state variables. The output is expressed in terms of these values and the input. Solution of the state equations can be obtained by time-domain or frequency-domain methods.
- The state-space model of a system can be derived from its difference equation, transfer function, or realization diagram.

- The state-space model is not unique, since there are infinite realizations of a system with the same input–output relationship.
- Since it is an internal description of the system, by using linear transformation of the state vector, we can obtain another realization of the system, although of the same input–output relationship, with different characteristics, such as amount of quantization noise, number of components required, sensitivity to parameter variations, etc.
- State-space models can be easily extended to the analysis of time-varying and non-linear systems, and systems with multiple inputs and multiple outputs.

Further Reading

1. Lathi, B. P., *Linear Systems and Signals*, Oxford University Press, New York, 2004.

Exercises

12.1 Given the difference equation governing a second-order system, with input $x(n)$ and output $y(n)$: (a) find the state-space model of the system realized as shown in Figure 12.1; and (b) find the state-space model of the system realized as shown in Figure 12.3. Find the first four values of the impulse response of the system, iteratively, using both the state-space models and verify that they are equal.

12.1.1 $y(n) - 5y(n - 1) + 3y(n - 2) = -6x(n) + 4x(n - 1) - 2x(n - 2)$

12.1.2 $y(n) + 5y(n - 1) + 4y(n - 2) = 5x(n) - 2x(n - 1) - 6x(n - 2)$

12.1.3 $y(n) + 3y(n - 1) + 2y(n - 2) = 4x(n) - 5x(n - 1) + 6x(n - 2)$

12.2 Given the difference equation governing a second-order system: (a) find the state-space model of the system realized as shown in Figure 12.1; and (b) find the state-space model of the system realized as shown in Figure 12.3. Find the outputs $y(0)$, $y(1)$, and $y(2)$ of the system for the input $x(n)$, iteratively, using both the state-space models and verify that they are equal. The initial conditions of the system are $y(-1) = 1$ and $y(-2) = 2$.

12.2.1 $y(n) - \frac{5}{4}y(n - 1) + \frac{3}{8}y(n - 2) = 3x(n) - 4x(n - 1) - 2x(n - 2)$, $x(n) = (\frac{1}{2})^n u(n)$.

12.2.2 $y(n) + 2y(n - 1) + 4y(n - 2) = 4x(n) - 2x(n - 1) - 6x(n - 2)$, $x(n) = u(n)$.

12.2.3 $y(n) - y(n - 1) + 2y(n - 2) = 2x(n) - 3x(n - 1) + 2x(n - 2)$, $x(n) = (-1)^n u(n)$.

12.3 Given the difference equation governing a second-order system, with input $x(n)$ and output $y(n)$, find the state-space model of the system realized as shown in Figure 12.1. Derive the closed-form expression of the impulse response of the system using the time-domain state-space method. Give the first four values of the impulse response.

* 12.3.1 $y(n) + y(n-1) + \frac{2}{9}y(n-2) = x(n) - 3x(n-1) + 2x(n-2)$.

  12.3.2 $y(n) - y(n-1) + y(n-2) = 2x(n) + 3x(n-1) + 4x(n-2)$.

  12.3.3 $y(n) + 3y(n-1) + 2y(n-2) = 3x(n) - 4x(n-1) + 2x(n-2)$.

12.4 Given the difference equation governing a second-order system, find the state-space model of the system realized as shown in Figure 12.1. Derive the closed-form expression of the zero-input and zero-state components of the state vector, zero-input and zero-state components of the response, and the total response of the system, using the time-domain state-space method, for the input $x(n)$. Give the first four values of the zero-input, zero-state, and total responses. The initial conditions of the system are $y(-1) = -1$ and $y(-2) = 2$.

  12.4.1 $y(n) + \frac{5}{6}y(n-1) + \frac{1}{6}y(n-2) = 4x(n) + 2x(n-1) - x(n-2)$,
  $x(n) = (\frac{1}{2})^n u(n)$.

* 12.4.2 $y(n) + y(n-1) + \frac{1}{4}y(n-2) = 2x(n) - x(n-1) + x(n-2)$,
  $x(n) = u(n)$.

  12.4.3 $y(n) + 3y(n-1) + 2y(n-2) = -2x(n) - x(n-1) + 3x(n-2)$,
  $x(n) = \cos(\frac{2\pi}{4}n)u(n)$.

12.5 Given the difference equation governing a second-order system, with input $x(n)$ and output $y(n)$, find the state-space model of the system realized as shown in Figure 12.1. Derive the closed-form expression of the impulse response of the system using the frequency-domain state-space method. Give the first four values of the impulse response.

  12.5.1 $y(n) + y(n-1) + \frac{2}{9}y(n-2) = x(n) - 2x(n-1) - 2x(n-2)$.

  12.5.2 $y(n) - \frac{3}{4}y(n-1) + \frac{1}{8}y(n-2) = 3x(n) - 2x(n-1) + x(n-2)$.

  *12.5.3 $y(n) + \frac{2}{3}y(n-1) + \frac{1}{9}y(n-2) = 2x(n) + x(n-1) + x(n-2)$.

  12.5.4 $y(n) + \sqrt{2}y(n-1) + y(n-2) = x(n-1)$.

12.6 Given the difference equation governing a second-order system, find the state-space model of the system realized as shown in Figure 12.1. Derive the closed-form expression of the zero-input and zero-state components of the state vector, zero-input and zero-state components of the response, and the total response of the system, using the frequency-domain state-space method, for the given input $x(n)$ and the initial conditions $y(-1)$ and $y(-2)$. Give the first four values of the zero-input, zero-state, and total responses.

  12.6.1 $y(n) - \frac{5}{6}y(n-1) + \frac{1}{6}y(n-2) = x(n-1)$, $x(n) = \sin(\frac{2\pi}{4}n)u(n)$,
  $y(-1) = 0$ and $y(-2) = 0$.

  *12.6.2 $y(n) + y(n-1) + \frac{1}{4}y(n-2) = x(n) + x(n-1) + x(n-2)$,
  $x(n) = (\frac{1}{3})^n u(n)$, $y(-1) = 1$ and $y(-2) = 1$.

  12.6.3 $y(n) + y(n-1) + y(n-2) = x(n) - 2x(n-1) + x(n-2)$,
  $x(n) = u(n)$, $y(-1) = 2$ and $y(-2) = 1$.

12.7 The state-space model of a system is given. Derive another state-space model of the system using the given transformation matrix P. Verify that the transfer function remains the same using either state-space model. Further verify that:

(i) the traces and determinants of matrices A and $\overline{A}$ are equal; and (ii) the determinants of $(zI - A)$ and $(zI - \overline{A})$ are the same.

12.7.1 $A = \begin{bmatrix} 1 & 2 \\ 1 & 3 \end{bmatrix}$ $B = \begin{bmatrix} 1 \\ 2 \end{bmatrix}$ $C = \begin{bmatrix} 2 & 2 \end{bmatrix}$ $D = 1$

$\quad\quad P = \begin{bmatrix} 1 & 1 \\ 1 & -1 \end{bmatrix}$

12.7.2 $A = \begin{bmatrix} 3 & -1 \\ 2 & 3 \end{bmatrix}$ $B = \begin{bmatrix} 1 \\ 2 \end{bmatrix}$ $C = \begin{bmatrix} -2 & 1 \end{bmatrix}$ $D = 3$

$\quad\quad P = \begin{bmatrix} 2 & 3 \\ 1 & 1 \end{bmatrix}$

12.7.3 $A = \begin{bmatrix} 2 & -1 \\ 2 & 1 \end{bmatrix}$ $B = \begin{bmatrix} 1 \\ -1 \end{bmatrix}$ $C = \begin{bmatrix} 2 & 3 \end{bmatrix}$ $D = 3$

$\quad\quad P = \begin{bmatrix} 0 & 1 \\ 1 & 0 \end{bmatrix}$

13

State-space Analysis of Continuous Systems

The state-space analysis of continuous systems is similar to that of discrete systems. The realization diagrams are the same, with the delay elements replaced by integrators. Therefore, we concentrate, in this chapter, on time-domain and frequency-domain solutions of the state equation. The state-space model is presented in Section 13.1. Time-domain and frequency-domain solutions of the state equation are presented, respectively, in Sections 13.2 and 13.3. The linear transformation of state vector to obtain different realizations of systems is described in Section 13.4.

13.1 The State-space Model

Consider the state-space model, shown in Figure 13.1, of a second-order continuous system, characterized by the differential equation

$$\ddot{y}(t) + a_1 \dot{y}(t) + a_0 y(t) = b_2 \ddot{x}(t) + b_1 \dot{x}(t) + b_0 x(t)$$

(In this chapter, a dot over a variable indicates its first derivative and two dots indicates its second derivative. For example, $\dot{y}(t) = \mathrm{d}y(t)/\mathrm{d}t$ and $\ddot{y}(t) = \mathrm{d}^2 y(t)/\mathrm{d}t^2$.) In addition to the input $x(t)$ and the output $y(t)$, we have shown two internal variables (called the state variables), $q_1(t)$ and $q_2(t)$, of the system. State variables are a minimal set of variables (N for an Nth-order system) of a system so that a knowledge of the values of these variables (the state of the system) at $t = t_0$ and those of the input for $t \geq t_0$ will enable the determination of the values of the state variables for $t > t_0$ and the output for $t \geq t_0$. An infinite number of different sets, each of N state variables, are possible for a particular Nth-order system.

A Practical Approach to Signals and Systems   D. Sundararajan
© 2008 John Wiley & Sons (Asia) Pte Ltd

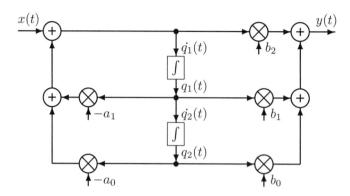

Figure 13.1   A state-space model of the canonical form I of a second-order continuous system

From Figure 13.1, we can write down the following state equations defining the state variables $q_1(t)$ and $q_2(t)$.

$$\dot{q}_1(t) = -a_1 q_1(t) - a_0 q_2(t) + x(t)$$

$$\dot{q}_2(t) = q_1(t)$$

The first derivative of each state variable is expressed in terms of all the state variables and the input. No derivatives of either the state variables or the input is permitted, on the right-hand side, to have the equation in a standard form. A second-order differential equation characterizing the system, shown in Figure 13.1, has been decomposed into a set of two simultaneous first-order differential equations. Selecting state variables as the output of the integrators is a natural choice, since an integrator is characterized by a first-order differential equation. With that choice, we can write down a state equation at the input of each integrator. However, the state variables need not correspond to quantities those are physically observable in a system. In the state-space model of a system, in general, an Nth-order differential equation characterizing a system is decomposed into a set of N simultaneous first-order differential equations of a standard form. With a set of N simultaneous differential equations, we can solve for N unknowns. These are the N internal variables, called the state variables, of the system. The output is expressed as a linear combination of the state variables and the input. The concepts of impulse response, convolution, and transform analysis are all equally applicable to the state-space model. The difference is that, as the system is modeled using matrix and vector quantities, the system analysis involves matrix and vector quantities. One of the advantages of the state-space model is the easier extension to multiple inputs and outputs. For simplicity, we describe systems with single input and single output only. The output $y(t)$ of the system, shown in Figure 13.1, is given by

$$y(t) = -b_2 a_1 q_1(t) - b_2 a_0 q_2(t)$$
$$+b_1 q_1(t) + b_0 q_2(t) + b_2 x(t)$$

The output equation is an algebraic (not a differential) equation. We can write the state and output equations, using vectors and matrices, as

$$\begin{bmatrix} \dot{q}_1(t) \\ \dot{q}_2(t) \end{bmatrix} = \begin{bmatrix} -a_1 & -a_0 \\ 1 & 0 \end{bmatrix} \begin{bmatrix} q_1(t) \\ q_2(t) \end{bmatrix} + \begin{bmatrix} 1 \\ 0 \end{bmatrix} x(t)$$

$$y(t) = \begin{bmatrix} b_1 - b_2 a_1 & b_0 - b_2 a_0 \end{bmatrix} \begin{bmatrix} q_1(t) \\ q_2(t) \end{bmatrix} + b_2 x(t)$$

Let us define the state vector $q(t)$ as

$$q(t) = \begin{bmatrix} q_1(t) \\ q_2(t) \end{bmatrix}$$

Then, with

$$A = \begin{bmatrix} -a_1 & -a_0 \\ 1 & 0 \end{bmatrix} \qquad B = \begin{bmatrix} 1 \\ 0 \end{bmatrix}$$

$$C = \begin{bmatrix} b_1 - b_2 a_1 & b_0 - b_2 a_0 \end{bmatrix} \qquad D = b_2,$$

the general state-space model description for continuous systems is given as

$$\dot{q}(t) = Aq(t) + Bx(t)$$
$$y(t) = Cq(t) + Dx(t)$$

The block diagram representation of the state-space model of an Nth-order continuous system, with single input and single output, is shown in Figure 13.2. Parallel lines terminating with an arrowhead indicate that the signal is a vector quantity.

Example 13.1. Consider the RLC circuit analyzed in Example 11.5, shown in Figure 13.3. It is a series circuit with a resistor of $5/3$ Ω, an inductance of two henries, and a capacitor of 3 F. The initial current through the inductor is three amperes and the initial voltage across the capacitor is two volts. This circuit is excited with a voltage

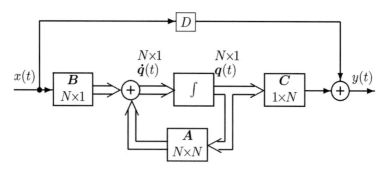

Figure 13.2 Block diagram representation of the state-space model of an Nth-order continuous system, with single input and single output

source of $3u(t)$ V. Assuming the capacitor voltage as the output of the circuit, find a state space model of the circuit.

Solution
The first step in finding the model is to write down the equations governing the circuit, using the circuit theorems and the input–output behavior of the components. Let the current through the circuit be $i(t)$, the voltage across the capacitor be $v_c(t)$, the input voltage be $x(t)$, and the output voltage be $y(t)$. The sum of the voltages across the components of the circuit must be equal to the input voltage. Therefore, we get

$$2\dot{i}(t) + \frac{5}{3}i(t) + v_c(t) = x(t)$$

The current in the circuit is given by $i(t) = 3\dot{v}_c(t)$. The next step is to select the minimum set of state variables required. Let the current through the inductor, $i(t)$, be the first state variable $q_1(t)$. Let the capacitor voltage, $v_c(t)$, be the second state variable $q_2(t)$. The next step is to substitute the state variables for the variables in the circuit differential equations. After substituting, these equations are rearranged such that only the first derivatives of the state variables appear on the left side and no

Figure 13.3 A RLC circuit

derivatives appear on the right side. For this example, we get

$$\dot{q}_1(t) = -\frac{5}{6}q_1(t) - \frac{1}{2}q_2(t) + \frac{1}{2}x(t)$$

$$\dot{q}_2(t) = \frac{1}{3}q_1(t)$$

These are the state equations of the circuit. The output equation of the circuit is $y(t) = v_c(t) = q_2(t)$. Using matrices, we get the state space model as

$$\dot{q}(t) = \begin{bmatrix} \dot{q}_1(t) \\ \dot{q}_2(t) \end{bmatrix} = \begin{bmatrix} -\frac{5}{6} & -\frac{1}{2} \\ \frac{1}{3} & 0 \end{bmatrix} \begin{bmatrix} q_1(t) \\ q_2(t) \end{bmatrix} + \begin{bmatrix} \frac{1}{2} \\ 0 \end{bmatrix} x(t)$$

$$y(t) = \begin{bmatrix} 0 & 1 \end{bmatrix} \begin{bmatrix} q_1(t) \\ q_2(t) \end{bmatrix}$$

Therefore,

$$A = \begin{bmatrix} -\frac{5}{6} & -\frac{1}{2} \\ \frac{1}{3} & 0 \end{bmatrix} \qquad B = \begin{bmatrix} \frac{1}{2} \\ 0 \end{bmatrix} \qquad C = \begin{bmatrix} 0 & 1 \end{bmatrix} \qquad D = 0$$

The state-space model of the RCL circuit is shown in Figure 13.4. It is similar to that shown in Figure 13.1, except that there is a multiplier with coefficient 1/3 between the two integrators. □

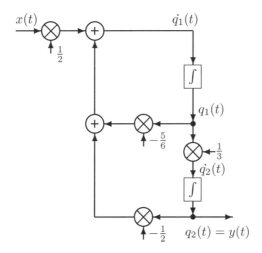

Figure 13.4 State-space model of the RCL circuit shown in Figure 13.3

13.2 Time-domain Solution of the State Equation

We have to find the solution to the state equation. For this purpose, we need the exponential of a matrix e^{At} and its derivative. Similar to the infinite series defining an exponential of a scalar,

$$e^{At} = I + At + A^2 \frac{t^2}{2!} + A^3 \frac{t^3}{3!} + \cdots$$

This series is absolutely and uniformly convergent for all values of t. Therefore, it can be differentiated or integrated term by term.

$$\frac{\mathrm{d}(e^{At})}{\mathrm{d}t} = A + A^2 t + A^3 \frac{t^2}{2!} + A^4 \frac{t^3}{3!} + \cdots = Ae^{At} = e^{At} A$$

By premultiplying both sides of state equation by e^{-At}, we get

$$e^{-At} \dot{q}(t) = e^{-At} Aq(t) + e^{-At} Bx(t)$$

By shifting the first term on the right-hand side to the left, we get

$$e^{-At} \dot{q}(t) - e^{-At} Aq(t) = e^{-At} Bx(t)$$

Since

$$\frac{\mathrm{d}(e^{-At} q(t))}{\mathrm{d}t} = e^{-At} \dot{q}(t) - e^{-At} Aq(t),$$

we can write the previous equation as

$$\frac{\mathrm{d}(e^{-At} q(t))}{\mathrm{d}t} = e^{-At} Bx(t)$$

Integrating both sides of this equation from 0^- to t, we get

$$e^{-At} q(t) \Big|_{0^-}^{t} = \int_{0^-}^{t} e^{-A\tau} Bx(\tau) \mathrm{d}\tau$$

Applying the limit and then premultiplying both sides by e^{At}, we get

$$q(t) = \overbrace{e^{At} q(0^-)}^{q_{zi}(t)} + \overbrace{\int_{0^-}^{t} e^{A(t-\tau)} Bx(\tau) \mathrm{d}\tau}^{q_{zs}(t)}$$

The first and second expressions on the right-hand side are, respectively, the zero-input and zero-state components of the state vector $q(t)$. Note that the part of the expression

$$\int_{0^-}^{t} e^{A(t-\tau)} Bx(\tau) d\tau$$

is the convolution of the matrices e^{At} and $Bx(t)$, $e^{At} * Bx(t)$. Convolution of matrices is the same as the multiplication of two matrices, except that the product of two elements is replaced by their convolution. If the initial state vector values are given at $t = t_0^-$, rather than at $t = 0^-$, the state equation is modified as

$$q(t) = e^{A(t-t_0)} q(t_0^-) + \int_{t_0^-}^{t} e^{A(t-\tau)} Bx(\tau) d\tau$$

The matrix e^{At} is called the state-transition matrix or the fundamental matrix of the system.

Once we know the state vector, we get the output of the system using the output equation as

$$y(t) = C \left(e^{At} q(0^-) + \int_{0^-}^{t} e^{A(t-\tau)} Bx(\tau) d\tau \right) + Dx(t)$$

$$= \overbrace{C e^{At} q(0^-)}^{y_{zi}(t)} + \overbrace{C \int_{0^-}^{t} e^{A(t-\tau)} Bx(\tau) d\tau + Dx(t)}^{y_{zs}(t)}$$

The first expression on the right-hand side is the zero-input component of the system response $y(t)$ and the other two expressions yield the zero-state component. The zero-input response of the system depends solely on the state-transition matrix e^{At}. This matrix, for an Nth-order system, is evaluated, using the Cayley–Hamilton theorem, as

$$e^{At} = c_0 I + c_1 A + c_2 A^2 + \cdots + c_{N-1} A^{(N-1)}$$

where

$$
\begin{bmatrix} c_0 \\ c_1 \\ \cdots \\ c_{N-1} \end{bmatrix}
=
\begin{bmatrix}
1 & \lambda_1 & \lambda_1^2 & \cdots & \lambda_1^{N-1} \\
1 & \lambda_2 & \lambda_2^2 & \cdots & \lambda_2^{N-1} \\
& & \cdots & & \\
1 & \lambda_N & \lambda_N^2 & \cdots & \lambda_N^{N-1}
\end{bmatrix}^{-1}
\begin{bmatrix} e^{\lambda_1 t} \\ e^{\lambda_2 t} \\ \cdots \\ e^{\lambda_N t} \end{bmatrix}
$$

and $\lambda_1, \lambda_2, \ldots, \lambda_N$ are the N distinct characteristic roots of A. For a root λ_r repeated m times, the first row corresponding to that root will remain the same as for a distinct root and the $m - 1$ successive rows will be successive derivatives of the first row with respect to λ_r. For example, with the first root of a fourth-order system repeating two times, we get

$$
\begin{bmatrix} c_0 \\ c_1 \\ c_2 \\ c_3 \end{bmatrix} = \begin{bmatrix} 1 & \lambda_1 & \lambda_1^2 & \lambda_1^3 \\ 0 & 1 & 2\lambda_1 & 3\lambda_1^2 \\ 1 & \lambda_2 & \lambda_2^2 & \lambda_2^3 \\ 1 & \lambda_3 & \lambda_3^2 & \lambda_3^3 \end{bmatrix}^{-1} \begin{bmatrix} e^{\lambda_1 t} \\ te^{\lambda_1 t} \\ e^{\lambda_2 t} \\ e^{\lambda_3 t} \end{bmatrix}
$$

Example 13.2. Find a closed-form expression for the output $y(t)$ of the system, described by the differential equation

$$
\ddot{y}(t) + 4\dot{y}(t) + 4y(t) = \ddot{x}(t) + \dot{x}(t) + 2x(t)
$$

using the time-domain method, with the initial conditions $y(0^-) = 2$ and $\dot{y}(0^-) = 3$ and the input $u(t)$, the unit-step function. Assume canonical form I realization of the system as shown in Figure 13.1.

Solution

$$
A = \begin{bmatrix} -4 & -4 \\ 1 & 0 \end{bmatrix} \qquad B = \begin{bmatrix} 1 \\ 0 \end{bmatrix} \qquad C = \begin{bmatrix} -3 & -2 \end{bmatrix} \qquad D = 1
$$

The initial state vector has to be found from the given initial output conditions using the state and output equations. From the state equation, we get

$$
\dot{q}_1(0^-) = -4q_1(0^-) - 4q_2(0^-)
$$
$$
\dot{q}_2(0^-) = q_1(0^-)
$$

Note that the input $x(t)$ is zero at $t = 0^-$. From the output equation, we get

$$
-3\dot{q}_1(0^-) - 2\dot{q}_2(0^-) = 3
$$
$$
-3q_1(0^-) - 2q_2(0^-) = 2
$$

Solving these equations, we get the initial state vector as

$$
q_1(0^-) = -\frac{15}{8} \qquad q_2(0^-) = \frac{29}{16}
$$

The characteristic polynomial of a system is given by the determinant of the matrix $(sI - A)$, where I is the identity matrix of the same size as A. While we can write down the characteristic polynomial from the differential equation, we just show how it can be found using the matrix A. For this example,

$$(sI - A) = s \begin{bmatrix} 1 & 0 \\ 0 & 1 \end{bmatrix} - \begin{bmatrix} -4 & -4 \\ 1 & 0 \end{bmatrix} = \begin{bmatrix} s+4 & 4 \\ -1 & s \end{bmatrix}$$

The characteristic polynomial of the system, given by the determinant of this matrix, is

$$s^2 + 4s + 4$$

With each of the infinite different realizations of a system, we get the A matrix with different values. However, as the system is the same, its characteristic polynomial, given by the determinant of $(sI - A)$, will be the same for any valid A. The characteristic roots, which are the roots of this polynomial, are $\lambda_1 = -2$ and $\lambda_2 = -2$. The transition matrix is given by

$$e^{At} = c_0 I + c_1 A$$

where

$$\begin{bmatrix} c_0 \\ c_1 \end{bmatrix} = \begin{bmatrix} 1 & \lambda_1 \\ 0 & 1 \end{bmatrix}^{-1} \begin{bmatrix} e^{\lambda_1 t} \\ te^{\lambda_1 t} \end{bmatrix} = \begin{bmatrix} 1 & 2 \\ 0 & 1 \end{bmatrix} \begin{bmatrix} e^{-2t} \\ te^{-2t} \end{bmatrix}$$

$$= \begin{bmatrix} e^{-2t} + 2te^{-2t} \\ te^{-2t} \end{bmatrix}$$

$$e^{At} = c_0 \begin{bmatrix} 1 & 0 \\ 0 & 1 \end{bmatrix} + c_1 \begin{bmatrix} -4 & -4 \\ 1 & 0 \end{bmatrix} = \begin{bmatrix} c_0 - 4c_1 & -4c_1 \\ c_1 & c_0 \end{bmatrix}$$

$$= \begin{bmatrix} e^{-2t} - 2te^{-2t} & -4te^{-2t} \\ te^{-2t} & e^{-2t} + 2te^{-2t} \end{bmatrix}$$

Since $q(t) = e^{At}q(0)$, with $t = 0$, we get $q(0) = e^{A0}q(0)$. That is, $I = e^{A0}$. This result, which can be used to check the state transition matrix, is also obvious from the infinite series for e^{At}.

The state vector $q(t)$ can be computed as follows.

$$q(t) = \begin{bmatrix} e^{-2t} - 2te^{-2t} & -4te^{-2t} \\ te^{-2t} & e^{-2t} + 2te^{-2t} \end{bmatrix} \begin{bmatrix} -\frac{15}{8} \\ \frac{29}{16} \end{bmatrix}$$

$$+ \begin{bmatrix} e^{-2t} - 2te^{-2t} & -4te^{-2t} \\ te^{-2t} & e^{-2t} + 2te^{-2t} \end{bmatrix} * \left(\begin{bmatrix} 1 \\ 0 \end{bmatrix} u(t) \right)$$

$$= \begin{bmatrix} e^{-2t} - 2te^{-2t} & -4te^{-2t} \\ te^{-2t} & e^{-2t} + 2te^{-2t} \end{bmatrix} \begin{bmatrix} -\frac{15}{8} \\ \frac{29}{16} \end{bmatrix}$$

$$+ \begin{bmatrix} \int_0^t (e^{-2\tau} - 2\tau e^{-2\tau})d\tau \\ \int_0^t (\tau e^{-2\tau})d\tau \end{bmatrix}$$

$$= \begin{bmatrix} -\frac{15}{8}e^{-2t} - \frac{7}{2}te^{-2t} \\ \frac{29}{16}e^{-2t} + \frac{7}{4}te^{-2t} \end{bmatrix} + \begin{bmatrix} te^{-2t} \\ -\frac{1}{4}e^{-2t} - \frac{1}{2}te^{-2t} + \frac{1}{4} \end{bmatrix}$$

$$= \begin{bmatrix} -\frac{15}{8}e^{-2t} - \frac{5}{2}te^{-2t} \\ \frac{1}{4} + \frac{25}{16}e^{-2t} + \frac{5}{4}te^{-2t} \end{bmatrix}$$

The output $y(t)$ can be computed using the output equation. The zero-input component of the output is given by

$$y_{zi}(t) = \begin{bmatrix} -3 & -2 \end{bmatrix} \begin{bmatrix} -\frac{15}{8}e^{-2t} - \frac{7}{2}te^{-2t} \\ \frac{29}{16}e^{-2t} + \frac{7}{4}te^{-2t} \end{bmatrix}$$

$$= 2e^{-2t} + 7te^{-2t}$$

The zero-state component of the output is given by

$$y_{zs}(t) = \begin{bmatrix} -3 & -2 \end{bmatrix} \begin{bmatrix} te^{-2t} \\ -\frac{1}{4}e^{-2t} - \frac{1}{2}te^{-2t} + \frac{1}{4} \end{bmatrix} + 1$$

$$= 0.5 + 0.5e^{-2t} - 2te^{-2t}$$

The total response of the system is the sum of the zero-input and zero-state components of the response and is given as

$$y(t) = (0.5 + 2.5e^{-2t} + 5te^{-2t})u(t) \qquad \square$$

Example 13.3. Find the closed-form expression for the impulse response of the system, described by the state-space model given in Example 13.2, using the time-domain method.

Solution

$$Ce^{At} = \begin{bmatrix} -3 & -2 \end{bmatrix} \begin{bmatrix} c_0 - 4c_1 & -4c_1 \\ c_1 & c_0 \end{bmatrix}$$

$$= \begin{bmatrix} -3c_0 + 10c_1 & -2c_0 + 12c_1 \end{bmatrix}$$

Since the convolution output of a function with the unit-impulse is itself and the vector B is a constant, the impulse response is given by

$$h(t) = Ce^{At}B + D\delta(t) = -3c_0 + 10c_1 + \delta(t)$$

$$h(t) = (\delta(t) - 3e^{-2t} + 4te^{-2t})u(t) \qquad \square$$

13.3 Frequency-domain Solution of the State Equation

The Laplace transform of a vector function, such as $q(t)$, is defined to be the vector function $Q(s)$, where the elements are the transforms of the corresponding elements of $q(t)$. Taking the Laplace transform of the state equation, we get

$$sQ(s) - q(0^-) = AQ(s) + BX(s)$$

We have used the time-differentiation property of the Laplace transform and $q(0^-)$ is the initial state vector. Since $IQ(s) = Q(s)$, where I is the identity matrix of the same size as the matrix A, and collecting the terms involving $Q(s)$, we get

$$(sI - A)Q(s) = q(0^-) + BX(s)$$

The inclusion of the identity matrix is necessary to combine the terms involving $Q(s)$. Premultiplying both sides by $(sI - A)^{-1}$, which is the inverse of $(sI - A)$, we get

$$Q(s) = \overbrace{(sI - A)^{-1}q(0^-)}^{q_{zi}(s)} + \overbrace{(sI - A)^{-1}BX(s)}^{q_{zs}(s)}$$

The inverse Laplace transforms of the first and second expressions on the right-hand side are, respectively, the zero-input and zero-state components of the state vector $q(t)$. Taking the Laplace transform of the output equation, we get

$$Y(s) = CQ(s) + DX(s)$$

Substituting for $Q(s)$, we get

$$Y(s) = \overbrace{C(sI - A)^{-1}q(0^-)}^{y_{zi}(s)} + \overbrace{(C(sI - A)^{-1}B + D)X(s)}^{y_{zs}(s)}$$

The inverse Laplace transforms of the first and second expressions on the right-hand side are, respectively, the zero-input and zero-state components of the system response $y(t)$. Comparing with the expression for $Q(t)$, we find that the inverse Laplace transform of $((sI - A)^{-1})$ is e^{At}, the transition or fundamental matrix of the system. With the system initial conditions zero, the transfer function is given by

$$H(s) = \frac{Y(s)}{X(s)} = (C(sI - A)^{-1}B + D)$$

Example 13.4. Solve the problem of Example 13.2 using the frequency-domain method.

Solution
The initial state vector is

$$q(0^-) = \begin{bmatrix} -\frac{15}{8} \\ \frac{29}{16} \end{bmatrix}$$

as derived in Example 13.2 from the given initial output conditions.

$$(sI - A) = s\begin{bmatrix} 1 & 0 \\ 0 & 1 \end{bmatrix} - \begin{bmatrix} -4 & -4 \\ 1 & 0 \end{bmatrix} = \begin{bmatrix} s+4 & 4 \\ -1 & s \end{bmatrix}$$

$$(sI - A)^{-1} = \frac{1}{s^2 + 4s + 4}\begin{bmatrix} s & -4 \\ 1 & s+4 \end{bmatrix} = \begin{bmatrix} \frac{s}{s^2+4s+4} & \frac{-4}{s^2+4s+4} \\ \frac{1}{s^2+4s+4} & \frac{s+4}{s^2+4s+4} \end{bmatrix}$$

We used the fact that $I = e^{A0}$ to check the computation of e^{At}. In the frequency-domain, the corresponding check, using the initial value theorem of the Laplace transform, is $\lim_{s\to\infty} s(sI - A)^{-1} = I$.

The zero-input component of the state vector is

$$q_{zi}(s) = (sI - A)^{-1}q(0^-) = \begin{bmatrix} \frac{s}{s^2+4s+4} & \frac{-4}{s^2+4s+4} \\ \frac{1}{s^2+4s+4} & \frac{s+4}{s^2+4s+4} \end{bmatrix}\begin{bmatrix} -\frac{15}{8} \\ \frac{29}{16} \end{bmatrix}$$

$$= \begin{bmatrix} \frac{-\frac{15}{8}s - \frac{29}{4}}{s^2+4s+4} \\ \frac{\frac{29}{16}s + \frac{43}{8}}{s^2+4s+4} \end{bmatrix} = \begin{bmatrix} -\frac{\frac{7}{2}}{(s+2)^2} - \frac{\frac{15}{8}}{s+2} \\ \frac{\frac{7}{4}}{(s+2)^2} + \frac{\frac{29}{16}}{s+2} \end{bmatrix}$$

Taking the inverse Laplace transform, we get

$$q_{zi}(t) = \begin{bmatrix} -\frac{15}{8}e^{-2t} - \frac{7}{2}te^{-2t} \\ \frac{29}{16}e^{-2t} + \frac{7}{4}te^{-2t} \end{bmatrix}$$

The zero-state component of the state vector is

$$q_{zs}(s) = (sI - A)^{-1}B)X(s) = \begin{bmatrix} \frac{s}{s^2+4s+4} & \frac{-4}{s^2+4s+4} \\ \frac{1}{s^2+4s+4} & \frac{s+4}{s^2+4s+4} \end{bmatrix} \begin{bmatrix} 1 \\ 0 \end{bmatrix} \frac{1}{s}$$

$$= \begin{bmatrix} \frac{1}{s^2+4s+4} \\ \frac{1}{s(s^2+4s+4)} \end{bmatrix} = \begin{bmatrix} \frac{1}{(s+2)^2} \\ \frac{1}{4}{s} - \frac{\frac{1}{2}}{(s+2)^2} - \frac{\frac{1}{4}}{s+2} \end{bmatrix}$$

Taking the inverse Laplace transform, we get

$$q_{zs}(t) = \begin{bmatrix} te^{-2t} \\ \frac{1}{4} - \frac{1}{4}e^{-2t} - \frac{1}{2}te^{-2t} \end{bmatrix}$$

Using the output equation, the output can be computed as given in Example 13.2.
□

Example 13.5. Find the closed-form expression for the impulse response of the system, described by the state-space model given in Example 13.2, using the frequency-domain method.

Solution
The transfer function is given by

$$H(s) = \frac{Y(s)}{X(s)} = (C(sI - A)^{-1}B + D)$$

$$H(s) = \begin{bmatrix} -3 & -2 \end{bmatrix} \begin{bmatrix} \frac{s}{s^2+4s+4} & \frac{-4}{s^2+4s+4} \\ \frac{1}{s^2+4s+4} & \frac{s+4}{s^2+4s+4} \end{bmatrix} \begin{bmatrix} 1 \\ 0 \end{bmatrix} + 1$$

$$= \frac{-3s - 2}{(s^2 + 4s + 4)} + 1 = 1 - \frac{3}{s + 2} + \frac{4}{(s + 2)^2}$$

Finding the inverse Laplace transform, we get

$$h(t) = (\delta(t) - 3e^{-2t} + 4te^{-2t})u(t)$$
□

13.4 Linear Transformation of State Vectors

In common with discrete systems, for a specific input–output relationship of a continuous system, the system can have different internal structures. By a linear transformation of a state vector, we can obtain another vector, implying different internal structure of the system. Let us find the state-space model of a system with state vector q using another state vector $\bar{q}$ such that $\bar{q} = Pq$ and $q = P^{-1}\bar{q}$, where P is the $N \times N$ transformation matrix and P^{-1} exists. With the new state vector, the state equation can be written as

$$P^{-1}\dot{\bar{q}}(t) = AP^{-1}\bar{q}(t) + Bx(t)$$

Premultiplying by P, we get

$$\dot{\bar{q}}(t) = PAP^{-1}\bar{q}(t) + PBx(t)$$

With $\bar{A} = PAP^{-1}$ and $\bar{B} = PB$, the state equation can be written as

$$\dot{\bar{q}}(t) = \bar{A}\bar{q}(t) + \bar{B}x(t)$$

With $\bar{C} = CP^{-1}$, the output equation can be written as

$$y(t) = \bar{C}\bar{q}(t) + Dx(t)$$

Some properties of A and $\bar{A}$ matrices can be used to check the computation of $\bar{A}$. The determinants of A and $\bar{A}$ are equal. The determinants of $(sI - A)$ and $(sI - \bar{A})$ are the same. The traces (sum of the diagonal elements) of A and $\bar{A}$ are equal.

Example 13.6. Derive the state-space model of the system in Example 13.1 with the new state vector that is related to old state vector as

$$\bar{q}_1(t) = q_1(t) + 2q_2(t)$$
$$\bar{q}_2(t) = -3q_1(t) + 4q_2(t)$$

Verify that the transfer function remains the same using either state-space model.

Solution

$$P = \begin{bmatrix} 1 & 2 \\ -3 & 4 \end{bmatrix} \qquad P^{-1} = \begin{bmatrix} \frac{4}{10} & -\frac{2}{10} \\ \frac{3}{10} & \frac{1}{10} \end{bmatrix}$$

$$\bar{A} = PAP^{-1} = \begin{bmatrix} 1 & 2 \\ -3 & 4 \end{bmatrix} \begin{bmatrix} -\frac{5}{6} & -\frac{1}{2} \\ \frac{1}{3} & 0 \end{bmatrix} \begin{bmatrix} \frac{4}{10} & -\frac{2}{10} \\ \frac{3}{10} & \frac{1}{10} \end{bmatrix} = \begin{bmatrix} -\frac{13}{60} & -\frac{1}{60} \\ \frac{119}{60} & -\frac{37}{60} \end{bmatrix}$$

Figure 13.5   The state-space model of a second-order continuous system with the new state vector

$$\overline{B} = PB = \begin{bmatrix} 1 & 2 \\ -3 & 4 \end{bmatrix} \begin{bmatrix} \frac{1}{2} \\ 0 \end{bmatrix} = \begin{bmatrix} \frac{1}{2} \\ -\frac{3}{2} \end{bmatrix}$$

$$\overline{C} = CP^{-1} = \begin{bmatrix} 0 & 1 \end{bmatrix} \begin{bmatrix} \frac{4}{10} & -\frac{2}{10} \\ \frac{3}{10} & \frac{1}{10} \end{bmatrix} = \begin{bmatrix} \frac{3}{10} & \frac{1}{10} \end{bmatrix}$$

The state-space model of a second-order continuous system with the new state vector is shown in Figure 13.5. This realization requires more components than that shown in Figure 13.4. However, it must be noted that, while minimum number of components is of great importance, there are other criteria, such as less coefficient sensitivity, that could decide which of the realizations of a system is suitable for a particular application.

The transfer function, using the new state-space model, is computed as follows.

$$(sI - \overline{A}) = s\begin{bmatrix} 1 & 0 \\ 0 & 1 \end{bmatrix} - \begin{bmatrix} -\frac{13}{60} & -\frac{1}{60} \\ \frac{119}{60} & -\frac{37}{60} \end{bmatrix} = \begin{bmatrix} s + \frac{13}{60} & \frac{1}{60} \\ -\frac{119}{60} & s + \frac{37}{60} \end{bmatrix}$$

$$(sI - \overline{A})^{-1} = \frac{1}{s^2 + \frac{5}{6}s + \frac{1}{6}} \begin{bmatrix} s + \frac{37}{60} & -\frac{1}{60} \\ \frac{119}{60} & s + \frac{13}{60} \end{bmatrix} = \begin{bmatrix} \dfrac{s + \frac{37}{60}}{s^2 + \frac{5}{6}s + \frac{1}{6}} & \dfrac{-\frac{1}{60}}{s^2 + \frac{5}{6}s + \frac{1}{6}} \\ \dfrac{\frac{119}{60}}{s^2 + \frac{5}{6}s + \frac{1}{6}} & \dfrac{s + \frac{13}{60}}{s^2 + \frac{5}{6}s + \frac{1}{6}} \end{bmatrix}$$

$$H(s) = \begin{bmatrix} \frac{3}{10} & \frac{1}{10} \end{bmatrix} \begin{bmatrix} \dfrac{s + \frac{37}{60}}{s^2 + \frac{5}{6}s + \frac{1}{6}} & \dfrac{-\frac{1}{60}}{s^2 + \frac{5}{6}s + \frac{1}{6}} \\ \dfrac{\frac{119}{60}}{s^2 + \frac{5}{6}s + \frac{1}{6}} & \dfrac{s + \frac{13}{60}}{s^2 + \frac{5}{6}s + \frac{1}{6}} \end{bmatrix} \begin{bmatrix} \frac{1}{2} \\ -\frac{3}{2} \end{bmatrix} = \dfrac{\frac{1}{6}}{(s^2 + \frac{5}{6}s + \frac{1}{6})}$$

The transfer function, using the old state-space model, is computed as follows.

$$(s\mathbf{I} - \mathbf{A}) = s\begin{bmatrix} 1 & 0 \\ 0 & 1 \end{bmatrix} - \begin{bmatrix} -\frac{5}{6} & -\frac{1}{2} \\ \frac{1}{3} & 0 \end{bmatrix} = \begin{bmatrix} s + \frac{5}{6} & \frac{1}{2} \\ -\frac{1}{3} & s \end{bmatrix}$$

$$(s\mathbf{I} - \mathbf{A})^{-1} = \frac{1}{s^2 + \frac{5}{6}s + \frac{1}{6}}\begin{bmatrix} s & -\frac{1}{2} \\ \frac{1}{3} & s + \frac{5}{6} \end{bmatrix} = \begin{bmatrix} \frac{s}{s^2 + \frac{5}{6}s + \frac{1}{6}} & \frac{-\frac{1}{2}}{s^2 + \frac{5}{6}s + \frac{1}{6}} \\ \frac{\frac{1}{3}}{s^2 + \frac{5}{6}s + \frac{1}{6}} & \frac{s + \frac{5}{6}}{s^2 + \frac{5}{6}s + \frac{1}{6}} \end{bmatrix}$$

$$H(s) = \begin{bmatrix} 0 & 1 \end{bmatrix}\begin{bmatrix} \frac{s}{s^2 + \frac{5}{6}s + \frac{1}{6}} & \frac{-\frac{1}{2}}{s^2 + \frac{5}{6}s + \frac{1}{6}} \\ \frac{\frac{1}{3}}{s^2 + \frac{5}{6}s + \frac{1}{6}} & \frac{s + \frac{5}{6}}{s^2 + \frac{5}{6}s + \frac{1}{6}} \end{bmatrix}\begin{bmatrix} \frac{1}{2} \\ 0 \end{bmatrix} = \frac{\frac{1}{6}}{(s^2 + \frac{5}{6}s + \frac{1}{6})},$$

which is the same as that obtained above.   □

13.5 Summary

- In this chapter, state-space analysis of continuous systems has been presented.
- The state-space model of a system is a generalization of input–output models, such as the transfer function.
- The state-space model, in addition to the input and the output, includes N internal variables of the system, called state variables, for an Nth-order system. All the outputs of the system are expressed in terms of the state variables and the input.
- A system, characterized by an Nth-order differential equation, is characterized, in the state-space model, by a set of N simultaneous first-order differential equations and a set of output equations.
- Solution of the N first-order differential equations yields the values of the state variables. The output is expressed in terms of these values and the input. Solution of the state equations can be obtained by time-domain or frequency-domain methods.
- The state-space model of a system can be derived from its differential equation, transfer function, or realization diagram.
- The state-space model is not unique, since there are infinite realizations of a system with the same input–output relationship.
- Since it is an internal description of the system, by using linear transformation of the state vector, we can obtain another realization of the system, although of the same input–output relationship, with different characteristics, such as sensitivity to parameter variations, number of components required, etc.
- State-space models can be easily extended to the analysis of time-varying and non-linear systems, and systems with multiple inputs and multiple outputs.

Further Reading

1. Lathi, B. P., *Linear Systems and Signals*, Oxford University Press, New York, 2004.

Exercises

13.1   Find the zero-input and zero-state components of the output of the circuit, described in Example 13.1, using the time-domain state-space method. Find the total output also.

13.2   Consider the series RLC circuit with a resistor of 9 Ω, an inductance of 3 H, and a capacitor of 1/6 F. The initial current through the inductor is 2 A and the initial voltage across the capacitor is 3 V. This circuit is excited with a voltage source $x(t) = 2e^{-3t}u(t)$ V. Assuming the current in the circuit as the output and the current through the inductor, $q1$, and the voltage across the capacitor, $q2$, as the state variables, find the state space model of the circuit. Find the zero-input and zero-state components of the output of the circuit using the frequency-domain state-space method. Find the total output also.

13.3   Consider the series RLC circuit with a resistor of 8 Ω, an inductance of 2 H, and a capacitor of 1/6 F. The initial current through the inductor is 4 A and the initial voltage across the capacitor is 3 V. This circuit is excited with a voltage source $x(t) = 3u(t)$ V. Assuming the inductor voltage as the output and the current through the inductor, $q1$, and the voltage across the capacitor, $q2$, as the state variables, find the state space model of the circuit. Find the zero-input and zero-state components of the output of the circuit using the time-domain state-space method. Find the total output also.

*13.4   Consider the series RLC circuit with a resistor of 2 Ω, an inductance of 1 H, and a capacitor of 1 F. The initial current through the inductor is 0 A and the initial voltage across the capacitor is 0 V. This circuit is excited with a voltage source $x(t) = 4e^{-t}u(t)$ V. Assuming the voltage across the resistor as the output and the current through the inductor, $q1$, and the voltage across the capacitor, $q2$, as the state variables, find the state space model of the circuit. Find the output of the circuit using the frequency-domain state-space method.

13.5   Consider the system described by the differential equation

$$\ddot{y}(t) + 5\dot{y}(t) + 6y(t) = 2\ddot{x}(t) - 3\dot{x}(t) + 4x(t)$$

with the initial conditions $y(0^-) = 2$ and $\dot{y}(0^-) = 1$ and the input $x(t) = 2u(t)$. Assign two state variables to the output of each integrator and assume canonical form I realization of the system as shown in Figure 13.1. Find the zero-input and zero-state components of the output of the system using the time-domain state-space method. Find the total output also.

*13.6   Consider the system described by the differential equation

$$\ddot{y}(t) + 4\dot{y}(t) + 3y(t) = \ddot{x}(t) - 2\dot{x}(t) + 3x(t)$$

with the initial conditions $y(0^-) = 3$ and $\dot{y}(0^-) = 1$ and the input $x(t) = 3e^{-2t}u(t)$. Assign two state variables to the output of each integrator and assume canonical form I realization of the system as shown in Figure 13.1. Find the zero-input and zero-state components of the output of the system using the time-domain state-space method. Find the total output also.

13.7   Consider the system described by the differential equation

$$\ddot{y}(t) + 5\dot{y}(t) + 4y(t) = x(t)$$

with the input $x(t) = \sin(t + \frac{\pi}{3})u(t)$. Assign two state variables to the output of each integrator and assume canonical form I realization of the system as shown in Figure 13.1. Find the zero-state output of the system using the time-domain state-space method.

13.8   Find the impulse response of the system characterized by the differential equation, with input $x(t)$ and output $y(t)$,

$$\ddot{y}(t) + 2\dot{y}(t) + y(t) = \ddot{x}(t) + \dot{x}(t) + 2x(t)$$

using the time-domain state-space method. Assign two state variables to the output of each integrator and assume canonical form I realization of the system as shown in Figure 13.1.

*13.9   Find the impulse response of the system characterized by the differential equation, with input $x(t)$ and output $y(t)$,

$$\ddot{y}(t) + \frac{5}{6}\dot{y}(t) + \frac{1}{6}y(t) = \dot{x}(t) + x(t)$$

using the time-domain state-space method. Assign two state variables to the output of each integrator and assume canonical form I realization of the system as shown in Figure 13.1.

13.10   Find the impulse response of the system characterized by the differential equation, with input $x(t)$ and output $y(t)$,

$$\ddot{y}(t) + 6\dot{y}(t) + 5y(t) = 2\ddot{x}(t)$$

using the time-domain state-space method. Assign two state variables to the output of each integrator and assume canonical form I realization of the system as shown in Figure 13.1.

*13.11 Consider the system described by the differential equation

$$\ddot{y}(t) + 3\dot{y}(t) + 2y(t) = 3\ddot{x}(t) - \dot{x}(t) + 4x(t)$$

with the initial conditions $y(0^-) = 2$ and $\dot{y}(0^-) = 3$ and the input $x(t) = 3u(t)$. Assign two state variables to the output of each integrator and assume canonical form I realization of the system as shown in Figure 13.1. Find the zero-input and zero-state components of the output of the system using the frequency-domain state-space method. Find the total output also.

13.12 Consider the system described by the differential equation

$$\ddot{y}(t) + 6\dot{y}(t) + 9y(t) = -2\ddot{x}(t) + \dot{x}(t) - 3x(t)$$

with the initial conditions $y(0^-) = -2$ and $\dot{y}(0^-) = -3$ and the input $x(t) = 2e^{-4t}u(t)$. Assign two state variables to the output of each integrator and assume canonical form I realization of the system as shown in Figure 13.1. Find the zero-input and zero-state components of the output of the system using the frequency-domain state-space method. Find the total output also.

13.13 Consider the system described by the differential equation

$$\ddot{y}(t) + 6\dot{y}(t) + 8y(t) = x(t)$$

with the input $x(t) = \cos(2t - \frac{\pi}{6})u(t)$. Assign two state variables to the output of each integrator and assume canonical form I realization of the system as shown in Figure 13.1. Find the zero-state output of the system using the frequency-domain state-space method.

13.14 Find the impulse response of the system characterized by the differential equation, with input $x(t)$ and output $y(t)$,

$$\ddot{y}(t) + 4\dot{y}(t) + 3y(t) = 3\ddot{x}(t) - 2\dot{x}(t) + x(t)$$

using the frequency-domain state-space method. Assign two state variables to the output of each integrator and assume canonical form I realization of the system as shown in Figure 13.1.

*13.15 Find the impulse response of the system characterized by the differential equation, with input $x(t)$ and output $y(t)$,

$$\ddot{y}(t) + 2\dot{y}(t) + y(t) = -2\ddot{x}(t) + 3\dot{x}(t) - 4x(t)$$

using the frequency-domain state-space method. Assign two state variables to the output of each integrator and assume canonical form I realization of the system as shown in Figure 13.1.

13.16   Find the impulse response of the system characterized by the differential
        equation, with input $x(t)$ and output $y(t)$,

$$\ddot{y}(t) + 7\dot{y}(t) + 12y(t) = -3\dot{x}(t) + 2x(t)$$

using the frequency-domain state-space method. Assign two state variables to
the output of each integrator and assume canonical form I realization of the
system as shown in Figure 13.1.

13.17   Derive the state-space model of the system in Example 13.1 with the new state
        vector $\bar{q}$ that is related to old state vector q as

$$\bar{q}_1(t) = q_2(t)$$
$$\bar{q}_2(t) = q_1(t)$$

Verify that the transfer function remains the same using either state-space
model. Further verify that: (i) the traces and determinants of matrices A and
$\overline{A}$ are equal; and (ii) the determinants of $(s\boldsymbol{I} - \boldsymbol{A})$ and $(s\boldsymbol{I} - \overline{\boldsymbol{A}})$ are the same.

13.18   Derive the state-space model of the system in Example 13.1 with the new state
        vector $\bar{q}$ that is related to old state vector q as

$$\bar{q}_1(t) = q_1(t) + q_2(t)$$
$$\bar{q}_2(t) = q_1(t) - q_2(t)$$

Verify that the transfer function remains the same using either state-space
model. Further verify that: (i) the traces and determinants of matrices A and
$\overline{A}$ are equal; and (ii) the determinants of $(s\boldsymbol{I} - \boldsymbol{A})$ and $(s\boldsymbol{I} - \overline{\boldsymbol{A}})$ are the same.

13.19   Derive the state-space model of the system in Example 13.1 with the new state
        vector $\bar{q}$ that is related to old state vector q as

$$\bar{q}_1(t) = q_1(t)$$
$$\bar{q}_2(t) = q_1(t) + q_2(t)$$

Verify that the transfer function remains the same using either state-space
model. Further verify that: (i) the traces and determinants of matrices A and
$\overline{A}$ are equal; and (ii) the determinants of $(s\boldsymbol{I} - \boldsymbol{A})$ and $(s\boldsymbol{I} - \overline{\boldsymbol{A}})$ are the same.

Appendix A:
Transform Pairs and Properties

Table A.1. DFT pairs

$x(n)$, period $= N$	$X(k)$, period $= N$
$\delta(n)$	1
1	$N\delta(k)$
$e^{j(\frac{2\pi}{N}mn)}$	$N\delta(k-m)$
$\cos\left(\frac{2\pi}{N}mn\right)$	$\frac{N}{2}(\delta(k-m)+\delta(k-(N-m)))$
$\sin\left(\frac{2\pi}{N}mn\right)$	$\frac{N}{2}(-j\delta(k-m)+j\delta(k-(N-m)))$
$x(n) = \begin{cases} 1 & \text{for } n = 0, 1, \ldots, L-1 \\ 0 & \text{for } n = L, L+1, \ldots, N-1 \end{cases}$	$e^{(-j\frac{\pi}{N}(L-1)k)}\frac{\sin(\frac{\pi}{N}kL)}{\sin(\frac{\pi}{N}k)}$

A Practical Approach to Signals and Systems   D. Sundararajan
© 2008 John Wiley & Sons (Asia) Pte Ltd

Table A.2. DFT properties

Property	$x(n), h(n)$, period $= N$	$X(k), H(k)$, period $= N$
Linearity	$ax(n) + bh(n)$	$aX(k) + bH(k)$
Duality	$\dfrac{1}{N}X(N \mp n)$	$x(N \pm k)$
Time-shifting	$x(n \pm m)$	$e^{\pm j\frac{2\pi}{N}mk}X(k)$
Frequency-shifting	$e^{\mp j\frac{2\pi}{N}mn}x(n)$	$X(k \pm m)$
Time-convolution	$\sum_{m=0}^{N-1}x(m)h(n-m)$	$X(k)H(k)$
Frequency-convolution	$x(n)h(n)$	$\dfrac{1}{N}\sum_{m=0}^{N-1}X(m)H(k-m)$
Time-expansion	$h(mn) = \begin{cases} x(n) & \text{for } n = 0, 1, \ldots, N-1 \\ 0 & \text{otherwise} \end{cases}$ where m is any positive integer	$H(k) = X(k \bmod N),$ $k = 0, 1, \ldots, mN - 1$
Time-reversal	$x(N - n)$	$X(N - k)$
Conjugation	$x^*(N \pm n)$	$X^*(N \mp k)$
Parseval's theorem	$\sum_{n=0}^{N-1}|x(n)|^2$	$\dfrac{1}{N}\sum_{k=0}^{N-1}|X(k)|^2$

Table A.3. FS pairs

$x(t)$, period $= T$	$X_{cs}(k)$, $\omega_0 = 2\pi/T$
$\begin{cases} 1 & \text{for } |t| < a \\ 0 & \text{for } a < |t| \leq \dfrac{T}{2} \end{cases}$	$\dfrac{\sin(k\omega_0 a)}{k\pi}$
$\sum_{n=-\infty}^{\infty}\delta(t - nT)$	$\dfrac{1}{T}$
$e^{jk_0\omega_0 t}$	$\delta(k - k_0)$
$\cos(k_0\omega_0 t)$	$0.5(\delta(k + k_0) + \delta(k - k_0))$
$\sin(k_0\omega_0 t)$	$0.5j(\delta(k + k_0) - \delta(k - k_0))$

Table A.4. FS properties

Property	$x(t), h(t),$ period $= T$	$X_{cs}(k), H_{cs}(k), \omega_0 = 2\pi/T$
Linearity	$ax(t) + bh(t)$	$aX_{cs}(k) + bH_{cs}(k)$
Time-shifting	$x(t \pm t_0)$	$e^{\pm jk\omega_0 t_0} X_{cs}(k)$
Frequency-shifting	$x(t)e^{\pm jk_0\omega_0 t}$	$X_{cs}(k \mp k_0)$
Time-convolution	$\int_0^T x(\tau)h(t - \tau)d\tau$	$TX_{cs}(k)H_{cs}(k)$
Frequency-convolution	$x(t)h(t)$	$\sum_{l=-\infty}^{\infty} X_{cs}(l)H_{cs}(k - l)$
Time-scaling	$x(at), a > 0, \ \text{Period} = \dfrac{T}{a}$	$X_{cs}(k), \ \omega_0 = a\dfrac{2\pi}{T}$
Time-reversal	$x(-t)$	$X_{cs}(-k)$
Time-differentiation	$\dfrac{d^n x(t)}{dt^n}$	$(jk\omega_0)^n X_{cs}(k)$
Time-integration	$\int_{-\infty}^{t} x(\tau)d\tau$	$\dfrac{X_{cs}(k)}{jk\omega_0}, \text{if } (X_{cs}(0) = 0)$
Parseval's theorem	$\dfrac{1}{T}\int_0^T |x(t)|^2 dt$	$\sum_{k=-\infty}^{\infty} |X_{cs}(k)|^2$
Conjugate symmetry	$x(t)$ real	$X_{cs}(k) = X_{cs}^*(-k)$
Even symmetry	$x(t)$ real and even	$X_{cs}(k)$ real and even
Odd symmetry	$x(t)$ real and odd	$X_{cs}(k)$ imaginary and odd

Table A.5. DTFT pairs

$x(n)$	$X(e^{j\omega})$, period= 2π
$\begin{cases} 1 & \text{for } -N \leq n \leq N \\ 0 & \text{otherwise} \end{cases}$	$\dfrac{\sin\left(\omega\dfrac{(2N+1)}{2}\right)}{\sin\left(\dfrac{\omega}{2}\right)}$
$\dfrac{\sin(an)}{\pi n}, \ 0 < a \leq \pi$	$\begin{cases} 1 & \text{for } |\omega| < a \\ 0 & \text{for } a < |\omega| \leq \pi \end{cases}$
$a^n u(n), \ |a| < 1$	$\dfrac{1}{1 - ae^{-j\omega}}$
$(n+1)a^n u(n), \ |a| < 1$	$\dfrac{1}{(1 - ae^{-j\omega})^2}$
$a^{|n|}, \ |a| < 1$	$\dfrac{1 - a^2}{1 - 2a\cos(\omega) + a^2}$
$a^n \sin(\omega_0 n)u(n), \ |a| < 1$	$\dfrac{(a)e^{-j\omega}\sin(\omega_0)}{1 - 2(a)e^{-j\omega}\cos(\omega_0) + (a)^2 e^{-j2\omega}}$
$a^n \cos(\omega_0 n)u(n), \ |a| < 1$	$\dfrac{1 - (a)e^{-j\omega}\cos(\omega_0)}{1 - 2(a)e^{-j\omega}\cos(\omega_0) + (a)^2 e^{-j2\omega}}$
$\delta(n)$	1
$\displaystyle\sum_{k=-\infty}^{\infty} \delta(n - kN)$	$\dfrac{2\pi}{N} \displaystyle\sum_{k=-\infty}^{\infty} \delta\left(\omega - \dfrac{2\pi}{N}k\right)$
$u(n)$	$\pi\delta(\omega) + \dfrac{1}{1 - e^{-j\omega}}$
1	$2\pi\delta(\omega)$
$e^{j\omega_0 n}$	$2\pi\delta(\omega - \omega_0)$
$\cos(\omega_0 n)$	$\pi(\delta(\omega + \omega_0) + \delta(\omega - \omega_0))$
$\sin(\omega_0 n)$	$j\pi(\delta(\omega + \omega_0) - \delta(\omega - \omega_0))$

Table A.6. DTFT properties

Property	$x(n), h(n)$	$X(e^{j\omega}), H(e^{j\omega})$
Linearity	$ax(n) + bh(n)$	$aX(e^{j\omega}) + bH(e^{j\omega})$
Time-shifting	$x(n \pm n_0)$	$e^{\pm j\omega n_0}X(e^{j\omega})$
Frequency-shifting	$x(n)e^{\pm j\omega_0 n}$	$X(e^{j(\omega \mp \omega_0)})$
Time-convolution	$\sum_{m=-\infty}^{\infty} x(m)h(n-m)$	$X(e^{j\omega})H(e^{j\omega})$
Frequency-convolution	$x(n)h(n)$	$\dfrac{1}{2\pi}\displaystyle\int_0^{2\pi} X(e^{jv})H(e^{j(\omega-v)})\mathrm{d}v$
Time-expansion	$h(n)$ $h(an) = x(n), \quad a > 0$ is a positive integer and $h(n) = 0$ zero otherwise	$H(e^{j\omega}) = X(e^{ja\omega})$
Time-reversal	$x(-n)$	$X(e^{-j\omega})$
Conjugation	$x^*(\pm n)$	$X^*(e^{\mp j\omega})$
Difference	$x(n) - x(n-1)$	$(1 - e^{-j\omega})X(e^{j\omega})$
Summation	$\sum_{l=-\infty}^{n} x(l)$	$\dfrac{X(e^{j\omega})}{(1 - e^{-j\omega})} + \pi X(e^{j0})\delta(\omega)$
Frequency-differentiation	$(n)^m x(n)$	$(j)^m \dfrac{\mathrm{d}^m X(e^{j\omega})}{\mathrm{d}\omega^m}$
Parseval's theorem	$\sum_{n=-\infty}^{\infty} \|x(n)\|^2$	$\dfrac{1}{2\pi}\displaystyle\int_0^{2\pi} \|X(e^{j\omega})\|^2\mathrm{d}\omega$
Conjugate symmetry	$x(n)$ real	$X(e^{j\omega}) = X^*(e^{-j\omega})$
Even symmetry	$x(n)$ real and even	$X(e^{j\omega})$ real and even
Odd symmetry	$x(n)$ real and odd	$X(e^{j\omega})$ imaginary and odd

Table A.7. FT pairs

$x(t)$	$X(j\omega)$
$u(t+a) - u(t-a)$	$2\dfrac{\sin(\omega a)}{\omega}$
$\dfrac{\sin(\omega_0 t)}{\pi t}$	$u(\omega + \omega_0) - u(\omega - \omega_0)$
$e^{-at}u(t), \ \mathrm{Re}(a) > 0$	$\dfrac{1}{a + j\omega}$
$te^{-at}u(t), \ \mathrm{Re}(a) > 0$	$\dfrac{1}{(a + j\omega)^2}$
$e^{-a|t|}, \ \mathrm{Re}(a) > 0$	$\dfrac{2a}{a^2 + \omega^2}$
$\dfrac{1}{a}((t+a)u(t+a) - 2tu(t) + (t-a)u(t-a))$	$a\left(\dfrac{\sin(\omega\frac{a}{2})}{\omega\frac{a}{2}}\right)^2$
$e^{-at}\sin(\omega_0 t)u(t), \ \mathrm{Re}(a) > 0$	$\dfrac{\omega_0}{(a + j\omega)^2 + \omega_0^2}$
$e^{-at}\cos(\omega_0 t)u(t), \ \mathrm{Re}(a) > 0$	$\dfrac{a + j\omega}{(a + j\omega)^2 + \omega_0^2}$
$\delta(t)$	1
$\sum_{n=-\infty}^{\infty} \delta(t - nT)$	$\dfrac{2\pi}{T}\sum_{k=-\infty}^{\infty} \delta\left(\omega - k\dfrac{2\pi}{T}\right)$
$u(t)$	$\pi\delta(\omega) + \dfrac{1}{j\omega}$
1	$2\pi\delta(\omega)$
$e^{j\omega_0 t}$	$2\pi\delta(\omega - \omega_0)$
$\cos(\omega_0 t)$	$\pi(\delta(\omega + \omega_0) + \delta(\omega - \omega_0))$
$\sin(\omega_0 t)$	$j\pi(\delta(\omega + \omega_0) - \delta(\omega - \omega_0))$

Table A.8. FT properties

Property	$x(t), h(t)$	$X(j\omega), H(j\omega)$
Linearity	$ax(t) + bh(t)$	$aX(j\omega) + bH(j\omega)$
Duality	$X(\pm t)$	$2\pi x(\mp j\omega)$
Time-shifting	$x(t \pm t_0)$	$X(j\omega)e^{\pm j\omega t_0}$
Frequency-shifting	$x(t)e^{\pm j\omega_0 t}$	$X(j(\omega \mp \omega_0))$
Time-convolution	$x(t) * h(t)$	$X(j\omega)H(j\omega)$
Frequency-convolution	$x(t)h(t)$	$\dfrac{1}{2\pi}(X(j\omega) * H(j\omega))$
Time-scaling	$x(at), a \neq 0$ and real	$\dfrac{1}{|a|}X(j\dfrac{\omega}{a})$
Time-reversal	$x(-t)$	$X(-j\omega)$
Conjugation	$x^*(\pm t)$	$X^*(\mp j\omega)$
Time-differentiation	$\dfrac{d^n x(t)}{dt^n}$	$(j\omega)^n X(j\omega)$
Time-integration	$\int_{-\infty}^{t} x(\tau)d\tau$	$\dfrac{X(j\omega)}{j\omega} + \pi X(j0)\delta(\omega)$
Frequency-differentiation	$t^n x(t)$	$(j)^n \dfrac{d^n X(j\omega)}{d\omega^n}$
Parseval's theorem	$\int_{-\infty}^{\infty} |x(t)|^2 dt$	$\dfrac{1}{2\pi}\int_{-\infty}^{\infty} |X(j\omega)|^2 d\omega$
Autocorrelation	$x(t) * x(-t) = \int_{-\infty}^{\infty} x(\tau)x(\tau - t)d\tau$	$|X(j\omega)|^2$
Conjugate symmetry	$x(t)$ real	$X(j\omega) = X^*(-j\omega)$
Even symmetry	$x(t)$ real and even	$X(j\omega)$ real and even
Odd symmetry	$x(t)$ real and odd	$X(j\omega)$ imaginary and odd

Table A.9. z-Transform pairs

$x(n)$	$X(z)$	ROC
$\delta(n)$	1	$\|z\| \geq 0$
$\delta(n-p), \; p > 0$	z^{-p}	$\|z\| > 0$
$u(n)$	$\dfrac{z}{z-1}$	$\|z\| > 1$
$a^n u(n)$	$\dfrac{z}{z-a}$	$\|z\| > \|a\|$
$n a^n u(n)$	$\dfrac{az}{(z-a)^2}$	$\|z\| > \|a\|$
$n u(n)$	$\dfrac{z}{(z-1)^2}$	$\|z\| > \|1\|$
$\cos(\omega_0 n) u(n)$	$\dfrac{z(z - \cos(\omega_0))}{z^2 - 2z \cos(\omega_0) + 1}$	$\|z\| > 1$
$\sin(\omega_0 n) u(n)$	$\dfrac{z \sin(\omega_0)}{z^2 - 2z \cos(\omega_0) + 1}$	$\|z\| > 1$
$a^n \cos(\omega_0 n) u(n)$	$\dfrac{z(z - a \cos(\omega_0))}{z^2 - 2az \cos(\omega_0) + a^2}$	$\|z\| > \|a\|$
$a^n \sin(\omega_0 n) u(n)$	$\dfrac{az \sin(\omega_0)}{z^2 - 2az \cos(\omega_0) + a^2}$	$\|z\| > \|a\|$

Table A.10. z-Transform properties

Property	$x(n)u(n),\ h(n)u(n)$	$X(z),\ H(z)$
Linearity	$ax(n)u(n) + bh(n)u(n)$	$aX(z) + bH(z)$
Left shift	$x(n+m)u(n),\ m > 0$	$z^m X(z) - z^m \sum_{n=0}^{m-1} x(n)z^{-n}$
Right shift	$x(n-m)u(n),\ m > 0$	$z^{-m} X(z) + z^{-m} \sum_{n=1}^{m} x(-n)z^{n}$
Multiplication by a^n	$a^n x(n)u(n)$	$X(\frac{z}{a})$
Time-convolution	$x(n)u(n) * h(n)u(n)$	$X(z)H(z)$
Summation	$\sum_{m=0}^{n} x(m)$	$\dfrac{z}{z-1} X(z)$
Multiplication by n	$nx(n)u(n)$	$-z\dfrac{\mathrm{d}X(z)}{\mathrm{d}z}$
Initial value	$x(0)$	$\lim_{z\to\infty} X(z)$
Final value	$\lim_{n\to\infty} x(n)$	$\lim_{z\to 1}((z-1)X(z))$
		ROC of $(z-1)X(z)$
		includes the unit-circle

Table A.11. Laplace transform pairs

$x(t)$	$X(s)$	ROC
$\delta(t)$	1	All s
$u(t)$	$\dfrac{1}{s}$	$\text{Re}(s) > 0$
$t^n u(t),\ n = 0, 1, 2, \ldots$	$\dfrac{n!}{s^{n+1}}$	$\text{Re}(s) > 0$
$e^{-at} u(t)$	$\dfrac{1}{s+a}$	$\text{Re}(s) > -a$
$t^n e^{-at} u(t),\ n = 0, 1, 2, \ldots$	$\dfrac{n!}{(s+a)^{n+1}}$	$\text{Re}(s) > -a$
$\cos(\omega_0 t)\, u(t)$	$\dfrac{s}{s^2 + \omega_0^2}$	$\text{Re}(s) > 0$
$\sin(\omega_0 t)\, u(t)$	$\dfrac{\omega_0}{s^2 + \omega_0^2}$	$\text{Re}(s) > 0$
$e^{-at} \cos(\omega_0 t)\, u(t)$	$\dfrac{s+a}{(s+a)^2 + \omega_0^2}$	$\text{Re}(s) > -a$
$e^{-at} \sin(\omega_0 t)\, u(t)$	$\dfrac{\omega_0}{(s+a)^2 + \omega_0^2}$	$\text{Re}(s) > -a$
$t \cos(\omega_0 t)\, u(t)$	$\dfrac{s^2 - \omega_0^2}{(s^2 + \omega_0^2)^2}$	$\text{Re}(s) > 0$
$t \sin(\omega_0 t)\, u(t)$	$\dfrac{2\omega_0 s}{(s^2 + \omega_0^2)^2}$	$\text{Re}(s) > 0$

Table A.12. Laplace transform properties

Property	$x(t)u(t), h(t)u(t)$	$X(s), H(s)$
Linearity	$ax(t) + bh(t)$	$aX(s) + bH(s)$
Time-shifting	$x(t - t_0)u(t - t_0), \ t_0 \geq 0$	$X(s)e^{-st_0}$
Frequency-shifting	$x(t)u(t)e^{s_0 t}$	$X(s - s_0)$
Time-convolution	$x(t) * h(t)$	$X(s)H(s)$
Time-scaling	$x(at), a > 0$ and real	$\dfrac{1}{a}X(\dfrac{s}{a})$
Time-differentiation	$\dfrac{dx(t)}{dt}$	$sX(s) - x(0^-)$
Time-differentiation	$\dfrac{d^2 x(t)}{dt^2}$	$s^2 X(s) - sx(0^-) - \dfrac{d\,x(t)}{dt}\big|_{t=0^-}$
Time-integration	$\int_{0^-}^{t} x(\tau)d\tau$	$\dfrac{X(s)}{s}$
Time-integration	$\int_{-\infty}^{t} x(\tau)d\tau$	$\dfrac{X(s)}{s} + \dfrac{1}{s}\int_{-\infty}^{0^-} x(\tau)\,d\tau$
Frequency-differentiation	$tx(t)u(t)$	$-\dfrac{dX(s)}{ds}$
Frequency-differentiation	$t^n x(t)u(t), \ n = 0, 1, 2, \ldots$	$(-1)^n \dfrac{d^n X(s)}{ds^n}$
Initial value	$x(0^+)$	$\lim_{s \to \infty} sX(s)$, if $X(s)$ is strictly proper
Final value	$\lim_{t \to \infty} x(t)$	$\lim_{s \to 0} sX(s)$, (ROC of $sX(s)$ includes the $j\omega$ axis)

Appendix B:
Useful Mathematical Formulas

Trigonometric Identities

Pythagorean identity

$$\sin^2 x + \cos^2 x = 1$$

Addition and subtraction formulas

$$\sin(x \pm y) = \sin x \cos y \pm \cos x \sin y$$

$$\cos(x \pm y) = \cos x \cos y \mp \sin x \sin y$$

Double-angle formulas

$$\cos 2x = \cos^2 x - \sin^2 x = 2\cos^2 x - 1 = 1 - 2\sin^2 x$$

$$\sin 2x = 2 \sin x \cos x$$

Product formulas

$$2 \sin x \cos y = \sin(x - y) + \sin(x + y)$$

$$2 \cos x \sin y = -\sin(x - y) + \sin(x + y)$$

$$2 \sin x \sin y = \cos(x - y) - \cos(x + y)$$

$$2 \cos x \cos y = \cos(x - y) + \cos(x + y)$$

A Practical Approach to Signals and Systems   D. Sundararajan
© 2008 John Wiley & Sons (Asia) Pte Ltd

Sum and difference formulas

$$\sin x \pm \sin y = 2 \sin \frac{x \pm y}{2} \cos \frac{x \mp y}{2}$$

$$\cos x + \cos y = 2 \cos \frac{x + y}{2} \cos \frac{x - y}{2}$$

$$\cos x - \cos y = -2 \sin \frac{x + y}{2} \sin \frac{x - y}{2}$$

Other formulas

$$\sin(-x) = \sin(2\pi - x) = -\sin x$$

$$\cos(-x) = \cos(2\pi - x) = \cos x$$

$$\sin(\pi \pm x) = \mp \sin x$$

$$\cos(\pi \pm x) = -\cos x$$

$$\cos\left(\frac{\pi}{2} \pm x\right) = \mp \sin x$$

$$\sin\left(\frac{\pi}{2} \pm x\right) = \cos x$$

$$\cos\left(\frac{3\pi}{2} \pm x\right) = \pm \sin x$$

$$\sin\left(\frac{3\pi}{2} \pm x\right) = -\cos x$$

$$e^{\pm jx} = \cos x \pm j \sin x$$

$$\cos x = \frac{e^{jx} + e^{-jx}}{2}$$

$$\sin x = \frac{e^{jx} - e^{-jx}}{2j}$$

Series Expansions

$$e^{jx} = 1 + (jx) + \frac{(jx)^2}{2!} + \frac{(jx)^3}{3!} + \frac{(jx)^4}{4!} + \cdots + \frac{(jx)^r}{(r)!} + \cdots$$

$$\cos(x) = 1 - \frac{x^2}{2!} + \frac{x^4}{4!} - \cdots + (-1)^r \frac{x^{2r}}{(2r)!} - \cdots$$

$$\sin(x) = x - \frac{x^3}{3!} + \frac{x^5}{5!} - \cdots + (-1)^r \frac{x^{2r+1}}{(2r+1)!} - \cdots$$

$$\sin^{-1} x = x + \frac{1}{2} \frac{x^3}{3} + \frac{(1)(3)}{(2)(4)} \frac{x^5}{5} + \frac{(1)(3)(5)}{(2)(4)(6)} \frac{x^7}{7} + \cdots, \quad |x| < 1$$

$$\cos^{-1} x = \frac{\pi}{2} - \sin^{-1} x, \quad |x| < 1$$

Summation Formulas

$$\sum_{k=0}^{N-1} (a + kd) = \frac{N(2a + (N-1)d)}{2}$$

$$\sum_{k=0}^{N-1} ar^k = \frac{a(1 - r^N)}{1 - r}, \quad r \neq 1$$

$$\sum_{k=0}^{\infty} r^k = \frac{1}{1 - r}, \quad |r| < 1$$

$$\sum_{k=0}^{\infty} kr^k = \frac{r}{(1 - r)^2}, \quad |r| < 1$$

$$1 + \cos(t) + \cos(2t) + \cdots + \cos(Nt) = \frac{1}{2} + \frac{\sin(0.5(2N + 1)t)}{2\sin(0.5t)}$$

Indefinite Integrals

$$\int u\,dv = uv - \int v\,du$$

$$\int e^{at}\,dt = \frac{e^{at}}{a}$$

$$\int te^{at} dt = \frac{e^{at}}{a^2}(at - 1)$$

$$\int e^{bt} \sin(at) dt = \frac{e^{bt}}{a^2 + b^2}(b\sin(at) - a\cos(at))$$

$$\int e^{bt} \cos(at) dt = \frac{e^{bt}}{a^2 + b^2}(b\cos(at) + a\sin(at))$$

$$\int \sin(at) dt = -\frac{1}{a}\cos(at)$$

$$\int \cos(at) dt = \frac{1}{a}\sin(at)$$

$$\int t\sin(at) dt = \frac{1}{a^2}(\sin(at) - at\cos(at))$$

$$\int t\cos(at) dt = \frac{1}{a^2}(\cos(at) + at\sin(at))$$

$$\int \sin^2(at) dt = \frac{t}{2} - \frac{1}{4a}\sin(2at)$$

$$\int \cos^2(at) dt = \frac{t}{2} + \frac{1}{4a}\sin(2at)$$

Differentiation Formulas

$$\frac{d(uv)}{dt} = u\frac{dv}{dt} + v\frac{du}{dt}$$

$$\frac{d(\frac{u}{v})}{dt} = \frac{v\frac{du}{dt} - u\frac{dv}{dt}}{v^2}$$

$$\frac{d(x^n)}{dt} = nx^{n-1}$$

$$\frac{d(e^{at})}{dt} = ae^{at}$$

$$\frac{d(\sin(at))}{dt} = a\cos(at)$$

$$\frac{d(\cos(at))}{dt} = -a\sin(at)$$

L'Hôpital's Rule

If $\lim_{x \to a} f(x) = 0$ and $\lim_{x \to a} g(x) = 0$, or

If $\lim_{x \to a} f(x) = \infty$ and $\lim_{x \to a} g(x) = \infty$, then

$$\lim_{x \to a} \frac{f(x)}{g(x)} = \lim_{x \to a} \frac{df(x)/dx}{dg(x)/dx}$$

The rule can be applied as many times as necessary.

Matrix Inversion

The inverse, A^{-1}, of a 2×2 matrix

$$A = \begin{bmatrix} a & b \\ c & d \end{bmatrix}$$

is defined as

$$A^{-1} = \frac{1}{ad - bc} \begin{bmatrix} d & -b \\ -c & a \end{bmatrix}$$

provided $ad - bc \neq 0$.

Answers to Selected Exercises

Chapter 2

2.1.2 Energy 100/9.

2.3.3 $x_e(n) = \begin{cases} \frac{(0.4)^n}{2} & \text{for } n > 0 \\ 1 & \text{for } n = 0 \\ \frac{(0.4)^{-n}}{2} & \text{for } n < 0 \end{cases}$

$x_e(-3) = 0.032 \qquad x_e(-2) = 0.08 \qquad x_e(-1) = 0.20 \qquad x_e(0) = 1,$

$x_e(1) = 0.20 \qquad x_e(2) = 0.08 \qquad x_e(3) = 0.032$

$x_o(n) = \begin{cases} \frac{(0.4)^n}{2} & \text{for } n > 0 \\ 0 & \text{for } n = 0 \\ -\frac{(0.4)^{-n}}{2} & \text{for } n < 0 \end{cases}$

$x_o(-3) = -0.032 \qquad x_o(-2) = -0.08 \qquad x_o(-1) = -0.20 \qquad x_o(0) = 0$

$x_o(1) = 0.20 \qquad x_o(2) = 0.08 \qquad x_o(3) = 0.032$

$x(n) = x_e(n) + x_o(n)$

$x(-3) = 0 \qquad x(-2) = 0 \qquad x(-1) = 0 \qquad x(0) = 1.0000$

$x(1) = 0.4000 \qquad x(2) = 0.16 \qquad x(3) = 0.064$

The sum of the values of the even component is 1.624 and that of the signal is also 1.624.

A Practical Approach to Signals and Systems   D. Sundararajan
© 2008 John Wiley & Sons (Asia) Pte Ltd

2.4.2 0

2.6.5 Periodic with period 9.

2.7.6 $x(n) = 2\sqrt{3}\cos\left(\frac{\pi}{6}n\right) + 2\sin\left(\frac{\pi}{6}n\right)$

  $3.4641, 4, 3.4641, 2, 0, -2, -3.4641, -4, -3.4641, -2, 0, 2$

2.8.3 $x(n) = 2\sqrt{3}\cos\left(\frac{\pi}{6}n - \frac{\pi}{6}\right)$

  $3, 3.4641, 3, 1.7321, 0, -1.7321, -3, -3.4641, -3, -1.7321, 0, 1.7321$

2.10.5 $x(n) = 5.9544e^{j(\frac{\pi}{3}n+0.6984)}$.

  $1.7321 + j1, j2, -1.7321 + j1, -1.7321 - j1, -j2, 1.7321 - j1$

  $2.8284 + j2.8284, -1.0353 + j3.8637, -3.8637 + j1.0353, -2.8284$

  $-j2.8284, 1.0353 - j3.8637, 3.8637 - j1.0353$

  $4.5605 + j3.8284, -1.0353 + j5.8637, -5.5958 + j2.0353, -4.5605$

  $-j3.8284, 1.0353 - j5.8637, 5.5958 - j2.0353$

2.11.3 $x(n) = (0.5)^n$.

  $x(0) = 1 \qquad x(1) = 0.5 \qquad x(2) = 0.25$

  $x(3) = 0.125 \qquad x(4) = 0.0625 \qquad x(5) = 0.0313$

2.13.4 $-3\sin\left(5\frac{2\pi}{8}n + \frac{\pi}{3}\right), 3\sin\left(11\frac{2\pi}{8}n - \frac{\pi}{3}\right), -3\sin\left(13\frac{2\pi}{8}n + \frac{\pi}{3}\right)$.

2.14.3 11 samples per second.

2.15.5 $0, -\frac{\sqrt{3}}{2}, -\frac{\sqrt{3}}{2}, 0, \frac{\sqrt{3}}{2}, \frac{\sqrt{3}}{2}$

  $x(n-7) = \cos\left(\frac{2\pi}{6}n + \frac{\pi}{6}\right)$

$$\frac{\sqrt{3}}{2}, 0, -\frac{\sqrt{3}}{2}, -\frac{\sqrt{3}}{2}, 0, \frac{\sqrt{3}}{2}$$

2.16.3 $\dfrac{\sqrt{3}}{2}, \dfrac{\sqrt{3}}{2}, 0, -\dfrac{\sqrt{3}}{2}, -\dfrac{\sqrt{3}}{2}, 0$

$$x(-n+1) = \cos\left(\frac{2\pi}{6}n - \frac{\pi}{6}\right) = x(n)$$

$$\frac{\sqrt{3}}{2}, \frac{\sqrt{3}}{2}, 0, -\frac{\sqrt{3}}{2}, -\frac{\sqrt{3}}{2}, 0$$

2.17.3 $0.5, 1, 0.5, -0.5, -1, -0.5$

$$x(-n+1) = \cos\left(\frac{2\pi}{6}n\right)$$

$$1, 0.5, -0.5, -1, -0.5, 0.5$$

2.18.8 $x(-3) = 0$ $\qquad$ $x(-2) = 0$ $\qquad$ $x(-1) = 0$ $\qquad$ $x(0) = -1$

$x(1) = -2$ $\qquad$ $x(2) = -1$ $\qquad$ $x(3) = 1$

$x(-3) = -1$ $\qquad$ $x(-2) = -1$ $\qquad$ $x(-1) = 2$ $\qquad$ $x(0) = -1$

$x(1) = -1$ $\qquad$ $x(2) = 0$ $\qquad$ $x(3) = 0$

Chapter 3

3.1.9 Energy 4.

3.3.4 $x_e(t) = \begin{cases} \frac{3}{2}|t|, & |t| < 1 \\ 0 & \text{otherwise} \end{cases}$

$x_o(t) = \begin{cases} \frac{3}{2}t, & -1 < t < 1 \\ 0 & \text{otherwise} \end{cases}$

The integral of the odd component is zero. The integral of the even component is 1.5 and that of the signal is also 1.5.

3.4.4 3.

3.5.3 $-1.2622, -1.4975, -1.5000, -1.5000,$ and -1.5000

3.6.3 7.3891.

3.7.2 $x(t) \approx \sum_{n=0}^{3} \cos\left(\frac{\pi}{6}(n)(1)\right) \delta_q(t - (n)(1))(1)$

$x(t) \approx \sum_{n=0}^{7} \cos\left(\frac{\pi}{6}(n)(0.5)\right) \delta_q(t - (n)(0.5))(0.5)$

3.8.3 $-6e^{-3t}u(t) + 2\delta(t)$

3.9.4 0.

3.10.3 $x(t) = -\frac{5}{\sqrt{2}}\cos(2\pi t) - \frac{5}{\sqrt{2}}\sin(2\pi t)$

$\frac{5}{8}, \frac{9}{8}, \frac{13}{8}$

3.11.4 $x(t) = \sqrt{2}\cos\left(\frac{2\pi}{6}t - \frac{\pi}{4}\right)$

$\frac{9}{4}, \frac{21}{4}, \frac{33}{4}$

3.12.3 $x(t) = 1.3483\cos(\frac{2\pi}{6}t - 2.9699).$

$1.5, -1.5, -3$

$2.8284, -1.0353, -3.8637$

$-1.3284, -0.4647, 0.8637$

3.14.2 1, 2

The shift of $x(t)$ to the right by 1 s makes the positive peak of the shifted waveform, $\sin(\frac{2\pi}{6}t - \frac{\pi}{6})$, occurs after one second of the occurrence of that of the given sinusoid.

3.15.2 $\dfrac{3}{4}, \dfrac{5}{4}$

3.16.4 $\dfrac{10}{3}, 10$

3.17.5 $x(-3) = 0 \qquad x(-2) = 0 \qquad x(-1) = 0 \qquad x(0) = -1.7321$

$x(1) = -1.7321 \qquad x(2) = 0 \qquad x(3) = 1.7321$

$x(-3) = -1.7321 \qquad x(-2) = -1.7321 \qquad x(-1) = 0 \qquad x(0) = 1.7321$

$x(1) = 1.7321 \qquad x(2) = 0 \qquad x(3) = -1.7321$

$x(-3) = 0 \qquad x(-2) = 0 \qquad x(-1) = -1.7321 \qquad x(0) = 1.7321$

$x(1) = 0 \qquad x(2) = -1.7321 \qquad x(3) = 1.7321$

Chapter 4

4.1.3 $h(n) = 3(-\frac{1}{3})^n u(n)$

$h(0) = 3 \qquad h(1) = -1 \qquad h(2) = \frac{1}{3} \qquad h(3) = -\frac{1}{9}$

$h(4) = \frac{1}{27} \qquad h(5) = -\frac{1}{81}$

4.2.4 $h(n) = 3\delta(n) + (-1)^n u(n) \qquad n = 0, 1, 2, \ldots$

$h(0) = 4 \quad h(1) = -1 \quad h(2) = 1 \quad h(3) = -1 \quad h(4) = 1 \quad h(5) = -1$

4.3.3 Linear.

4.4.3 Time-invariant.

4.5.3 $\{y(n), n = -1, 0, 1, 2, 3, 4, 5\} = \{6, 10, 13, 28, 19, 16, 16\}$.

4.6.5 $y(n) = 0.9(1 - (0.6)^{n-2})u(n - 3)$

$y(0) = 0 \qquad y(1) = 0 \qquad y(2) = 0 \qquad y(3) = 0.36$

$y(4) = 0.576 \qquad y(5) = 0.7056$

4.10 $8.4276 \cos(\frac{2\pi}{5}n + \frac{\pi}{4} - 0.9964)$.

$4.2138 e^{j(\frac{2\pi}{5}n - 0.9964)}$.

4.13 The zero-state response is

$$y(n) = \frac{20}{9} + \frac{4}{3}n - \frac{20}{9}\left(\frac{1}{4}\right)^n$$

The zero-input response is

$$\left(\frac{1}{4}\right)^{(n+1)}$$

The complete response is

$$y(n) = \frac{20}{9} + \frac{4}{3}n - \frac{71}{36}\left(\frac{1}{4}\right)^n \qquad n = 0, 1, 2, \ldots$$

$y(0) = 0.2500 \qquad y(1) = 3.0625 \qquad y(2) = 4.7656 \qquad y(3) = 6.1914$

$y(4) = 7.5479 \qquad y(5) = 8.8870$

The transient response is

$$-\frac{71}{36}\left(\frac{1}{4}\right)^n \qquad n = 0, 1, 2, \ldots$$

The steady-state response is $(\frac{20}{9} + \frac{4}{3}n)u(n)$.

4.17.2 (i)

$$h(n) = \left(\frac{25}{3}\right)\delta(n) - 4\left(\frac{1}{5}\right)^n - \left(\frac{7}{3}\right)\left(\frac{3}{5}\right)^n$$

The first four values of $h(n)$ are

$\{2, 0.6, -1, 0.472\}$

(ii)

$$h(n) = \left(\frac{50}{3}\right)\delta(n) - \left(\frac{14}{3}\right)\left(-\frac{3}{5}\right)^n - 11\left(\frac{1}{5}\right)^n, \qquad n = 0, 1, 2, \ldots$$

The first four values of $h(n)$ are

$\{1, 0.6, -2.12, 0.92\}$

Chapter 5

5.1.9 Nonlinear.

5.2.4 Time-invariant.

5.3.4 $y(t) = tu(t) - 2(t-3)u(t-3) + (t-6)u(t-6)$

$y(0) = 0$ $y(1) = 1$ $y(2) = 2$ $y(3) = 3$ $y(4) = 2,$ $y(5) = 1$

5.7.2 $h(t) = 2\delta(t) + 5e^t u(t)$

$y(t) = (-3 + 5e^t)u(t)$

5.11 The zero-input response is $3e^{-t}u(t)$. The zero-state response is $(2\sin(t) - 2e^{-t})u(t)$. The complete response is

$y(t) = (2\sin(t) + e^{-t})u(t)$

The transient response is $e^{-t}u(t)$.
The steady-state response is $(2\sin(t))u(t)$.

5.14 $y(t) = 4\sin(\frac{2\pi}{6}t - \frac{\pi}{6} + 1.5247)$.

$y(t) = 2e^{j(\frac{2\pi}{6}t + 1.5247)}$.

Chapter 6

6.1.3 $x(n) = \dfrac{1}{4}\left(1 + 8\cos\left(\dfrac{2\pi}{4}n - \dfrac{\pi}{3}\right) - 3\cos(\pi n)\right)$

6.2.2 $\{x(0) = 0 \qquad x(1) = -3 + \sqrt{3} \qquad x(2) = 2 \qquad x(3) = -3 - \sqrt{3}\}$

$\{X(0) = -4 \qquad X(1) = -2 - j2\sqrt{3} \qquad X(2) = 8 \qquad X(3) = -2 + j2\sqrt{3}\}$

6.3.4 $\left\{x(0) = -2 - \dfrac{3}{\sqrt{2}} \qquad x(1) = -6 + \dfrac{3}{\sqrt{2}}\right.$

$\left. x(2) = -2 + \dfrac{3}{\sqrt{2}} \qquad x(3) = -6 - \dfrac{3}{\sqrt{2}}\right\}$

6.5.1 $\{x(0) = 2.25 \qquad x(1) = 0.25 \qquad x(2) = 0.25 \qquad x(3) = -1.75 - j1\}$

6.6.2 $X(k) = \{6 - j3, -j1, j11, -2 + j1\}$

$x(-14) = 2 + j2, \ x(43) = 1 - j4$

$X(12) = 6 - j3, \ X(-7) = -j1$

6.9.2 $\{208, 224, 208, 224\}$

6.11.3 $\{-4, -13, 24, -9\}$

Chapter 7

7.2.4 $\omega_0 = 2$.

$$X_c(0) = \frac{3}{8} \qquad X_c(1) = \frac{1}{2} \qquad X_c(2) = \frac{1}{8}$$

$$X_p(0) = \frac{3}{8} \qquad X_p(1) = \frac{1}{2} \qquad \theta(1) = 0 \qquad X_p(2) = \frac{1}{8} \qquad \theta(2) = 0$$

$$X_{cs}(0) = \frac{3}{8} \qquad X_{cs}(\pm1) = \frac{1}{4} \qquad X_{cs}(\pm2) = \frac{1}{16}$$

7.3.5 $\omega_0 = 2\pi$.

$$X_c(0) = 1 \qquad X_c(1) = \frac{\sqrt{3}}{2} \qquad X_s(1) = \frac{1}{2} \qquad X_c(3) = \sqrt{3} \qquad X_s(3) = 1$$

$$X_p(0) = 1 \qquad X_p(1) = 1 \qquad \theta(1) = -\frac{\pi}{6} \qquad X_p(3) = 2 \qquad \theta(3) = -\frac{\pi}{6}$$

$$X_{cs}(0) = 1 \qquad X_{cs}(1) = \frac{1}{4}(\sqrt{3} - j1) \qquad X_{cs}(3) = \frac{1}{2}(\sqrt{3} - j1),$$

$$X_{cs}(-1) = \frac{1}{4}(\sqrt{3} + j1) \qquad X_{cs}(-3) = \frac{1}{2}(\sqrt{3} + j1)$$

7.4.2 $\omega_0 = \frac{1}{63}$.

$$X_c(0) = 2 \qquad X_s(14) = -2 \qquad X_c(27) = -5$$

7.9.6 $X_{cs}(0) = 2/\pi$ and $X_{cs}(k) = 2/[\pi(1 - 4k^2)] \qquad k \neq 0$.

7.11 $x(t) = \frac{1}{2} - \frac{4}{\pi^2}\left(\cos\frac{\pi}{2}t + \frac{1}{9}\cos 3\frac{\pi}{2}t + \frac{1}{25}\cos 5\frac{\pi}{2}t\cdots\right)$

$$x(t+2) = \frac{1}{2} + \frac{4}{\pi^2}\left(\cos\frac{\pi}{2}t + \frac{1}{9}\cos 3\frac{\pi}{2}t + \frac{1}{25}\cos 5\frac{\pi}{2}t\cdots\right)$$

$$3x(t) - 2 = -\frac{1}{2} - \frac{12}{\pi^2}\left(\cos\frac{\pi}{2}t + \frac{1}{9}\cos 3\frac{\pi}{2}t + \frac{1}{25}\cos 5\frac{\pi}{2}t\cdots\right)$$

$$\frac{\pi^2}{8} = 1 + \frac{1}{9} + \frac{1}{25} + \frac{1}{49} + \cdots$$

The DFT approximation of the trigonometric FS coefficients are
$$\{X_c(0) = \tfrac{1}{2} \qquad X_c(1) = -\tfrac{1}{2} \qquad X_c(2) = 0\}.$$
The power of the signal is $\frac{1}{3}$.
The power of the signal, up to the third harmonic, is 0.3331.
The power of the signal, up to the fifth harmonic, is 0.3333.

7.15 $y(t) = \dfrac{1}{\sqrt{5}} e^{j(2t + \tan^{-1}(\frac{-2}{1}))} + \dfrac{1}{\sqrt{10}} e^{j(3t + \tan^{-1}(\frac{-3}{1}))}$

Chapter 8

8.2 $X(e^{j\omega}) = \dfrac{\sin\left(\frac{2N+1}{2}\omega\right)}{\sin\left(\frac{\omega}{2}\right)}$

$X(e^{j0}) = 11$ and $X(e^{j\pi}) = -1$.

8.4 $X(e^{j\omega}) = \dfrac{(a)e^{-j\omega}\sin(\omega_0)}{1 - 2(a)e^{-j\omega}\cos(\omega_0) + (a)^2 e^{-j2\omega}}$

$X(e^{j0}) = 0.7/1.49 \text{ and } X(e^{j\pi}) = -0.7/1.49$.

8.9.5 $X(e^{j\omega}) = \dfrac{0.36e^{-j2\omega}}{1 - 0.6e^{-j\omega}}$

8.10.2 $\left\{ X(e^{j0}) = 10\left(\dfrac{2\pi}{4}\right)\delta(\omega) \qquad X(e^{j\frac{2\pi}{4}}) = (2 + j2)\left(\dfrac{2\pi}{4}\right)\delta\left(\omega - \dfrac{2\pi}{4}\right) \right.$

$X(e^{j2\frac{2\pi}{4}}) = 2\left(\dfrac{2\pi}{4}\right)\delta\left(\omega - 2\dfrac{2\pi}{4}\right),$

$\left. X(e^{j3\frac{2\pi}{4}}) = (2 - j2)\left(\dfrac{2\pi}{4}\right)\delta\left(\omega - 3\dfrac{2\pi}{4}\right) \right\}.$

8.16.2 $2(0.5)^n u(n) - (0.25)^n u(n)$.

8.21.1 $y(n) = \{x(n) * h(n) \qquad n = 1, 2, \ldots, 6\} = \{-2, 1, -8, -4, -5, -12\}$.

8.24.2 $\dfrac{1}{1-(0.6)e^{j4\omega}}$ $-\pi < \omega \le \pi$.

8.26.4 $Y(e^{j\omega}) = \dfrac{1}{(1-0.6e^{-j\omega})(1-e^{-j\omega})} + 2.5\pi\delta(\omega)$.

8.29 The IDFT values are {1.0082, 0.3024, 0.0907, 0.0272}.
The exact values of $x(n)$ are {$x(0) = 1$ $x(1) = 0.3$ $x(2) = 0.09$ $x(3) = 0.027$}.

8.31 $h(n) = \left(\dfrac{56}{5}\left(-\dfrac{2}{3}\right)^n - \dfrac{51}{5}\left(-\dfrac{1}{4}\right)^n\right) u(n)$

The first four values of the impulse response are 1, -4.9167, 4.3403, -3.1591.

8.35

$$y(n) = 55 \left(\frac{1}{2}\right)^n - 90 \left(\frac{1}{3}\right)^n + 38 \left(\frac{1}{5}\right)^n \qquad n = 0, 1, \ldots$$

8.36.2 $x_H(n) = -0.5 \sin(n)$.

Chapter 9

9.7 $X(j\omega) = \dfrac{1}{(2+j\omega)^2}$ $X(j0) = \dfrac{1}{4}$

9.14.2 $X(j\omega) = \begin{cases} 2\pi & \text{for } |\omega| < 3 \\ 0 & \text{for } |\omega| > 3 \end{cases}$

9.15.3 $X(j\omega) = \dfrac{\sin((\omega - \omega_0)a)}{(\omega - \omega_0)} + \dfrac{\sin((\omega + \omega_0)a)}{(\omega + \omega_0)}$

9.16 $x(t) = \begin{cases} -2 & \text{for } t < 0 \\ (t-2) & \text{for } 0 < t < 4 \\ 2 & \text{for } t > 4 \end{cases}$

9.18.2 $\dfrac{-j\pi}{2}(\delta(\omega - \omega_0) - \delta(\omega + \omega_0)) - \dfrac{\omega_0}{(\omega^2 - \omega_0^2)}$

9.20.9 $\pi\delta(\omega) - \dfrac{1}{j\omega}$

9.22.4 $X(j\omega) = \dfrac{-1 + e^{-j\omega} + e^{-j2\omega} - e^{-j3\omega}}{\omega^2}$

9.23.6 $Y(j\omega) = \pi\delta(\omega) + \dfrac{1}{j\omega} - \dfrac{1}{1+j\omega}$

9.24.4 $\dfrac{(1+j\omega)e^{-j\omega} - 1}{\omega^2}$

9.28.3 $\{X(0) = -1 \quad X(3) = \dfrac{j3}{2} \quad X(-3) = -\dfrac{j3}{2} \quad X(5) = 1$

$X(-5) = 1 \quad X(-7) = 6\}$

$X(j\omega) = \pi(-2\delta(\omega) + j3(\delta(\omega - 3) - \delta(\omega + 3)) + 2(\delta(\omega - 5)$
$+ \delta(\omega + 5)) + 12\delta(\omega + 7))$

9.29.4 $X_{cs}(k) = \dfrac{j}{k\pi} \quad k \neq 0 \quad \text{and} \quad X_{cs}(0) = 1$

9.31.2 $X(j\omega) = \dfrac{2}{1+\omega^2} \quad \text{and} \quad X_s(j\omega) = \dfrac{1}{T_s} \displaystyle\sum_{k=-\infty}^{\infty} \dfrac{2}{1+(\omega - k\omega_s)^2} \qquad \omega_s = \dfrac{2\pi}{T_s}$

$X(j0) = 2 \quad X_s(j0) = 200.0017 \quad T_s = 0.01$

$X_s(j0) = 20.0167 \quad T_s = 0.1,$

$$X_s(j0) = 2.1640 \qquad T_s = 1 \qquad X_s(j0) = 1.0001 \qquad T_s = 10$$

9.33 The exact values of the FT are $X(j0) = 1$ and $X(j\pi) = 4/\pi^2 = 0.4053$. The four samples of the signal are $\{x(0) = 1, x(1) = 0.5, x(2) = 0, x(3) = 0.5\}$ and the DFT is $\{X(0) = 2, X(1) = 1, X(2) = 0, X(3) = 1\}$. As the sampling interval is 0.5 s, the first two samples of the spectrum obtained by the DFT are $0.5\{2, 1\} = \{1, 0.5\}$.

Chapter 10

10.1.3 $X(z) = 2 - 4z^{-3}$.

10.2.4 $\{x(0) = 1, x(1) = 1, x(2) = -1\}$.

10.3.4 $X(z) = \left(\dfrac{-2z^2 + 3z}{(z-1)^2} \right)$

10.4.1 The nonzero values of $y(n)$ are $\{y(1) = 4, y(3) = -2, y(5) = -16, y(7) = 8\}$.

10.5.2 $X(z) = \dfrac{2z}{(z-2)^2}$

10.6.2 $X(z) = \dfrac{4z}{(z-4)^2}$

10.7.2 $y(n) = (0.5 + 0.5\sin(\frac{2\pi}{4}n) - 0.5\cos(\frac{2\pi}{4}n))u(n)$.

10.8.3 $x(0) = 2. \ x(\infty) = 16$.

10.9.4 $X(z) = \dfrac{z(z+1)}{(z^2+1)}$

10.13 $x(n) = (\sqrt{2})^{n+1}\cos\left(\dfrac{3\pi}{4}n - \dfrac{\pi}{4}\right)u(n)$

$$x(0) = 1 \qquad x(1) = 0 \qquad x(2) = -2 \qquad x(3) = 4$$

10.19.2 $y(n) = (0.8192)(3)\cos(\frac{2\pi}{4}n - \frac{\pi}{3} - 0.6107)u(n)$.

10.21 $y(n) = -\dfrac{16}{21}(-1)^n + \dfrac{325}{56}\left(\dfrac{3}{4}\right)^n - \dfrac{23}{12}\left(\dfrac{1}{2}\right)^n \qquad n = 0, 1, 2, \ldots$

The first four values of $y(n)$ are

$\{3.1250, 4.1563, 2.0234, 2.9707\}$

The zero-input response is

$$\dfrac{27}{8}\left(\dfrac{3}{4}\right)^n - \dfrac{5}{4}\left(\dfrac{1}{2}\right)^n$$

The zero-state response is

$$-\dfrac{16}{21}(-1)^n + \dfrac{17}{7}\left(\dfrac{3}{4}\right)^n - \dfrac{2}{3}\left(\dfrac{1}{2}\right)^n$$

The transient response is

$$\frac{325}{56}\left(\frac{3}{4}\right)^n - \frac{23}{12}\left(\frac{1}{2}\right)^n$$

The steady-state response is $-\frac{16}{21}(-1)^n u(n)$.

10.25.1 $h(n) = 12\delta(n) - \left(6\left(\frac{1}{4}\right)^n\right)u(n), \ n = 0, 1, 2, \ldots$

The first four values of the impulse response are

$\{6, -1.5, -0.3750, -0.0938\}$

10.26.3 $h(n) = -10\delta(n) + \left(7\left(\frac{1}{3}\right)^n + 7\left(\frac{1}{2}\right)^n\right)u(n), \ n = 0, 1, 2, \ldots$

The first four values of the impulse response are

$\{4, 5.8333, 2.5278, 1.1343\}$

Chapter 11

11.2.4 $X(s) = \left(\dfrac{e^{-2s}}{s^2} + \dfrac{2e^{-2s}}{s}\right)$

11.5.2 The poles of $X(s)$ are located at $s = -1$ and $s = -2$. The zero is located at $s = 1$. $x(t) = (-2e^{-t} + 3e^{-2t})u(t)$.

The transform of the scaled signal is

$$\frac{s-2}{(s+2)(s+4)}$$

The poles are located at $s = -2$ and $s = -4$. The zero is located at $s = 2$. $x(at) = (-2e^{-2t} + 3e^{-4t})u(t)$.

11.6.5 $y(t) = \dfrac{1}{2}\left(t^2 e^{-t}\right)u(t)$

11.8.4 $x(0^+) = 3$. $x(\infty) = 1$.

11.9.3 $X(s) = \dfrac{(1-e^{-2s})}{s^2(1+e^{-2s})}$

11.15 $x(t) = \left(2t - 1 + e^{-t}\right)u(t)$

11.18.3 $h(t) = (-0.5e^t + 0.5e^{3t})u(t)$

11.20 $y(t) = (\overbrace{e^{-2t} + te^{-t} - e^{-t}}^{\text{zero-state}} + \overbrace{te^{-t} + 3e^{-t}}^{\text{zero-input}})u(t)$
$= (e^{-2t} + 2te^{-t} + 2e^{-t})u(t)$

The steady state response is $e^{-2t}u(t)$ and the transient response is $(2te^{-t} + 2e^{-t})u(t)$. The initial and final values of $y(t)$ are 3 and 0 , respectively. The initial and final values of the zero-state response are 0 and 0, respectively.

11.23.1 $y(t) = 3\sqrt{2}\cos(0.5t - \frac{\pi}{3} - \frac{\pi}{4})u(t)$.

11.24.2 $h(t) = \left(e^t - 1\right)u(t)$

11.25.3 $h(t) = 5\delta(t) - \left(7e^{-3t} + 4e^{-2t}\right)u(t)$

11.29 $v(t) = \frac{25}{8}e^{-\frac{1}{8}t}u(t)$.

Chapter 12

12.3.1 $h(n) = \delta(n) + \left(\dfrac{28}{3}\left(-\dfrac{1}{3}\right)^{n-1} - \dfrac{40}{3}\left(-\dfrac{2}{3}\right)^{n-1}\right)u(n-1)$ $n = 0, 1, 2, \ldots$

$$h(0) = 1 \qquad h(1) = -4 \qquad h(2) = \frac{52}{9} \qquad h(3) = -\frac{44}{9}$$

12.4.2 The zero-input component of the state-vector is given by

$$\begin{bmatrix} -\frac{1}{8}\left(-\frac{1}{2}\right)^n \\ \frac{1}{4}\left(-\frac{1}{2}\right)^n \end{bmatrix}$$

The zero-input response is given by

$$\left(\frac{1}{2}\left(-\frac{1}{2}\right)^n\right)u(n)$$

The first four values of the zero-input response $y(n)$ are

$$y(0) = \frac{1}{2} \qquad y(1) = -\frac{1}{4} \qquad y(2) = \frac{1}{8} \qquad y(3) = -\frac{1}{16}$$

The zero-state component of the state-vector is given by

$$\begin{bmatrix} \frac{2}{3}\left(1 - (-\frac{1}{2})^n\right) + \left(-\frac{2}{9} + \frac{2}{9}(-\frac{1}{2})^n - \frac{2}{3}n(-\frac{1}{2})^n\right) \\ -2\left(-\frac{2}{9} + \frac{2}{9}(-\frac{1}{2})^n - \frac{2}{3}n(-\frac{1}{2})^n\right) \end{bmatrix}u(n)$$

The zero-state response is given by

$$\left(\frac{8}{9} + \frac{10}{9}\left(-\frac{1}{2}\right)^n + \frac{8}{3}n\left(-\frac{1}{2}\right)^n\right)u(n)$$

The first four values of the zero-state response $y(n)$ are

$$y(0) = 2 \qquad y(1) = -1 \qquad y(2) = \frac{5}{2} \qquad y(3) = -\frac{1}{4}$$

The total response is

$$y(n) = \left(\frac{8}{9} + \frac{29}{18}\left(-\frac{1}{2}\right)^n + \frac{8}{3}n\left(-\frac{1}{2}\right)^n\right)u(n) \qquad n = 0, 1, 2, \ldots$$

The first four values of the total response $y(n)$ are

$$y(0) = \frac{5}{2} \qquad y(1) = -\frac{5}{4} \qquad y(2) = \frac{21}{8} \qquad y(3) = -\frac{5}{16}$$

12.5.3 $h(n) = (9\delta(n) - 7(-\frac{1}{3})^n + 8n(-\frac{1}{3})^n)u(n) \qquad n = 0, 1, 2, \ldots$

The first four values of the sequence $h(n)$ are

$$h(0) = 2 \qquad h(1) = -\frac{1}{3} \qquad h(2) = 1 \qquad h(3) = -\frac{17}{27}$$

12.6.2 The zero-input component of the state vector is

$$q(n) = \left[\begin{array}{c} -\frac{1}{4}n(-\frac{1}{2})^{n-1} + \frac{4}{3}(-\frac{1}{2})^n \\ \frac{1}{2}n(-\frac{1}{2})^{n-1} - \frac{5}{3}(-\frac{1}{2})^n \end{array} \right] u(n)$$

The zero-input response is given by

$$\left(-\frac{3}{4}n\left(-\frac{1}{2}\right)^n - \frac{5}{4}\left(-\frac{1}{2}\right)^n \right) u(n)$$

The first four values of the zero-input response $y(n)$ are

$$y(0) = -\frac{5}{4} \qquad y(1) = 1 \qquad y(2) = -\frac{11}{16} \qquad y(3) = \frac{7}{16}$$

The zero-state component of the state vector is

$$q(n) = \left[\begin{array}{c} \frac{3}{5}n(-\frac{1}{2})^{n-1} - \frac{12}{25}(-\frac{1}{2})^n + \frac{12}{25}(\frac{1}{3})^n \\ -\frac{6}{5}n(-\frac{1}{2})^{n-1} - \frac{36}{25}(-\frac{1}{2})^n + \frac{36}{25}(\frac{1}{3})^n \end{array} \right] u(n)$$

The zero-state response is given by

$$\left(\frac{9}{5}n\left(-\frac{1}{2}\right)^n - \frac{27}{25}\left(-\frac{1}{2}\right)^n + \frac{52}{25}\left(\frac{1}{3}\right)^n \right) u(n)$$

The first four values of the zero-state response $y(n)$ are

$$y(0) = 1 \qquad y(1) = \frac{1}{3} \qquad y(2) = \frac{31}{36} \qquad y(3) = -\frac{25}{54}$$

The total response is

$$y(n) = \left(\frac{21}{20}n\left(-\frac{1}{2}\right)^n - \frac{233}{100}\left(-\frac{1}{2}\right)^n + \frac{52}{25}\left(\frac{1}{3}\right)^n \right) u(n) \qquad n = 0, 1, 2, \ldots$$

The first four values of the total response $y(n)$ are

$$y(0) = -\frac{1}{4} \qquad y(1) = \frac{4}{3} \qquad y(2) = \frac{25}{144} \qquad y(3) = -\frac{11}{432}$$

Chapter 13

13.4 $A = \begin{bmatrix} -2 & -1 \\ 1 & 0 \end{bmatrix}$ $B = \begin{bmatrix} 1 \\ 0 \end{bmatrix}$ $C = [2\ 0]$, $D = 0$

$(-4t^2 e^{-t} + 8te^{-t})u(t)$

13.6 The zero-input component of the output is given by

$5e^{-t} - 2e^{-3t}$

The zero-state component of the output is given by

$9e^{-t} - 33e^{-2t} + 27e^{-3t}$

The total response of the system is

$y(t) = (14e^{-t} - 33e^{-2t} + 25e^{-3t})u(t)$

13.9 $h(t) = (-3e^{-\frac{1}{2}t} + 4e^{-\frac{1}{3}t})u(t)$

13.11 The zero-input component of the output is given by

$7e^{-t} - 5e^{-2t}$

The zero-state component of the output is given by

$6 - 24e^{-t} + 27e^{-2t}$

The total response of the system is

$y(t) = (6 - 17e^{-t} + 22e^{-2t})u(t)$

13.15 $h(t) = (-2\delta(t) + 7e^{-t} - 9te^{-t})u(t)$

Index

A Practical Approach to Signals and Systems   D. Sundararajan
© 2008 John Wiley & Sons (Asia) Pte Ltd